MW01170126

From the Flight Deck

Also by Peter B. Mersky

The Naval Air War in Vietnam (with Norman Polmar)

The Grim Reapers: Fighting Squadron Ten in WWII

Amphibious Warfare: An Illustrated History (with Norman Polmar)

Vought F-8 Crusader

Time of the Aces: Marine Pilots in the Solomons, 1942–1944

Israeli Fighter Aces

U.S. Marine Corps Aviation: 1912 to the Present

From the Flight Deck

AN ANTHOLOGY OF THE BEST WRITING ON CARRIER WARFARE

Edited and Introduced by
Peter B. Mersky

Brassey's, Inc.
Washington, D.C.

Library of Congress Cataloging-in-Publication Data

From the flight deck : an anthology of the best writing on carrier warfare / edited by Peter B. Mersky.—1st ed.
p. cm.
Includes index.
ISBN 1-57488-433-6 (hardcover : alk. paper) — ISBN 1-57488-612-6 (pbk. : alk. paper)
1. Aircraft carriers. 2. Naval aviation. 3. Air warfare. I. Mersky, Peter B., 1945–

VG90.F76 2003
359.9'435—dc21 2003007351

Printed in the United States of America on acid-free paper that meets the American National Standards Institute Z39-48 Standard.

Brassey's, Inc.
22841 Quicksilver Drive
Dulles, Virginia 20166

First Edition

10 9 8 7 6 5 4 3 2 1

For Rear Admiral W. F. "Bud" Flagg and his wife, Dee. After living through combat deployments to Vietnam and more than 3,200 hours of Crusader time, they were on American Flight 77 when terrorists flew it into the Pentagon on September 11, 2001.

CONTENTS

CHAPTER 3
The Early Cold War, 1946–1953

CHAPTER 6
A New Age, 1976–Present

ILLUSTRATIONS

PREFACE

No other type of military flying is as colorful, exciting, or dangerous as flying from an aircraft carrier. In a profession filled with danger, glory, and satisfaction, nothing compares with launching from a huge yet confined warship, completing your mission, and returning with exacting precision to that same flight deck.

Although "air capable" ships have existed since the late 19th century, with occasional balloon launches, powered flight from a deck occurred little more than seven years after the Wright Brothers' December 1903 flight at Kitty Hawk. It took another seven years for more formal, more technically evolved operations to begin during World War I. By the end of the 1920s, carrier aviation had become an established part of naval operations for the United States, Great Britain, and Japan. Other nations, including France, Germany, and Italy, were planning their own ambitious programs.

Each year brought exciting developments in flight-deck equipment and procedures, new aircraft, and grudging acceptance from entrenched cadres of "ship drivers," who saw the upstart carrier and its growing community as threats to their traditional position as the capital ships of the fleet.

World War II brought tremendous responsibilities and fame to carrier aviators. Carrier task forces contributed greatly to the ultimate Allied victory, particularly in the Pacific. At the beginning of the war, the Japanese carrier fleet possessed a much higher degree of combat experience. For the first six months of the war, Imperial Japanese carrier aviators were on what we would call today "the tip of the spear."

After the war, with many military services reducing size and operational commitment, the U.S. Navy continued testing and developing aircraft, carriers, and procedures, aided immeasurably by the British, whose considerable number of aircraft carrier experiments produced the carrier as we know it today with angled deck, steam-powered catapults, and landing-light systems.

USS *Forrestal* (CVA-59), launched in 1954, was the world's first carrier built to incorporate these three important British developments, as well as being the first "super" carrier.

A new conflict in Korea in 1950, and then in Vietnam 15 years later, confirmed the role of the carrier as the capital ship of any navy rich enough to operate such an expensive weapons system. During this time, many changes occurred—larger, more capable carriers, some nuclear powered, newer aircraft to fly from them, and even the career paths of the people who served in these ships.

There were also basic social changes, not the least of which was the introduction of women, first as naval aviators in the mid-1970s, eventually followed in the early 1990s by their assignments to carrier combat squadrons. The 1991 Tailhook scandal ratcheted up the intensity of female participation in front-line activities traditionally reserved for men. During the 1991 Gulf War, before the Tailhook scandal broke, several female pilots served in Navy helicopter squadrons that flew back and forth from carriers to rear-area airfields. The first women aviators who flew fixed-wing aircraft went to S-3 Viking antisubmarine warfare squadrons and C-2 Greyhound carrier onboard delivery (COD) squadrons. However, by the mid-1990s, with a push from the highest levels of the U.S. government, women took their seats in F-14 Tomcat and FA-18 Hornet squadrons. (Note: although many writers, even some specializing in military subjects, use F/A-18, the official form dictated by the Naval Air Systems Command is without the slash to denote the aircraft's dual-mission capability.) The reception was chilly, to say the least, but the momentum was irresistible. And when U.S. carriers sailed into the Adriatic Sea in the late 1990s to fight in the war against the former Yugoslavia, several female Hornet pilots added green ink, signifying combat missions, to their logbooks. By October 2001, with the opening salvos of the War on Terrorism, and the campaign against Afghanistan's Taliban regime, female naval aviators were regular members of the carrier squadrons that flew daily missions over that desolate, hostile country.

The question of types and numbers of aircraft carriers reared its head in the 1980s, and by the end of the century, political and economic infighting had reduced the American carrier fleet to 12, with a pair of nuclear ships building at Newport News, Virginia. But these won't add to the fleet; they will replace the three remaining fossil-fueled ships within the next five years. After the 1991 Gulf War, six veteran carriers were retired or quickly scrapped. By 1999, the four ships of the pioneering *Forrestal* class had been decommissioned, leaving only *Kitty Hawk*, *Constellation*, and *Kennedy* to continue serving.

Of the many other countries that have had at least one carrier since World War II, only France, and with the purchase of the French carrier *Clemenceau* in 2000, Brazil, maintain conventional ships. Other nations, such as Great Britain, Spain, India, Italy, and Thailand, have carriers that operate Harrier jump jets and various helicopters. After the fall of the Soviet Union in 1991, Russia took over the USSR's fleet of carriers, including the *Kuznetsov*, the only Soviet carrier to attain operation with a functioning air wing of fixed-wing jets, Su-27 Flankers. (Because of political ramifications, the ship's original name of *Riga* was subsequently changed to the *Leonid Brezhnev* [the Soviet Secretary General of the late 1970s], then *Tbilisi*, then to its current name.) The ship has not been to sea since mid-1999.

The naval aviator of the 1940s would probably raise his eyebrows in surprise—perhaps disbelief—but also in appreciation of where his community has gone in the last 55 years, how far it has come. His heritage is proud, deep, and full of history and stories, some of which are found in this collection. There is a lot I have not included, mainly because we would need two or three more books to adequately present the remaining stories. And these excerpts are just from books, not the myriad thousands of fine magazine articles that have appeared over the years all over the world.

As we start a new century, already with incredible challenges facing everyone around the globe, the world's military forces are changing, often reducing or consolidating available resources. In the United States, economic forces have generated a steady exodus of trained aviators who want to fly for the airlines, earn a lot of money, and not deploy overseas for six months, leaving family and friends far behind. It takes a lot of courage to leave home in that fashion. Many relationships do not survive even the first such separation. Yet, for those who do keep the faith and make perhaps six or seven deployments in the course of a 20-year naval career, the satisfaction is usually great and the bonding with squadron mates is steel-strong. I hope the reader finds some of that emotion in these pages.

Peter B. Mersky
Alexandria, Virginia

ACKNOWLEDGMENTS

I could not have accomplished my task without the help of many of the authors and their representatives whose work appears in these pages, and the people who follow.

Rick Russell, my editor; Jim Sutton, negotiator supreme; Ellen Lawson; and Marilyn Elkins.

Research aides Dan Crawford, Robert Cressman, Jim Farmer, Fred Freeman, Hill Goodspeed, Tony Holmes, Richard Knott, Harold Livingston, Robert C. Mikesh, Dr. Frank Olynyk, Norman Polmar, Rosario Rausa, and Andrew Thomas

And for translations and research, Captain Robert Feuilloy, French Navy (Ret); and Stephane Nicolaou.

CREDITS

"Taking the First Chance: Eugene Ely, the Navy's First Carrier Aviator," from *Wings for the Fleet*, by Rear Admiral George van Deurs, copyright 1966. Reprinted by permission of the U.S. Naval Institute Press.

"The Aircraft Carrier Goes to War," and "The Navy Wants a Piece of the Action: Nuclear Bombers Aboard Carriers," from *Aircraft Carriers: A Graphic History*, by Norman Polmar, copyright 1969. Reprinted by permission of the author.

"The U.S. Navy's First Aircraft Carrier, USS *Langley*," from *The Golden Age Remembered: U.S. Naval Aviation 1919–1941*, edited by E. T. Wooldridge, copyright 1998. Reprinted by permission of the U.S. Naval Institute Press.

"Fighting the Germans in the North Atlantic," from *Wings on My Sleeve*, by Commander Eric Brown, copyright 1961. Reprinted by permission of the author.

" 'The Famous Victory': Swordfish at Taranto," from *With Naval Wings: The Autobiography of a Fleet Air Arm Pilot in World War II*, by John Wellham, copyright 1995. Reprinted by permission of Spellmount Publishers Ltd.

"Japan Attacks Pearl Harbor," from *Day of Infamy*, by Walter Lord, copyright 1957, 1991, 2001. Reprinted by permission of Henry Holt and Company, LLC.

"Payback Time: The Doolittle Raiders Bomb Tokyo," from *Thirty Seconds Over Tokyo*, by Captain Ted Lawson and Bob Considine, copyright 1943. Reprinted by permission of Mrs. Ellen Lawson.

"Something for the Home Folks: Taking Liberties with the Facts," by Richard Sale, copyright 1943, from *Teenage Aviation Stories*, edited by Don Samson, Grosset & Dunlap, 1948.

"The Midway Admiral: Raymond A. Spruance," from *Admiral Raymond A. Spruance*, by Vice Admiral E. P. Forrestel, 1966. Naval History Center, Washington, D.C.

"A Lost Aviator Disappears," from *The Last Flight of Ensign C. Martin Kelly, Jr.*, by Bowen P. Weisheit, copyright 1993. Reprinted by permission of the author.

"From the Japanese Viewpoint," from *Midway: the Battle That Doomed Japan*, by Mitsuo Fuchida and Masatake Okumiya, copyright 1959. Reprinted by permission of the U.S. Naval Institute Press.

"Fighting for the Solomons: the Carrier Navy Joins the Marines," from *Carrier Warfare in the Pacific: An Oral History*, edited by E. T. Wooldridge, copyright 1993 by the Smithsonian Institution. Used by permission of the publisher.

"There Were Heroes on Both Sides," from *The Last Tallyho*, by Richard Newhafer, 1964.

"So Young to Command," from *Bring Back My Stringbag*, by John Kilbracken, copyright 1979, 1996. Reprinted by permission of Pen & Sword.

"Marines on Carriers in Korea," from *Into the Jet Age: Conflict and Change in Naval Aviation 1945–1975*, edited by E. T. Wooldridge, copyright 1995. Reprinted by permission of the U.S. Naval Institute Press.

"A Whitehat's Viewpoint," from *Sailors in the Sky: Memoirs of a Navy Aircrewman in the Korean War*, by Jack Sauter, copyright 1996. Reprinted by permission of the author.

"Fiction As True As Fact," from *The Bridges at Toko-Ri*, by James Michener, copyright 1953. Reprinted by permission of Random House.

"Royal Navy Carriers in Korea," from *With the Carriers in Korea*, by John R. P. Lansdown, copyright 1997. Reprinted by permission of Crecy Publishing Ltd.

"The Rubber Deck," from *The Wrong Stuff: Flying on the Edge of Disaster*, by Commander John Moore, USN (Ret), copyright 1997. Courtesy of John Moore, author, and published by Specialty Press, 11605 Kost Dam Road, North Branch, MN 55056, 800-895-4585.

"Young Pups and Old ADs," from *Gold Wings, Blue Sea: A Naval Aviator's Story*, by Captain Rosario Rausa, USNR (Ret), copyright 1981. Reprinted by permission of the author and the U.S. Naval Institute Press.

"Carriers in the Middle East: The Suez Campaign," from *Wings Over Suez*, by Brian Cull, with David Nicolle and Shlomo Aloni, copyright 1997. Reprinted by permission of Grub Street Publishing Ltd.

"French Hellcats in Vietnam," from *Au-Dela du Pont d'Envol* [Beyond the Flight Deck], by Vice Admiral Roger Vercken, FN (Ret), Editions Alerion, copyright 1995. Reprinted by permission of the author.

"The Skyhawk's War," from *Dark Sky Black Sea: Aircraft Carrier Night and All-Weather Operations*, by Charles H. Brown, copyright 1999. Reprinted by permission of the U.S. Naval Institute Press.

"Paying the Price: Frank Elkins of VA-164," from *The Heart of a Man*, by Frank Elkins, copyright 1991. Reprinted by permission of Marilyn Elkins.

"Fire on the Oriskany," from *RF-8 Crusader Units Over Cuba and Vietnam*, by Peter Mersky, copyright 1999. Osprey Publishing. Reprinted by permission of author and publisher.

"*America* and *Liberty*: Passing Ships," from *Wings and Warriors: My Life As a Naval Aviator*, by Donald D. Engen, copyright 1997 by the Smithsonian Institution. Used by permission of the publisher.

"War in an Unknown Place," from *Sea Harrier Over the Falklands: A Maverick at War*, by Commander Sharkey Ward, DSC, AFC, RN, copyright 1993. Reprinted by permission of U.S. Naval Institute Press.

"Managing the Store: What It Takes to Run an Aircraft Carrier," from *Supercarrier: An Inside Account of Life Aboard the World's Most Powerful Ship, the USS* John F. Kennedy, by George C. Wilson, copyright 1986. Reprinted by permission of the author.

"A Helo Crew's War," from *Weapons Free: The Story of a Gulf War Helicopter Pilot*, by Richard Boswell, copyright 1998. Reprintedby permission of Crecy Publishing, Ltd.

"Prowlers in the Gulf," from *Ironclaw: A Navy Carrier Pilot's Gulf War Experience*, by Sherman Baldwin, copyright 1996. Reprinted by permission of HarperCollins Publishers.

"Bailout in the Balkans," from *No Escape Zone*, by Nick Richardson, copyright 2000. Reprinted by permission of Little, Brown UK, and Lucas Alexander Whitley Ltd.

"Recce Over Kosovo," from *Le Porte-Avions Foch*, by Alexandre Paringaux, Prestige Aeronautique, copyright 2000. Used by permission of the author.
"Women in the Cockpits," from *She's Just Another Navy Pilot: An Aviator's Sea Journal*, by Loree Draude Hirschman and Dave Hirschman, copyright 2000. Reprinted by permission of the U.S. Naval Institute Press.

CHAPTER 1

The Early Years, 1910–1938

Taking the First Chance: Eugene Ely, the First Carrier Aviator

Someone has to be first in any endeavor. The Wright Brothers had succeeded in making the first powered flight on December 17, 1903, and from that milestone, there were plenty of other "firsts" to come in the succeeding decade. With communications technology less immediate than today's real-time television cable coverage, the world had difficulty in coming to grips with the fact that two American bicycle mechanics from Ohio had accomplished one of man's oldest dreams. The European capitals had the most trouble, particularly in France and England. Some people simply refused to believe the Wrights' claim—until the businesslike brothers journeyed to France in 1908 and put on several months' of flying displays.

While the Old World came to understand what had happened on the desolate dunes of coastal North Carolina, another young American was preparing to make what proved to be nearly as important a pair of aviation milestones. The U.S. Navy had been forced to consider an aviation section along with the Army. It had established a small department headed by Captain Washington Irving Chambers, an experienced sailor. A member of the Annapolis class of 1876, he had risen to command the battleship *Louisiana* before being reassigned to special duties in Washington in 1909, cutting short his command tour.

Although disappointed by the brevity of his command, Chambers went to work handling the business of the Navy's newly created aviation section. He had watched the Wrights fly along the Hudson River to Grant's Tomb from the bridge of his battleship, but like many ship drivers, he considered the new vehicle as no more than an interesting diversion, with little promise for the military, especially the Navy.

As part of his office's coverage, Chambers occasionally attended air shows, which were becoming the rage in the U.S. and Europe. In October 1910, he visited a display at Belmont Park on Long Island, where he met Glenn Curtiss, one of America's best-known aircraft designers and pilots. He also met a young daredevil, Eugene Ely, one of Curtiss's demonstrator pilots. By the time Chambers had spent a week inspecting the more than 40 aircraft on display and seeing them go through their paces, he had become a convert and knew the Navy needed airplanes in some way.

A month later, Captain Chambers again met Eugene Ely at another show in Baltimore. It was a fortunate encounter for all concerned. However, Ely's flying career lasted only 18 months. He died on October 19, 1911 in the crash of his aircraft before more than 8,000 air show spectators at Macon, Georgia. He was only twenty-five. He received a posthumous Distinguished Flying Cross in 1933 and had also been designated a lieutenant in the Naval Reserve.

George van Deurs wrote a splendid account of early naval aviation. After graduating from Annapolis with the class of 1921, he earned his wings as Naval Aviator No. 3109 in 1924 and flew in several squadrons of the period. He served in ship's company aboard USS Saratoga *(CV-3), and commanded Patrol Squadron 23 at Pearl Harbor. His World War II service included assignments in the South Pacific, working with the Royal New Zealand Air Force. He eventually commanded the naval occupation forces in western Japan. He commanded the carrier USS* Philippine Sea, *before retiring in 1951. Admiral Van Deurs died in August 1984.*

From *Wings for the Fleet,* by Rear Admiral George Van Deurs

On 3 November 1910, another air show opened at Halethorpe Field near Baltimore. Again, Chambers left his desk in Washington to attend. Many of the Belmont pilots sent their planes to the Baltimore show, but an expressmen's strike held up delivery of most of the planes and, on opening afternoon, only two Curtiss planes flew. One of these was flown by Eugene Ely. Ely was feeling pretty confident. He liked to fly. He was making big money and a good name in an exciting business, which promised an unlimited future. His wife, Mabel, was also an aviation fan.

A rain storm stopped the show; by evening tent hangars were blown down and Ely's planes were smashed. Since nothing could be done at the field in such weather, Gene and Mabel went shopping in Baltimore. When they returned to the hotel at the end of the wet afternoon, they met Captain Chambers.

During their conversation, Chambers mentioned he had just asked Wilbur Wright for a pilot and a plane to fly from a ship. Wright had flatly refused all help, saying it was too dangerous. He would not even meet Chambers to talk it over. Chambers was taken aback because it was Orville's suggestion in 1908 that had given him the idea. "I had hoped it would get the Navy interested in planes," he said.

Eugene Ely poses by his Curtiss biplane on January 18, 1911, after landing aboard USS *Pennsylvania* in San Francisco. He wears a football helmet, jacket, and inner-tube life preservers. (Courtesy of the Naval Historical Center)

Gene Ely quickly asked for the job. "I've wanted to do that for some time," he told the surprised captain. Ely would furnish his own plane and he asked for no fee. He had three reasons for his eagerness. He had argued shipboard takeoffs with other fliers and he wanted to show them it could be done, he wanted the publicity, and he wanted to do a patriotic service.

Chambers wanted to get Curtiss' consent. "Not necessary," Ely assured him. "I make my own dates under our contract." That was a happy chance. As a matter of fact, Curtiss did his best to talk Ely out of it. Maybe he agreed with Wilbur Wright and thought it too dangerous. He argued that a failure would hurt plane sales. Mabel Ely believed he feared success even more. It might detract from the naval value of his hydroaeroplane.

Back in Washington, Wainwright turned down Chambers' proposal to let Ely fly from a cruiser. Chambers' boss, Captain Fletcher, told reporters the Navy had no money for such things. Meyer returned to Washington; in

Baltimore that same day, Chambers asked Curtiss and Ely to back his appeal with technical arguments. Only Ely went to Washington with him to confer with Meyer.

Secretary Meyer was back at his desk after his long inspection trip. Undoubtedly Wainwright had coached him before the conference. Ely never forgot how the Secretary covered his technical ignorance of aircraft and ships with an imperious coldness, and he never forgave him for calling Ely's plane a mere carnival toy when he turned down the proposed shipboard takeoff.

Then John Barry Ryan, a millionaire publisher and politician, got into the act. Two months earlier he had organized financiers, investors, and scientists interested in aeronautics, with a few pilots, as the U.S. Aeronautical Reserve. He furnished this organization with a Fifth Avenue clubhouse, provided several cash prizes for aeronautical achievements by its members, and made himself commodore of the organization. One of these prizes was $1,000 for the first ship-to-shore flight of a mile or more. Ryan was in Washington to pledge the club's pilots and their planes to the Army and Navy in case of war, when he heard of the Chambers-Ely plan and reopened the subject with Secretary Meyer.

When Ryan urged Chambers' proposal, Secretary Meyer responded that the Navy had no funds for such experiments. Ryan then offered to withdraw the $1,000 prize, which the non-member Ely could not win anyway, and use it to pay the costs of the test. Meyer had little interest in planes, but he was an accomplished politician, and he knew Ryan could swing votes in both Baltimore and New York. After consulting the White House, he agreed that the Navy would furnish a ship, but no money. Thereupon he left town.

Winthrop, "acting," acted in a hurry. He rushed the *Birmingham*, commanded by Captain W. B. Fletcher, to the Norfolk Navy Yard and told the yard commandant to help equip her with the ramp which Constructor McEntree had designed. The ship was a scout cruiser, with four tall stacks. Her open bridge was but one level about the flush main deck. On her forecastle, sailors sawed and nailed until they finished an 83-foot ramp, which sloped at five degrees from the bridge rail to the main deck at the bow. The forward edge was 37 feet above water.

Meanwhile, Henning and Callen, Ely's mechanics, worked at Piny Beach, where later the Hampton Roads Naval Base would be built. Using bits shipped from Hammondsport and pieces salvaged in Baltimore, they built a plane. Ely got there on a Sunday in foul weather. He added cigar-shaped aluminum floats under the wings and a splash-board on the landing gear. Late in the day he saw the plane—without its engine—aboard the Navy tug *Alice*, headed for the Navy Yard. The engine had been shipped; no one knew when it might arrive.

Gene Ely was not a worrying man. But the storm at Baltimore had cost him money. Shortly before Belmont, a speck of paint in a gas tank vent had robbed him of fame and a $50,000 prize. In previous months other crack-ups had bruised his body and damaged his pocketbook. These mishaps taught him how tiny, unexpected flaws could foul up a flight. Each time he charged it off to experience and tried again. Since his interview with Mr. Meyer, the cruiser flight had become a must. To his original motives, Ely had added an intense desire to show Secretary Meyer the error of his ways. At the same time he knew that, if he failed in his first try, Meyer would never give him another chance. And so Ely was worried when he joined his wife and Chambers.

At the old Monticello Hotel in Norfolk, Ely told reporters, "Everything is ready. If the weather is favorable, I expect to make the flight tomorrow without difficulty." Mabel knew that her husband was whistling in the dark. He had not seen the platform. The plane was untested. He hoped his engine would come on the night boat. But she had complete confidence in Gene, so she enjoyed a seafood dinner and untroubled sleep. Ely ate little, turned in early, and slept poorly.

In the morning, as he worried into his clothes, the clouds looked level with the hotel roof. He skipped breakfast and took the Portsmouth ferry.

Callen and Henning had hoisted the plane aboard the *Birmingham*, pushed it to the after end of the platform, and secured it with its tail nearly over the ship's wheel. Only 57 feet of ramp remained in front of the plane. Henning was worried. But Callen reassured him. "Old Gene can fly anywhere," he said. Then Ely's chief mechanic, Harrington, arrived with the engine. The three were getting it out of the crate when Ely and Chambers boarded the ship.

At 1130, sooty, black coal smoke rolled from the *Birmingham*'s stacks as she backed clear and headed down river. Two destroyers cleared the next dock. One followed the cruiser; the other headed for Norfolk to pick up Mabel Ely and the Norfolk reporters.

Going down river, Ely helped his men install the engine. He wanted to double check everything to avoid another failure; besides, the familiar work eased his tensions. He blew out the gas tank vent twice. In spite of squalls, they had the plane ready before the ship rounded the last buoy off Piny Beach. They had almost reached the destroyers *Bailey* and *Stringham*, waiting with Winthrop and other Washington officials, when another squall closed in. A quarter mile off Old Point Comfort, Captain Fletcher anchored the *Birmingham.* Hail blotted out the Chamberlain Hotel.

It was nearly two o'clock when that squall moved off to the north. Ely climbed to his plane's seat. Henning spun the propeller. Under the bridge the wireless operator tapped out a play-by-play account of the engine testing. When the warm-up came to an end, nobody liked the looks of the weather. Black clouds scudded just above the topmast. The cruiser *Washington* radioed that it was thick up the bay, and the Weather Bureau reported it would be worse the next day. Chambers nodded toward the torpedo boats. "If this weather holds till dark," he said, "a lot of those guys will go back to Washington shouting 'I told you so.' "

By 1430 the sky looked lighter to the south. Captains Fletcher and Chambers decided to get under way. Iowa-born Ely could not swim, feared the water, got seasick on ferryboats, and knew nothing about ships. He thought the cruiser would get under way as quickly as a San Francisco Bay ferry. He had no idea that the windlass he heard wheezing and clanking under the aeroplane platform might take half an hour to heave 90 fathoms of chain out of the mud. So he paced first the bridge, then the launching platform. Then he climbed into his seat and tried the controls. Sixty fathoms of chain were still out. Henning span the propeller. Ely opened the throttle and listened approvingly to the steady beat. Under the plane's tail, the helmsman at the wheel took the full force of the blast.

Ely was ready. He idled the engine and waited. Then he gunned the engine to clear it, twisted the wheel for the feel of the rudder, rechecked the setting of the elevator, and looked back at the captains on the bridge wing. They looked completely unhurried.

Then Ely noticed the horizon darkening with another squall and he began to wonder why the *Birmingham* did not start. He looked at Chambers, pointed at the approaching blackness. The captain nodded. He knew it would be close, but he could do nothing. Thirty fathoms of chain were still in the water.

Gene Ely checked everything again, and stared at the squall ahead. He seemed about to lose his chance because the Navy was too slow. At 1516 he decided he would wait no longer for the ship to start steaming into the wind. If ever he was going to fly off that ship, it had to be now. He gave the release signal.

Harrington, who knew the plan, hesitated. Ely emphatically repeated his signal. The mechanic yanked the toggle, watched the plane roll down the ramp and drop out of sight. Water splashed high in front of the ship. Then the plane came into sight, climbing slowly toward the dark clouds. Men on the platform and bridge let out the breath they had held. One of them spoke into a voice tube, and the wireless operator tapped out, "Ely just gone."

In 1910, Curtiss pilots steered with their rudder, balanced with their ailerons and kept the elevator set, by marks on its bamboo pushrod, either at a climb, level, or a glide position. In order to dip and pick up a bit more speed, Ely took off with his elevator set for glide. Off the bow he waited the fraction of an instant too long to shift to climb. The machine pointed up, but squashed down through the air.

Gene felt a sudden drag. Salt water whipped his face. A rattle, like hail on a tin roof, was louder than his engine. He tried to wipe the spray from his goggles but his gloved hand only smeared them, so he was blinded. Then the splashboard pulled the wheels free of the water. The rattle stopped. He snatched off his goggles and saw dirty, brown water just beyond his shoes.

The seat shook. The engine seemed to be trying to jump out of the plane. Ely's sense of direction left him. There were no landmarks, only shadows in the mist, and that terrifying dirty water below. He swung left toward the darkest misty shadow. He had to land quickly. On the ground he might stop the vibration, take off again, and find the Navy Yard. He wondered if the bulky life jacket that fouled his arms would keep him afloat if the plane splashed.

A strip of land bordered by gray, weathered beach houses loomed ahead. Five minutes after the mechanic had pulled the toggle, Ely landed on the beach at Willoughby Spit. "Where am I?" he asked Julia Smith, who had dashed out of the nearest house.

"Right between my house and the yacht club," she said.

It sounded funny but it wasn't. He knew the splintered propeller would not take him to the Navy Yard. He had failed. He blamed himself bitterly for the split second delay in shifting the elevator. Now he knew how to do it without hitting the water, but would he ever get another chance?

Boats full of people converged on the yacht club dock. Their enthusiastic congratulations confused him. "I'm glad you did not head for the Navy Yard," Chambers told him. "Nobody could find it in this weather." Captain Fletcher agreed. John Barry Ryan offered him $500 for the broken propeller. "A souvenir of this historic flight," he explained.

Ely figured that in not making the Navy Yard, he had failed, and Chambers and Ryan spent the evening trying to convince him that he had succeeded. His particular landing place was unimportant. It would soon be forgotten. The world would remember that he had shown that a plane could fly from a ship, and that navies could no longer ignore aeroplanes. Ely did not cheer up until Chambers promised to try to arrange a chance for him to do it again. "I could land aboard, too," was Ely's comment.

The next morning Ryan's valet wrapped the splintered propeller in a bathrobe and carried it into his pullman drawing room. There Ryan gave a champagne party until train time, presented Ely with a check for the propeller, and made him a lieutenant in his U.S. Aeronautical Reserve. After the train pulled out, Gene spent the check on a diamond for Mabel.

The morning of 15 November 1910, the *Birmingham* flight filled front pages all over the United States and Europe. Foreign editors speculated that the United States would probably build special aviation ships immediately. American editors, more familiar with naval conservatism, said the flight should at least lead Secretary Meyer to ask for appropriations for aviation. But Wainwright's friends belittled the performance. A ship could not fight with its guns boxed by a platform. A masthead lookout, they said, could see farther than Ely had flown.

And so it went and so it would go for a long time, this argument between the Navy's black shoe conservatives and the brown shoe visionaries.

The Aircraft Carrier Goes to War

Although two Americans had given the world its long-sought wings, enthusiasm and development in the United States waned in the years following the Wrights' accomplishments. While Europeans might have smarted after the imagined insult of having upstarts from the "colonies" show the way, they quickly took the lead in most aspects of aircraft development. France, Germany, and England quickly created aviation sections of their armies and navies, and found missions for their aircraft and crews, too.

Britain's Royal Flying Corps and Royal Naval Air Service were two distinct organizations that fought intense, bloody wars throughout World War I. Both groups boasted highly experienced aviators and flight crews and many of the world's leading aircraft. They were matched by their counterparts in France and Germany. (The RFC and RNAS combined on April 1, 1918, to form the Royal Air Force.)

The British had also been tinkering as far back as 1912 with putting aircraft aboard men-of-war. The vaunted Royal Navy had fought Germany's well-armed High Seas Fleet in the Battle of Jutland in May 1916. During the opening phase of the engagement, Short seaplanes had launched from converted seaplane carriers as long-range scouts. It was a portent of things to come during the next twenty years. Within the next twelve months, what we would today call the first carrier-based alpha strikes were launched from British carriers and struck German Zeppelin sheds; they also successfully engaged the lighter-than-air aerial monsters. It was a time of great expectations and developments, where only imagination limited what might be done.

Norman Polmar is a well-published authority on defense matters, especially naval interests. He has traveled to many exotic countries at the behest and sponsorship of those governments as well as that of the United States. He writes a regularly published compendium of American naval organizations and developments called Ships and Aircraft of the U.S. Fleet *(17th edition, Naval Institute Press, 2001). Among his many published books are* Code-Name Downfall: The Secret War to Invade Japan *(Simon & Schuster, New York, 1995),* World War II: America at War: 1941–1945 *(with Thomas B. Allen, Random House, New York, 1996), and* The Naval Air War in Vietnam *(with Peter B. Mersky, Nautical & Aviation Publishing Co. of America, South Carolina, 1981). He was a Ramsey Fellow at the National Air and Space Museum, and he has served as a consultant for several government officials, including the Chief of Naval Operations. This excerpt comes from his first book, originally published in 1969.*

From *Aircraft Carriers: A Graphic History,* by Norman Polmar

The year 1917 was the year of the first true *aircraft* carrier. The advantages of wheeled airplanes over seaplanes were by this time well known, especially for combat with Zeppelins. The problem was how to take airplanes to sea. Rutland, now a flight-commander, became a leader in the effort to equip ships of the Grand Fleet with airplanes. Posted as senior flying officer of the seaplane carrier *Manxman*, he first convinced his pilots that it was safer to be afloat in an airplane than a seaplane. Then, using the seaplane/trolley take-off deck of the *Manxman*, Commander Rutland was able to get a Sopwith Pup biplane into the air after a run of only 20 feet down the deck. The *Campania* and *Manxman* were both provided with Sopwith Pups, between them

HMS *Furious* in 1918 displays the initial deck layout of the first operational aircraft carriers. (Courtesy of the Royal Navy)

carrying 15 of the fighters plus fourteen reconnaissance planes. But more fighters were needed in the Fleet, and in ships that could steam with the faster battleships and cruisers. A technical committee studying the problem was amazed when Commander Rutland announced that he could fly a fighter off a cruiser with a run of only 15 feet with an air speed of 20 knots. Most other pilots were taking 35 to 45 feet to get a Pup off the *Campania* and *Manxman.* The light cruiser *Yarmouth* was ordered into the Rosyth Dockyard for installation of a flying-off platform. The ship's gunnery officer, Lieutenant-Commander C. H. B. Gowan, had designed the platform after Commander Rutland convinced him a fighter could be flown off in less than 45 feet.

While the cruiser was being modified, the *Manxman* went to sea and on May 17, Commander Rutland and another flier flew off airplanes on an anti-Zeppelin sweep. The weather was poor, and the second pilot came down almost immediately after taking off; he alleged his plane had engine trouble. He was picked up by a destroyer, but his plane was lost. Commander Rutland was forced to land on the Danish coast. He was able to leave neutral Denmark for Sweden and then Norway, finally returning to England in time for the *Yarmouth* tests in June of 1917.

Taking off from the *Yarmouth*, Rutland was able to get a Sopwith Pup into the air with a run of less than 20 feet. Another pilot duplicated the feat, although with considerable misgivings. Then tests were resumed aboard the *Manxman*, with all of her pilots taking off in just one-third of the ship's 60-foot flight deck.

Orders were given for all cruisers that could take the additional weight to be outfitted with 20-foot flying-off platforms and allocated fighter aircraft.

The *Yarmouth* went back to sea with her Sopwith Pup and on August 21 she was part of a cruiser squadron conducting a sweep of the Danish coast. Soon after daylight the Zeppelin *L.23* was sighted tracking the squadron from a "safe" distance of 12 miles. After luring the airship further away from its base, the *Yarmouth* turned into the wind and flew off her Sopwith Pup piloted by Flight-Sub-Lieutenant B. A. Smart. The fighter climbed above the Zeppelin and dived on the airship, spitting incendiary bullets. The first attack failed. On his second run Lieutenant Smart closed the range and his bullets sent the Zeppelin down in flames. The airplane then came down on the water and Lieutenant Smart was picked up by a boat from a British destroyer.

In all, twenty-two light cruisers were fitted with flying-off platforms. However, the Commander-in-Chief of the Grand Fleet refused to have the planes and their flying-off platforms installed in capital ships—battleships and battle cruisers—because of the necessity of turning the ship into the wind to launch planes. He foresaw such a maneuver being unacceptable for warships racing to close with an enemy. Commander Rutland, who had now been appointed senior flying officer in the hermaphrodite carrier *Furious*, turned his efforts to overcoming this problem. The solution lay in attaching the flying-off platform to the main battery turret and then turning only the turret into the wind produced by the true wind and the movement of the ship.

The battle cruiser *Repulse* had a platform affixed to her second (forward) 15-inch gun turret and, on October 1, Commander Rutland flew a Sopwith Pup from the platform, seemingly without effort, as the ship steamed underway. In the aviator's words, it was "absurdly simple."

Then to ensure that the ship's superstructure would not create air currents to prevent launchings from her after main turret, on October 9 he flew a fighter from the after turret on the *Repulse*. After these experiments all British battle cruisers were fitted to carry two fighter aircraft and smaller cruisers which could not take the weight of the *Yarmouth*-type installation were provided with the lighter, revolving platforms. In later tests two-place aircraft were flown off turret platforms, the first going aloft from a midships turret of the battle cruiser *Australia*.

The first ship to rightly be designated an "aircraft carrier" was H.M.S. *Furious*. Shortly after the start of World War I, Admiral Sir John Fisher, First Sea Lord of the Royal Navy, ordered the construction of three *light* battle cruisers, the *Glorious*, *Courageous*, and *Furious*. Begun in 1915, they were to be fast, lightly armored, and heavily gunned. Their draft was to be kept to an absolute minimum for operations in the relatively shallow Baltic Sea. Instead of the 30 to 34 feet of water "drawn" by standard battle cruisers, Admiral Fisher's ships would have a draft of only 25 or 26 feet.

The *Glorious* and *Courageous* were completed in late 1916, each mounting a main battery of four 15-inch guns and fourteen torpedo tubes of 21-inch diameter (probably more torpedo tubes than any other ship of their day). The *Furious* was delayed in order to give her a pair of 18-inch guns (the largest guns mounted in a warship until two Japanese ships appeared with 18.1-inch guns in World War II). She was also to mount 18 torpedo tubes. But on March 17, 1917, orders were issued for the *Furious* to be completed as a "partial" aircraft carrier at the expense of one 18-inch gun.

In place of the forward 18-inch turret a hangar was built on the forecastle of the *Furious* with its roof forming a slanted flight deck 228 feet long and 50 feet wide, ample for take-offs of two-seat aircraft. Behind her flight deck the *Furious* retained the bridge structure, masts, and funnel of a battle cruiser along with a giant 18-inch gun aft.

The hermaphrodite *Furious* joined the Fleet in July 1917. Forward she was an aircraft carrier; aft a battle cruiser. Over-all she was 786½ feet long, displaced 19,100 tons, and an 18 boiler-turbine power plant could move her at 31½ knots. Her 18-inch gun was the most powerful—and unsuitable—weapon afloat. Its maximum range was over 20 miles and it fired shells weighing 3320 pounds. The recoil was "tremendous" according to one officer who served in the ship. His cabin was beneath the big gun and "every time she fired it was like a snowstorm in my cabin, only instead of snowflakes sheared rivet-heads would come down from the deckhead and partition."

As originally completed the *Furious* carried an "air group" of four seaplanes and six wheeled aircraft. A hydraulic-operated lift transported them between the hangar and flight deck. Since she had only a flying-off deck, the planes had to fly to nearby shore bases after being launched or come down at sea. The latter practice presented a dangerous situation for the land planes.

After less than a month's operations with the new "carrier," Squadron-Commander E. H. Dunning, the ship's senior flying officer, resolved to solve the problem of landing on the ship. On August 2, Commander Dunning approached the stern of the *Furious* in a Sopwith Pup. The ship was steaming into a 21-knot wind at a speed of 26 knots, putting a 47-knot wind over the flight deck. This was about the landing speed of the fighter, meaning the plane would almost hover over the flight deck.

Commander Dunning flew up along the starboard side of the ship and when abreast her bridge he side-slipped to port and cut off his engine as he came over the flight deck. As the plane settled toward the deck several men grabbed straps attached to the plane and brought it to a stop. It was the first landing of an aircraft aboard a warship underway.

On August 7 the *Furious* again went to sea for landing experiments. Commander Dunning made the first approach. This time there was more wind over the flight deck and after the plane settled toward the deck and was grabbed by waiting crewmen the craft was blown back against the combing of the elevator hatch and damaged. Commander Dunning was not injured and immediately climbed into another plane which was waiting to take off. He flew off the Pup and came around for another landing. As the plane came over the deck Commander Dunning waved the crew away. He was not satisfied with the approach and wanted to come around again. But as he opened up his engine the airplane stalled. Men raced to grab the straps attached to the plane. They were too late; the Pup was blown over the side and crashed into the sea. It took 20 minutes for the *Furious* to slow, turn, and hoist out a boat. The smashed Pup remained afloat because of an air bag in its tail. Commander Dunning, apparently knocked unconscious in the crash, had drowned.

After Commander Dunning's fatal accident the landing experiments were halted and the *Furious* briefly joined her near sister ships *Courageous* and *Glorious* in the North Sea for operations against the Germans. In November 1917, the *Furious* entered a dockyard for installation of a landing deck. The ship's mainmast, 18-inch gun, and one of her secondary battery 5.5-inch guns were removed. Behind the ship's funnel was installed a landing deck 287 feet long and 70 feet wide. Under this deck, as under the forward take-off deck, a hangar was installed which could accommodate ten aircraft. Thus modified the ship could carry a total of twelve Sopwith Pups and eight Short seaplanes. At this stage her conventional armament consisted of ten 5.5-inch guns, five 3-inch anti-aircraft guns, and the 18 torpedo tubes. (The 18-inch guns built for the *Furious* were installed in British monitors but did not, as popularly recorded, end up at Singapore.)

The aircraft "equipment" in the *Furious* included trackways around the ship's funnel and bridge to connect the two flight decks so that after a plane landed it could be placed on a wheeled dolly and brought forward to the flying-off deck. A sandbag arresting gear, similar to the system used by Eugene Ely was provided to engage a hook under the aircraft. In addition, aircraft would be slowed by small V-shaped hooks on their wheel axles which would engage fore-and-aft wires. Together these devices would, hopefully, halt the aircraft before they reached a barrier of manila hawsers stretched across the end of the landing-on deck.

This form of arresting gear worked reasonably well when tested ashore. However, it was never satisfactory on carriers although it was retained for a number of years. When finally abandoned in 1925 there was no replacement system and for a number of years Royal Navy carriers had to rely solely on a strong head wind and slow landing speed to halt aircraft.

In March 1918 when the *Furious* emerged from the yard she was an aircraft carrier at both ends and a monstrosity amidships; as an aircraft carrier she was a horror. The hot boiler gases that poured from her centerline funnel affected the density of air over her stern. This coupled with the bridge structure cutting off the air streams over the landing deck and creating air eddies made landing on a most difficult proposition. Almost every attempt ended in an accident of varying severity. Even Commander Rutland, now senior flying officer in the *Furious*, went over the side while making a landing and narrowly escaped with his life. For a while the aircraft were fitted with twin skids in lieu of wheels, but these were no improvement; being rigid and unsprung they would not yield on impact as did tires and they tended to break on touching down on the hard deck of the *Furious*. In short order it was decided that aircraft would not be landed on the *Furious*, and the ship became operational as a "take-off-only" carrier.

On the morning of July 19, 1918, the *Furious*, in company with a force of light cruisers and destroyers, launched two flights of Camel fighters against the Zeppelin sheds at Tondern. The first flight of three aircraft bombed one of the large airship sheds. Of the four aircraft in the second flight, one was forced down at sea when its engine failed, another crashed shortly after taking off, and the third had to make a forced landing in Denmark. The last plane of this flight made it to Tondern and destroyed another Zeppelin shed with its bombs. Each of the destroyed sheds had contained an airship—the Zeppelins *L.54* and *L.60*.

However, only one plane from each flight was able to return to the *Furious* and they had to come down in the water alongside because of problems with the landing procedure on the ship's abbreviated landing deck. (Four of the seven pilots eventually reached Denmark.) Because of the aircraft losses the Admiralty decided against further carrier strikes from the *Furious*. For the remainder of the war she was relegated to service with captive balloons used for observation.

The effect on the Germans of the loss of two airships to bombs from the two 2F.1 Camels from the *Furious* was considerable. According to Douglas Robinson's major study of the air war: "Until the Armistice the [German] Naval Airship Division lived in constant fear of a similar attack on one of the other bases . . . because of its exposed position, Tondern was maintained on a standby basis as an emergency landing ground only."

Less than a month after the Tondern raid the Germans lost another Zeppelin to a "shipboard" fighter, this time the airship being destroyed in an aerial encounter. The ingenious Colonel Samson—the amalgamation of the British air services on April 1, 1918, had given him military rank—was behind the project. Upon returning from the eastern Mediterranean he was engaged in administrative work and was then given command of the Yarmouth air station on the eastern coast of England. From time to time British ships had towed seaplanes to sea, cutting them loose to take off when they were within range of their objectives. Then, to save wear and tear on the aircraft, the seaplanes were towed to sea on floats. The seaplanes were too slow to intercept Zeppelins and Samson's fighter planes lacked the range to engage the airships in their patrol areas. Samson proposed towing a fighter on a barge behind a destroyer. When the warship reached full speed, more than 30 knots, the fighter would be able to take off with a deck run of just a few feet, about the same distance as planes flying off platforms atop gun turrets.

A 40-foot barge was taken in hand and fitted with a "flight deck" and a device to lock the plane in place until the pilot had run its engine to maximum power. Troughs were fitted in the barge's deck to mate with the skilike skids on the aircraft to hold the fighter steady in rough water. A Camel biplane was then loaded on the modified barge and a destroyer towed the 40-foot "carrier" to sea. As the destroyer picked up speed Samson climbed into the Camel's cockpit and started the plane's engine. A crewman—tethered to the barge with a line so he would not be blown overboard—turned the plane's propeller and the engine started at once. When Samson had warmed up the plane's engine a signalman on the barge flagged the destroyer and the sleek ship held steady at 31 knots.

Samson open his throttle all the way and the Camel tugged at its restraining wire. He pulled the release toggle and the cable holding back the plane came loose. A moment later the plane was airborne . . . or almost so.

The biplane started to leave the deck and then faltered. The skids apparently left their grooves and fouled a cross-bar. The plane fell into the water in front of the on-rushing barge. The barge slammed into the airplane, pushing plane and pilot underwater. Colonel Samson escaped from the wrecked plane and was plucked from the water, wet and frustrated, but otherwise all right.

Photographs taken from the destroyer showed what had happened. The "flight deck" was modified so it would be horizontal when the barge was being pulled at high speeds, a tail guide was made to keep the tail of the airplane up and straight for the first four feet of run, and wheels were used instead of skids. For the second test Flight-Sub-Lieutenant Stuart Culley was selected as pilot. Culley was a serious-minded young man, born in the United States of a Canadian mother and an English officer.

On August 1 the destroyer *Redoubt* towed the barge to sea with Culley's biplane perched on it like a bird which had come to rest on a piece of flotsam. Soon the *Redoubt* was making 35 knots, just one knot short of the year-old destroyer's maximum speed. A final word of encouragement by Samson was shouted into Culley's ear above the drone of the plane's engine and sea and the young pilot was ready.

As the plane strained against the restraining wire Culley pulled the release toggle. He felt the plane rise and gave it left rudder to avoid hitting the destroyer or being run over by the barge should he crash. But he didn't crash. The Camel took to the air almost as if it had taken off from a long land strip. Culley circled the destroyer, dipped his wings in salute, and then headed for shore.

The evening of August 10 again found the *Redoubt* pulling a barge-borne Camel fighter to sea. This time there were four light cruisers and a dozen destroyers with her. On the decks of the cruisers were motor torpedo boats which would be launched if German ships were encountered. In addition to the Camel fighter, the destroyers also towed three seaplanes on barges for reconnaissance and rescue work. The next day three additional seaplanes would fly out to join the force.

The following morning, off the Dutch coast, the motor torpedo boats were put in the water. The three seaplanes that had been towed across the North Sea were unable to take off because of the long swell that was running, but the three which had flown across the Channel arrived over the force on schedule. Almost upon their arrival one of the trio sighted a Zeppelin. The seaplanes were unable to operate at 22,000 feet, the maximum altitude of the German airship, so no attempt was made to intercept. Instead, the task force headed out to sea in hope the Germans would follow. They did.

The Zeppelin *L.53* was lured out to sea and away from its base. The *Redoubt* turned up speed and at 8:41 A.M. Lieutenant Culley made a perfect take-off with a deck run of only five feet. According to some reports the Zeppelin sighted the Camel taking off from the barge, but it was no concern for the airship which could climb higher and faster than any British aircraft. For almost an hour the Camel climbed and made a circling search for the Zeppelin. As the Camel climbed above 14,000 feet its controls got sluggish; at 17,000 feet the engine coughed. At 18,000 feet Lieutenant Culley broke through a layer of clouds and found himself in bright sunshine and some 200 feet below the German airship. But his plane would go no higher. Pulling back on his control stick, Culley literally hung the Camel on its propeller until the plane was aimed straight at the airstrip. He triggered the plane's two machine guns. One gun jammed after seven rounds were fired. The other Lewis gun fired a double drum of incendiary bullets into the Zeppelin.

In a few seconds bursts of flame shot from the airship's envelope and the *L.53* erupted into a mass of flames; seconds later its charred metal frame dropped into the sea. Only one man escaped the death of the airship, having bailed out from 19,000 feet, possibly a record for the time.

But now the young pilot had problems. As Culley's guns stopped firing the Camel fell into a stall, dropping 2000 feet before he regained control. His fuel was almost gone and the British task force was nowhere to be seen. He headed toward the Dutch coast. In the distance appeared some fishing boats and Culley decided to come down near them. As he came near the "fishing boats" seemed to grow in size until Culley realized his mistake. They were destroyers! And there were cruisers . . . the task force which had taken him to sea. Culley brought his plane down on the water and a boat from the *Redoubt* picked him up. A derrick devised by Samson salvaged the fighter. For the action Culley was decorated with the Distinguished Service Order, high recognition of his feat.

The earlier raid of the *Furious* on the Zeppelin sheds at Tondern and Lieutenant Culley's kill of the *L.53* opened new vistas to "naval" aviation. Indeed, in the fall of 1918 the aircraft carriers *Vindictive* and *Argus* were completed and the carrier *Eagle* had been launched.

The *Vindictive* was begun in June of 1916 as the light cruiser *Cavendish.* She was launched in January of 1918 and completed as an "aircraft carrier" in October 1918, similar in appearance to the *Furious,* with her landing and flying-off decks separated by a cruiser superstructure. With her new name and configuration she was officially designated as a "light cruiser fitted as aircraft carrier." As such she displaced 9750 tons, was 605 feet over-all in length, and had a trial speed of 29 knots. Her aircraft capacity was rated at six land planes and she mounted four 7.5-inch guns, several smaller guns, and six torpedo tubes. The *Vindictive* was not particularly successful as an aircraft carrier and was reconstructed as a cruiser in 1923–1925.

Of considerably more significance, in 1916 the British government purchased the unfinished Italian liner *Conte Rosso* for conversion to an aircraft carrier with a flight deck running her entire length, uninterrupted by superstructure or funnels. She had been laid down in 1914 and work on her halted when the war began. Renamed *Argus*, she was launched in late 1917 and completed with a flight deck 550 feet long and 68 feet wide on a hull of 14,450 tons which was 565 feet over-all in length. To provide a clear flight deck, her chart house was installed on an elevator which lowered during flight operations and her exhaust gases were led aft through trunks and expelled over her stern.

The flat appearance of the *Argus* gave rise to her nickname of "flat iron." She could make 20.75 knots for short periods, adequate to get her twenty aircraft into the air. There were two elevators to move the planes between the hangar and flight decks. Normally the elevators fitted flush with the flight deck. However, in early flight operations an elevator would be left a few inches below the flight deck level to "trap" landing aircraft and prevent them from running down the deck and over the bow. The trap worked too well. Of five hundred landings, forty ended in crashes and ninety in seriously damaged aircraft. The lowered-elevator scheme was "dropped."

Ready for operation when the *Argus* was completed was the first aircraft designed to carry torpedoes and operate from carriers, the Sopwith Cuckoo. In October of 1918, a month after the *Argus* was completed, a squadron of these torpedo planes went aboard the ship. However, the war ended before the *Argus* could prove herself in combat.

Two other true aircraft carriers were under construction when the war ended, the *Eagle* and *Hermes*. The first had been laid down in February of 1913 as the battleship *Almirante Cochrane* intended for Chile. All work on her stopped in August of 1914 and in 1917 she too was purchased by the British government for conversion to a carrier. She was launched in June of 1918 as the *Eagle* and commissioned for trials on April 13, 1920. As completed the *Eagle* displaced 22,600 tons and was 667 feet over-all, had a maximum speed of 24 knots, and could handle twenty-one aircraft. Of significance, she introduced the "island" structure to aircraft carrier design with her bridge, mast, and a single funnel incorporated into a streamlined superstructure on the starboard side of the flight deck. This position was chosen because the natural torque of most pull-type propeller engines pull to the left. The *Eagle* also introduced the two-level aircraft hangar.

From November of 1920 until 1923 she was in a dockyard being modified as a result of her trials. When she rejoined the Fleet the *Eagle* had two funnels protruding from her island structure.

The *Hermes* was the world's first ship begun as an aircraft carrier, work on her having started in January 1918. She was launched on September 11, 1919. With the urgency of war past, she was not completed until July of 1923. The *Hermes* was somewhat similar in appearance to the *Eagle*, having a large island structure on her starboard side. She displaced 10,850 tons, was 598 feet over-all in length, and, although smaller than the *Eagle*, her improved design gave her a speed of 25 knots and a capacity of twenty-five aircraft.

Argus, *Eagle*, and *Hermes:* the first generation of "true" aircraft carriers; although too late for World War I, they would be invaluable experimental and training ships between the great wars and would have vital roles in the Second World War.

The U.S. Navy's First Aircraft Carrier, USS Langley

Alfred M. Pride came in on the ground floor, or first deck, of American naval aviation. A rarity in later flag ranks, he did not come from Annapolis, but from the enlisted community. A former machinist's mate, he received a commission in the Naval Reserve and served in England and France during World War I. Augmenting to the regular Navy, he saw many of naval aviation's early developments aboard its first carriers, the *Langley* and *Lexington*. He served throughout World War II and eventually rose to four-star rank. He died in December 1988 at the age of 91.

The two decades following World War I witnessed tremendous expansion and development in all areas of military hardware, especially in aviation, and in particular, the field of naval aviation. The world's richest countries, and the ones who had been on the winning side of World War I, endowed themselves with great fleets that included the new capital ship, the aircraft carrier. Although the new ship struggled to find acceptance with senior officers who came from the entrenched battleship and cruiser communities, only the most myopic observer would deny that it held the future of naval combat on its huge, wooden deck.

Technology, however, as well as operating procedures, was still far from providing the nascent carrier fleets with dependable systems with which crews could launch and recover regularly—and safely. Mel Pride witnessed, and in some cases, participated in many of the first postwar experiments with aircraft aboard ships. Often, the planes used were seconded products from British and French manufacturers. The American aircraft industry, aside from the early Curtiss company, which eventually joined with its arch rivals, the Wrights, was still a long way from becoming the huge collection of monolithic organizations of the post–World War II period.

With the tentative, occasionally uncertain atmosphere that surrounded the aircraft and aviators of the time, the early naval aviators were a breed apart, with courage and fortitude far beyond that normally expected of military men. They were always ready to fly and to try new things.

E. T. Wooldridge enjoyed a full career as a Navy fighter pilot. A member of the Naval Academy's class of 1950, he flew props and jets as a test pilot, and eventually flew the F-4 Phantom as the commanding officer of VF-102. He also served as the executive officer of USS Forrestal *(CVA-59). After his retirement, Captain Wooldridge joined the staff of the National Air and Space Museum, becoming chairman of the aeronautics department, and the first Ramsey Fellow. He received the Admiral Arthur W. Radford award in 2001 for excellence in naval aviation writing.*

From *The Golden Age Remembered: U.S. Naval Aviation 1919–1941*, edited by E. T. Woolridge

I only got in one or two antisubmarine patrols off the French coast before World War I ended. By that time, I loved aviation. It was pretty hard to be in at all without being very enthusiastic about it. Being a youngster, it was a whole new environment, and there was a lot of satisfaction in just flying an aircraft. You felt rather egotistical about it, as if you were one of the privileged.

America's first carrier, USS *Langley* (CV-1) was basically a flight deck atop a collier's hull. Although not the best platform for early development, the "covered wagon," as she was occasionally known, served long and well. She was eventually sunk by the Japanese in 1942 while ferrying much-needed aircraft to the Pacific. Note the two funnels canted away from the landing area. (Courtesy of the Naval Historical Center)

When the war was over, I was still a reserve, and we were ordered to the stations nearest our homes in the United States until it was determined what should be done with us. The navy had a coastal air station at Chatham, Massachusetts, so that was the one that I was ordered to. While there, I learned that the navy was about to fly aircraft off of battleships for spotting gunfire. Chevalier, Whiting, and others who were influential in naval aviation had watched the British experiments in flying off ships to scout for the ships and spot their gunfire. Chevalier went over to Britain and observed the tests that they were making. Our Navy Department thought this was a pretty good idea because our aerial observation from battleships was confined to the use of kite balloons. They obviously couldn't leave the ship to go out on scouting trips since they were moored to the ship.

So the Navy Department made some tests on the USS *Texas* (BB-35) on which they built a platform on the guns of the number two turret, the high turret. The platforms were made of the wooden painting stages, which all the battleships carried. The framework could be assembled very quickly on top of the guns. In March 1919, Lt. Comdr. E. O. McDonnell, flying a Sopwith Camel, made the first successful flight from the number two turret of the *Texas*, lying at anchor at Guantánamo.

So I thought I'd like to try flying from battleships, and I wrote a letter via channels and asked to be assigned to this duty, which I was. I went to Carlstrom Field in Florida, where I joined what was to become the Atlantic Fleet Ship Plane Division. This group at Carlstrom Field, an Army Air Service field, was under the command of Lt. Comdr. Godfrey de Courcelles Chevalier (naval aviator no. 7). I think there were eighteen of us that were assigned to this duty for the air service to teach us to fly land planes, because we were to fly them off the battleships. We had no land planes in the navy, except for the few we had used in the Northern Bombing Group in France and a few here and there on special projects.

They started us right in through the air service's elementary training, as though we had never seen an airplane before. This was rather hard on our egos because very few of the instructors had been on duty in Europe during the war, and most all of us had been. They put us right through their elementary and advanced training. We got into their single-engine fighters, then were turned loose to go to our battleships. I was ordered to the USS *Arizona* (BB-39). I joined the ship in her home yard, the Brooklyn Navy Yard. Then we left for Guantánamo within a week or two, where we did most of the work.

On the forward turrets of the *Arizona* we were to use an aircraft called

LCdr. Marc Mitscher, flying a Chance Vought UO-1, makes the first landing aboard USS *Saratoga* (CV-3) in 1927. Note the fore-aft wires engaged by hooks on aircraft's main landing gear spreader bar. Fiddle bridges held the wires off the flight deck. This arrangement was abandoned for cross-deck wires. (Courtesy of the Naval Historical Center)

the Sopwith-Strutter, which was a two-seater British airplane. On the after turrets, we had French Nieuport 28s. The runway was fifty-two feet long, as I remember, which meant that you had to have a pretty good breeze blowing down the line of the runway to be sure to get you in the air before you struck the water. In fact, we had two or three airplanes go in the water. Then we arrived through experience at the notion that you had to have at least 22 knots of wind down the runway. That meant that with the number three turret—the battleship had a speed of about 20 to 22 knots—you had to get a fairly good breeze blowing. Then the ship would be steaming almost across the wind to make the apparent wind come down the deck.

I flew off number three turret because that was a single-seater, and I was the senior of the two aviators. The other pilot on the *Arizona* was Lt. (jg) Jacob F. Wolfer, who unfortunately was killed the following year when he spun in down in Guantánamo Bay. The single-seaters were always thought to be more desirable probably because they were more lively, more maneuverable.

We put the tail up on a wooden horse and restrained the plane with a pelican hook, a quick-release device shaped somewhat like a pelican's beak. Then you revved the engine up full, nodded your head, and somebody released the pelican hook. Down the platform you went, hoping to God that you'd get flying speed before you got to the water. You always barely made it, if you made it at all. I never went in the water that way. I've been in the water several times, but I never failed to take off from the turret.

Both types of airplanes had rotary engines, which sometimes were rather unreliable. The nonaviators didn't think much of all this. I caught hell because the rotary engine used castor oil, which would spew out, and drops of it would go down on the beautiful teak on the quarterdeck of the *Arizona*. When I reported in, the skipper, Capt. John H. Dayton—who was a very fine man—told me that he didn't believe in airplanes on ships, and that the only future for aviation in the fleet that he could see was small dirigibles towed by the battleships. They could cast them loose and go out and scout and come back to another ship. I think that they appreciated our spotting for long-range firing very much, but they regretted these damn dirty airplanes on their ships.

When we operated from the battleship, we landed on the beach, wherever we could find a field, and towed the airplane with anything we could since these planes weren't very heavy. I used an ox team in Cuba to tow the airplane down to the shore. We got them onto a fifty-foot motor launch and brought them back out and hoisted them back on the ship with a boat crane. Most of the time that we were around Guantánamo, we'd keep the airplanes flying to keep our hands in. We operated from Hicacal Beach over on the west side of Guantánamo Bay, where we established a camp.

There were other battleships with us: the *Nevada* (BB-36), the *Oklahoma* (BB-37), and the *Pennsylvania* (BB-38), all of which had their turrets rigged for aircraft. The pilots would gather together a great deal in port, and every day we were over at Hicacal Beach flying. In fact, we had to assemble our aircraft there. We didn't take any planes south with us in the battleships on the first cruise. We got down there and were met by a collier, which had our aircraft aboard in crates. We took those ashore, up over Conde Bluff and down onto the flats, and assembled them ourselves. Even though we had a ground crew with us, everybody had to work on the aircraft, despite the fact that most of us had never seen any of these aircraft before.

When the ship came back to the navy yard after the cruise to Guantánamo, we assembled out at Mitchel Field at Mineola, Long Island, New York, where we borrowed facilities from the air service. We spent that summer of

1921 flying around there to keep our hands in, then we went back to battleships in the autumn, at which time I was shifted to the *Nevada* for the same type of duty. After another cruise in the winter in the *Nevada* to Guantánamo Bay, we came back and assembled again at Norfolk Naval Air Station.

Just prior to that, the people who were interested in this business, Chevalier and Whiting particularly, had figured out that we'd better get a carrier in the navy. The collier *Jupiter* (AC-3) was sent to the navy yard at Norfolk to be converted into a carrier and named the *Langley*. Chevalier told me that I was to stay ashore at Norfolk the summer of 1921 and devise an arresting gear to stop the aircraft on the *Langley* deck. There was no provision for arresting in her original plans—in fact, nobody had figured what to do about that. So I stayed at Norfolk and worked on the arresting gear, designed it, and saw it installed. When the *Langley* went into commission in '22, I was in their commissioning detail.

There had been some previous work on arresting gear. Eugene Ely had made a landing on 18 January 1911 on the stern of the USS *Pennsylvania* (ACR-4) in San Francisco with an old pusher aircraft. They had built a long platform from her mainmast out over the stern. To stop the airplane, they hung some hooks on it and put lines across the deck with a sandbag on each end of each line. There were quite a lot of them, so that when Ely came down and landed, the hooks snagged the lines and dragged more and more of the sandbags, bringing him to rest. It seemed such an obvious way to stop anything that I started out to try and see how this thing was going to work out. I just put sandbags on wires across the roadway down there, to find out how I should design the hook for the planes that we then had.

In the meantime, the Navy Department had built a turntable on the field at Hampton Roads, about a hundred feet in diameter, flush with the ground. They mounted some gear on it that the British were then using. The British type of gear simply used some cables stretched real taut about nine inches apart and a little over a foot off the deck, running fore-and-aft on the flight deck. Our version of it had them fifteen inches off the deck. On the axle of the landing gear were hooks, like anchors, that went down between the longitudinal cables. Just the friction of the cables was supposed to stop the airplane.

It never did; the plane went up on its nose at the end of the run. The British called it a Harp or a B-gear. In their version, the wires were high enough so the planes actually rested on them and coasted along on the wires. The friction was much greater. It tore up the plane and busted the propeller in almost every landing. It was disastrous. They never put it into other than experimental use.

On the same turntable I mounted some crosswires that were going to drag along weights suspended in towers. Then it became much more of an engineering operation; you knew what the plane weighed, you knew about how fast it was coming in, so you knew its kinetic energy. You knew how much weight you had and how high the weights could go, so you knew how much potential energy they were going to have. You could balance those out and come up with a rational prediction of where the airplane was going to stop, how far it was going to run out. Actually, there was such a great loss of friction in the system that the potential energy of the weights at the end of a run was usually only a part of the kinetic energy of the aircraft as it landed. But at least you had a rational approach to the thing, whereas you never knew what you were going to get dragging the sandbags along.

I tested the designs that I came up with. I was not confident that it was going to work; every test was a question of whether it would work. We were shortsighted. We should have discarded the fore-and-aft cables at the very beginning, but we didn't, and they even put them into the *Lexington* and *Saratoga.* It was there that a naval constructor named Leslie C. Stevens said, "Let's do away with those fore-and-aft cables." They were breaking up more aircraft than they were saving, but there was a great fear of going over the side. The British had used these because they had had some very disastrous experiences in early experiments. Because they were going to have to take off from the ship, they built the platform forward. They used very small, light airplanes, Sopwith Pups and Camels. They would come in, have to make a very sharp turn in by the foremast and down onto the platform. There was one of the cruises when they lost some overboard. It was a bad place to go overboard, because the ship ran over you.

So the British were very strongly orientated toward the fore-and-aft wires, and we were too, but it began to dawn on us that these things were causing more breakups than they were worth. If you were low enough so that your axle hooks engaged the fore-and-aft wires, and your trailing hook did not engage a crosswire, then you came to the end of the fore-and-aft wires, and you went up on your nose.

There were usually two or three other people involved in my experiments at Norfolk, but I was pretty much on my own. It didn't take very long to come up with a good design, probably not more than five or six months. I had to because I understood the ship was going into commission in April 1922, and this was the start of the summer of '21. As I would get ideas, I

A Boeing F4B of VF-1B makes an inflight engagement while landing aboard the *Langley* in the early 1930s. Picking up the landing wire before the wheels are on the deck has always been something to avoid as it puts undue stress on the airframe and could result in a crash. Note a second F4B "in the groove." (Courtesy of the Naval Historical Center)

would have to go to Norfolk to tell the draftsmen over there what we'd better put in the ship. I had to work pretty fast.

The weights that were designed for the system obviously were heavy and cumbersome. They had a great advantage in that the system was self-contained. If the ship would lose its power, you could still fly airplanes after the wire had been pulled out. You had to get the wire back into battery, and the weights took care of that. I had a come-along arrangement. The wire would play out, but then you could control its coming back. The first weights were just blocks of cast iron on the towers that supported the flight deck. Later, much better-designed weights were put down in one of the holds of the ship. However, it was still a bulky and an awkward system.

The landing area was about 265 feet long. On the earliest design we graduated the weights; the top weight was the lightest, and the bottom was the heaviest. A light airplane would drag the arresting wire out aways. When the top weight had moved up a little bit, it picked up another weight, and so forth. So there was no adjustment when a different type of airplane landed on the deck. As planes became faster and larger and more diverse in their types, this became necessary.

Not the least of the problems on the *Langley* was the matter of power. She was, I guess, the first electric drive ship in the navy, and it wasn't too reliable. I remember when we left the navy yard at Norfolk, we went at about 6 knots out the channel. The maximum speed was about 13 knots, and we were lucky if we got that. But she was a very useful ship and taught us an awful lot about carrier technique.

We spent the summer, autumn, and part of the winter of 1922 shaking down, which was a very eventful, and rewarding, time since almost everything we were doing was a first-time experience. As an example, the business of using arm signals to show whether the pilot was high or low or fast or slow came about in an interesting way. We were at anchor in the York River when the *Langley* was being shaken down. The executive officer, Comdr. Kenneth Whiting—who had been largely responsible for our having a carrier—was in the netting, just below the flight deck level where the personnel go while aircraft are landing. He used to stand in the netting all the way aft on the port side. That was a good place to see what was going on.

We had one pilot who had not landed on the deck before but had had a lot of training and practice ashore. Up to that moment, it never occurred to any of us that anybody could know any more about handling the airplane than the fellow that was flying in it. It was a very parochial point of view, but it was one that all pilots had at that time. This chap came in, and apparently he was very reluctant to actually set his plane down. He kept coming in high, and then he'd give her the gun before he quite got to the deck and go around again. This had happened several times.

Whiting jumped up on the deck and grabbed the white hats from two bluejackets who were there. He held them up to indicate that this character was too high, then he put them down. He coached the fellow in, and that seemed like a good idea. So from then on, an officer was stationed aft there with flags to signal whether the plane was high or low or coming in too fast or too slow. It was a stroke of genius by Whiting, and out of that has grown the present, very sophisticated electrical signaling system.

Up to the time the *Langley* was commissioned, every naval air station had carrier pigeons that we used to take on flights. Before you started on your flight, you went over to the pigeon loft and got your little box with four pigeons in it. If you had a forced landing, of which we had quite a number, you wrote your message on the piece of paper, stuck it in the capsule that was fastened to the pigeon's leg, and let it go. It flew back to the air station, and the people there knew where you were, presumably. This had been going on for a long while in the very early days of aviation.

On the fantail of the *Langley* was a room that was the pigeon loft. The pigeon quartermaster—there was such a fellow—would let his pigeons out, one or two at a time, for exercise. They'd leave the ship and fly around, and they usually stayed in sight. Pretty soon they'd come back and land on a little platform outside the coop, the little alarm bell would ring, and the pigeon quartermaster opened the door, and in they'd go. While we were in the navy yard, after we were commissioned and before we went to shake down, the pigeon quartermaster would put the pigeons in a cage and put them on railway express and send them to Richmond or somewhere where the expressman would let them go, and they'd all come back to Norfolk.

CHAPTER 2

Carrier Aviation Comes of Age, 1939–1945

Fighting the Germans in the North Atlantic

It takes a great deal of courage and skill to fly aircraft from carriers. Control movements must be precise and measured. Anyone who accumulates several hundred "traps," as arrested landings are known in aviator slang, has amply demonstrated his supply of both attributes. To have tallied 2,407 traps in a wide variety of aircraft, props and jets, lifts an individual aviator to a higher level than the "average" carrier pilot. Commander Eric Brown holds the world's record for arrested landings. His closest rival is U.S. Navy Captain John Leenhouts, with an admirable total of 1,645.

Today, at age 80, "Winkle" Brown can look back on an extraordinary career as a naval aviator in peace and war. Arguably the world's most experienced test pilot—with due regard for such luminaries as America's Don Engen and Israel's Danny Shapira—he produced an equally extraordinary memoir. Here, he describes his experiences as a Martlet pilot fighting the notorious Focke-Wulf FW–200 Condor maritime raiders that constantly threatened the vital convoys from America as they entered the eastern Atlantic approaches to Britain in 1941.

At this early pivotal point of World War II, American supplies on the cold, water-logged lines of vulnerable ships were absolutely essential to Great Britain's remaining in the war after the victorious but draining Battle of Britain in the summer and fall of 1940. Without the massive resupply effort from the United States, the RAF's repulse of the Luftwaffe, and the resulting postponement and ultimate cancellation of the German invasion of England, would have been only a delaying action, and not the definite strategic victory that it became.

Thus, defense of the convoys became the main concern of the British in 1941 and on into 1942 as the combined prowler force of U-boats and long-range patrol aircraft like the Condor—a prewar airliner named Kurier—gave the Allies many anxious weeks. In a little reprise of the RAF's savior role nine months before, the Fleet Air Arm pilots in American-supplied F4F Wildcats (renamed Martlet in British service), offered the main airborne defense for the beleaguered convoys.

From *Wings on My Sleeve,* by Commander Eric Brown, Royal Navy

On the morning of the 19th two more Kuriers appeared and I quickly despatched one of them with the head-on attack technique which I had inadver-

Sub-Lt. Eric Brown in the cockpit of a Grumman F4F aboard the escort carrier HMS *Audacity* in 1941. (Courtesy of Eric Brown)

tently learned. Sheepy got a short burst in at the other one in a similar attack but found that by the time he had manœuvred round for his next run-in the Kurier had reached the safety of the cloud layer. She was certainly damaged, as he had seen bits fly off her in his first attack.

Soon after our return to the carrier our third and last serviceable aircraft sighted a U-boat on the port beam of the convoy. It dived and managed to avoid the escorts.

In the afternoon we had another Kurier visit. This one was sighted by *Stork*, who called up *Audacity*. Yellow Section took off at once and was directed towards the Focke-Wulf by bearings passed from *Stork* to *Audacity*. *Stork* also fired an occasional four-inch round in the direction of the enemy aircraft.

Our Martlets made a number of astern attacks without any visible effect on the heavily armoured German aircraft. Then Jimmy Sleigh got impatient and tried a head-on attack, with impressive results. The Kurier fell and as Jimmy pulled away to avoid it he hit its port wing tip, and landed back on *Audacity* with the Focke-Wulf's aileron hanging on his tail wheel. Shortly afterwards we sailed past the wreckage of my FW of the morning.

That night was unexpectedly quiet, but next morning at 1030 hours we had our routine visit from a Kurier. Red Section was off again, but we had to play a most exasperating game of hide and seek in the heavy cloud layer. This pilot was taking no chances of joining the long casualty list of his squadron, Kampfgeschwader 40.

At 1500 hours the Martlets reported two U-boats nearly ahead of the convoy. Commander Walker altered the convoy course but did not send a striking force as it was too late in the afternoon.

Again the night was uneventful. It was the calm before the next storm.

At 0910 hours on the morning of the 21st I was on my dawn patrol stint when I saw something I could hardly believe. Two U-boats were lying side by side on the surface some twenty-five miles astern of the convoy. There was a plank between them, and men were moving from one to the other.

As I approached they opened up on me with Oerlikon-type guns. Remembering Fletch's fate I circled round and climbed. I soon noticed that the guns did not seem capable of being elevated above about sixty degrees, so I got right overhead. By this time I had seen that one of the U-boats had a hole in the port bow. These were probably our two friends of the previous afternoon. This explained why they had not attacked us last night, I thought. From my position overhead I rolled over into a screeching dive and blazed away in the general direction of the two conning towers. Three men were shot off the plank, while the others vanished into the bowels of their submarines. Then slowly both U-boats made off on the surface away from the convoy.

Commander Walker got my radio report and deduced that the two U-boats had collided during the night and were transferring the whole of one crew, or perhaps only a working party, from one to the other. He detached four escorts to hunt them down, and I was ordered not to attack further but merely to keep the U-boats in sight until *Audacity* could get relief aircraft to the scene.

By 1126 hours I could see the striking force of four escorts some twelve miles from the U-boats, and then had to start back to the *Audacity* as I was getting low on petrol.

At 1130 hours I heard the other Martlet which had started the dawn patrol with me call that he could see two U-boats on the convoy's port quarter. These were probably the ones I had just left, as by now I was coming in from the port quarter also to join *Audacity*.

There was trouble getting a third Martlet serviceable to go out and keep an eye on these two U-boats, and when it did eventually get off it reported that the submarines had dived. They had probably sighted the approaching hunter force and decided to duck out of sight. Diving must have involved a grave risk for the damaged boat. It is possible that this was the U-567, a boat whose signals faded that day and were never heard again.

At 1300 hours *Stork* sighted another U-boat ten miles on the port bow, and at 1510 hours yet another, twelve miles away. Walker decided that no matter where the convoy turned it was for it, so adopted a direct course for home in the daylight, then a drastic alteration after dark.

Every pilot on *Audacity* had flown a long sortie that day, and it came the turn of Sheepy and I to do the dusk anti-submarine patrol again. This was to be a double sweep at ten miles and twenty miles range, and it took us longer than usual. By the time we got back to the ship it was getting quite dark. We had seen no U-boats. All heads were down for action.

Sheepy came in for his landing. The ship was rolling in a steady cross swell. I could see that they were using two hand torches to bat him in. He made it, and I came in hard on his heels. The ship fired two Very lights, a warning for me to go round again, as she was rolling too violently. She altered course slightly and steadied up. I watched those two twinkling eyes of light with grim concentration as Pat guided me in. I didn't want another wave-off, as it was getting darker every minute, and we had no deck lighting.

Finally, at 1920 hours, I was aboard, and the whole company wheeled into a turn to alter course. *Audacity* normally stayed in the middle of the convoy at night, but now she went off alone on the starboard side, zigzagging at her full speed of fourteen knots. The Captain was afraid that if the U-boats made a dead set at the *Audacity*, snug in the centre of the convoy, they would be very likely to hit the merchant ships around us. He did not have Walker's approval for this move, but as he was actually senior to the escort commander he could do as he liked with his ship.

Walker ordered four detached escorts to stage a mock battle astern of the convoy with depth charges and starshell to draw the U-boats away from the real position of the convoy. The plan failed because some of our merchant ships, believing it was the genuine thing, fired snowflake and lit us up.

At 2033 hours the balloon went up. The rear ship of the centre column was torpedoed and another shower of snowflake was fired. This illumination revealed to Oberleutnant Bigalk on the conning tower of U-751 a sight that made his heart leap. An aircraft carrier was passing him within torpedo range.

At 2037 hours I was sitting in the wardroom finishing my after-dinner coffee when a torpedo hit us aft right under the point where our mechanics were still working on the Martlets. The tremendous explosion and shock from aft spilled my coffee on my lap. We all rushed on deck.

The ship was down by the stern with half the gun platform awash and the gun useless. We were still under way, but the rudder was damaged and we could not steer. The Captain was afraid of ramming a merchantman or escort, so he ordered the engines stopped. There we lay, dead in the water, hoping that we would last till morning, then get a tow. We were about ten miles from the convoy.

Meanwhile U-751 was closing us. She stopped some little distance from us, and the Germans, for whom this was just too good to be true, worked frantically to reload their torpedo tubes.

Some twenty-five minutes after the first hit we saw the U-boat about 200 yards away from the ship on our port beam. She was covered in phosphorescence, glowing eerily.

For about five minutes nothing happened. We simply regarded each other across the water. Then one of our seamen leapt for an Oerlikon and began firing at the U-boat.

It must have been at about the same time that Bigalk's tubes were finally brought to the ready. Our lone gunner had only got off about two rounds when we all saw plainly the white bubbling tracks of two more torpedoes coming at us. Everyone rushed for the starboard side of the flight deck.

Then they struck, both in the bows. There was a tremendous explosion, probably of aviation petrol, and the whole of the for'ard quarter of the ship disappeared. The rest reared up in split seconds so steeply that we could not keep our feet. We were all scrambling about amongst the aircraft lashed down aft. Above the other noises I heard the frightening sound of the wire lashings whining under the impossible strain. . . .

They seemed to part together with a great twang. The Martlets plunged down the wildly tilting deck as if in formation. There was a jarring, broken crash as they hit each other, then splayed out over the deck. The cries and screams of men being mowed down by the monsters mingled with warning shouts. Many leapt off the flight deck into the sea, a long jump from the high-angled deck, and were badly hurt on impact.

I jumped down the ladder on to the catwalk which surrounded the flight deck, then down on to the promenade deck outside our cabins. From there the only step was overboard.

It was about twenty feet to the sea. Going over I was frightened of landing on someone in the water—the sea seemed thick with heads. When I landed I was scared of someone else jumping down on me.

I had on my leather Irvine flying jacket and my flying boots over my full uniform. Stuck into the opening of the jacket were the two things I had given priority in the scramble to salvage something before the second torpedo had hit—my log book and a pair of silk pyjamas I was bringing back for my fiancée. I tried to blow up my Mae West and found my flying boots dragging me down. I kicked them off. I tried to swim away but found the log book restricting my movements. Reluctantly I let it go. Then I struck out away from the ship, frightened of being jumped on or sucked under by the ship.

At a safe distance I turned round. The ship was tilted up now at a terrifying angle, her single propeller right out of the water. Suddenly she lurched farther up until she was almost vertical, then very quickly was gone. There was no suction, but a great series of explosions cracked through the water like depth charges.

I swam off, wondering if the U-boat was going to search for prisoners. The familiar voice of Leading Seaman Budge called out, 'Sir, we've got a dinghy here. We can't get into it but we can hang on.' I swam over and joined him and three other ratings clinging to the rubber dinghy. The water did not strike me as very cold, and we were all in that high-spirited mood of escape from death.

Then I heard Sheepy Lamb's cheerful voice call out from somewhere, 'Hi! You all right, Winkle?'

'Yes, I am.'

'Is your Mae West working?'

'Yes, fine.'

'Well, come on with me. We'll make much better progress.'

So I left the dinghy and the two of us struck purposefully out as if we had somewhere to go. We did manage to get well clear of the wreckage, in case the U-boat came round looking for likely evidence of success.

Presently we watched a corvette steam up and begin to pick up survivors out of the water. We struck out towards her and were within hailing distance when she suddenly turned about and made off again. She must have picked up a ping and gone off on a hunt.

To the best of my knowledge it was about three hours before she came back—it seemed more like twenty-three. There had been about twenty of us around in the sea earlier in the night, but only the two of us had survived the three hours, probably because of our pilot-type life jackets.

We yelled, and this time they took us aboard. She came as close as she could to us and lifebelts on the ends of heaving lines came plopping down from her decks. We were supposed to climb that wet heaving cliff. My legs were too weak and I got badly skinned going up.

We were taken below to the ship's tiny wardroom, where there were already some of our men lying about asleep. They had all been massaged and dosed with rum, and there was a terrible stench of the stuff everywhere. They poured a lot of it down us as well, and we began to come to life a little. The C.O. was there, awake, waiting to see which of his boys would come in. He managed a weary, 'Glad you're okay, Winkle.' Leading Seaman Budge was there as well. He said, 'You should have stayed with me, sir. You'd have got here earlier.' I thumbed at Sheepy. 'I may have to follow him in the air still, but this is the last time I let him lead me in the water.'

Next morning our ship, the *Convolvulus*, transferred us to the *Deptford.* Here we learned something of the aftermath of the *Audacity* sinking. Three corvettes had rushed to assist the carrier. Half an hour later a U-boat was sighted by *Deptford.* In the resulting confusion of the chase *Deptford* rammed *Stork*, her bows cutting into *Stork's* quarterdeck and killing two of the five German U-boat prisoners in *Stork.* The collision reduced both ships to a maximum speed of ten knots and disabled their asdic sets. The three corvettes sent to help *Audacity* now had to take over the asdic hunt for the contact and leave her.

Next morning a Liberator appeared and gave anti-submarine cover to the convoy for two and a half hours. This did not deter our usual morning Kurier, which came on the scene at 1115 hours. Then the relief Liberator spotted two U-boats on the surface at 1600 hours, both of which at once submerged.

That night Walker ordered a mock battle by *Deptford* and *Jonquil*, and altered the convoy course. I was aboard *Deptford* with the other *Audacity* pilots by then. We did not enjoy this invitation to lurking U-boats to have a go at us.

The *Deptford* made Liverpool, however, with no further trouble. I was to be married when we got home, and I reflected sadly that my first chosen best man, poor Fletch, was dead, and my second, Pat, missing.

Later we heard that the Captain was dead. The *Penstemon* had sighted him swimming exhaustedly in the gathering sea. Her first lieutenant dived in from the whaler which was going to pick him up and managed to get a lifebuoy on his limp body. He was being hauled in when a big sea jerked the lifeline into the water and Commander McKendrick drifted away out of reach. He was a great loss, a man of very high principles and deep understanding, who had managed to turn the strange hybrid *Empire Audacity* into a brave and useful little ship, a pattern and example for many new ships of her type which were to come.

Her epitaph was written by Grand-Admiral Doenitz himself. 'In HG 76,' he said, 'the worst feature from our point of view was the presence of the aircraft carrier *Audacity*. The year 1941 came to an end in an atmosphere of worry and anxiety for U-boat Command.'

"The Famous Victory": Swordfish at Taranto

It's interesting to consider that of the three main Axis partners—Germany, Japan, and Italy—only Japan kept her Navy in the fight for the entire war. While possessing fine examples of the naval architect's art, Germany's Kriegsmarine was pretty well out of the war by 1942, with the exception of the U-boat force. And even that potent weapon had begun a major decline by late 1942. Italy was never a serious naval threat; however, the potential was always there, especially in the early years of the conflict. With a 7,600-km (4,500-mile) coastline, the peninsular nation boasted many fine harbors, not the least of which was the important naval anchorage at Taranto, at the top of the Italian boot's heel.

Early on, an attack on Taranto was high on the Royal Navy's list. The carrier HMS *Illustrious*, with a squadron of Swordfish torpedo planes with special range-extending fuel tanks, was the primary attack ship, in company with another carrier, HMS *Eagle*. Intelligence placed a growing fleet of Italian battleships at Taranto, an increasing threat in the Mediterranean that could not long be ignored.

A raid was scheduled for October 21, 1940, but a fire on *Illustrious*, caused by a wayward spark that ignited gasoline dripping from a Swordfish's auxiliary fuel tanks, postponed the attack at first for 10 days, then until November 11. In the interim, replacement Swordfish and aircrews were brought aboard.

An early morning launch of a reconnaissance Fulmar, a two-place, single-engine fighter then in ragged use in the fleet, brought news of a sixth battleship at Taranto. It was a target too rich to ignore: all six Italian battleships were in one place, just right for attack. At 8:40 pm, 170 miles from

Taranto, *Illustrious* launched her 12 Swordfish, followed 50 minutes later by a second wave of 9 aircraft. The 21 raiders flew with varying combinations of torpedos, bombs, and flares.

One of the junior pilots in the second wave was Lieutenant (A) John A. G. Wellham. Flying behind him in the open-cockpit biplane was Lieutenant Pat Humphreys. Not many men would be able to tell their grandchildren about their participation in what came to be known as "the famous victory," but John Welham wrote about it as well.

A note of interest concerns Wellham's rank, "Lieutenant (A)." The "(A)" indicates assignment to the Royal Navy's Air Branch, normally an officer pilot or observer. The prewar arrangement between the Royal Air Force and the Royal Navy regarding naval aviation units could probably fill a book of explanation. Typical of military establishments, the two services were constantly at odds in various areas of mutual concern. Originally, the RAF had responsibility for naval aircraft, and this group of aircraft and people that flew from Navy carriers actually belonged to the RAF, and was loosely referred to as the Fleet Air Arm.

As John Kilbracken wrote in his wartime memoir, *Bring Back My Stringbag*, an excerpt of which appears in an upcoming chapter, "The RAF had neglected naval aviation: apart from being under strength, the FAA was equipped, as it would always be, with aircraft no self-respecting pilot in the Air Force proper would consider." When the Navy assumed control of its own aviation in 1938–1939, many of the former RAF personnel transferred to the Navy with short-term commissions. Kilbracken continues, "Many RN officers certainly held there had been no FAA since control of its aircraft and aircrews had been handed back to the Navy. We were sometimes described as the Air Branch of the RN, or the Branch for short . . . Right up to the end of the Second World War, there was still an occasional RAF sergeant or aircraftsman to be seen in the squadrons, probably forgotten through some administrative oversight . . ."

From *With Naval Wings: The Autobiography of a Fleet Air Arm Pilot in World War II*, by John Wellham

At 2115 we were told to man our aircraft. I climbed on to the flight deck to find E5H; my faithful E5B was languishing at Dekheila. I climbed in, settling myself on the parachute. My fitter and rigger strapped me in, gave me a pat on the helmet and said 'Good luck, Sir, see you in the morning.' I hoped that they would. Pat Humphreys struggled into his cramped cockpit behind me. In a few minutes the FDO's illuminated wand started to circle with the signal to start up our engines. The crew wound the inertia starter: as the revs. were building I set the throttle, then knocked on the two magneto switches as the clutch was engaged. With a cough, splutter and a cloud of exhaust the engine fired and I caught it on the throttle, setting it for warming up. I checked the engine instruments, which all read what they should, then ran the engine to full throttle, switching off and on each magneto switch. The old Pegasus was running perfectly so I reduced to a tick-over and waited.

I could feel the ship building up speed and turning slightly into the fitful wind. I heard the leader's aircraft open up and roar down the deck. The air-

This Fairey Swordfish Mk. 1 is on a training sortie early in the war. Only the pilot seems to be on board as both rear cockpits appear unoccupied. The aircraft carries an 18-inch training torpedo on centerline as well as underwing racks for bombs and flares. The big, slow, lightly armed "Stringbag" could barely make 100 knots, but fought throughout the war from Taranto to D-Day, and beyond. Note the canted, in-flight angle of the main landing gear wheels. (Courtesy of the Royal Navy)

craft handlers were doing their usual hair-raising act of ducking round the whirling propellors to pull chocks away. When my turn came I followed the FDO's signals to taxi into the centre line of the deck, held on the brakes while the wings were spread and locked, then, opening the throttle fully and easing in the boost override, I let the brakes go and started my take-off run. With the ship's speed of 30kts, air speed rose rapidly; as we passed the island a quick glance at the ASI showed it rising to 70kts. We rose smoothly into the air and I climbed away in a gentle turn to port and reduced the revs. to normal climbing power. I looked above for the other aircraft and Pat and I saw them at the same time. I slid into my slot in the formation as we continued to circle the ship. I asked Pat if he knew the reason for the delay. He said that he could only count seven of us, then corrected this to eight. As we passed again through north the leader, Lieut-Cdr. Hale, straightened up and set course to the north west. Later we found that L5F and L5Q had started to taxi at the same time and had collided; the latter had not been damaged and had taken off to join us but L5F with Lieuts. Clifford and Going had suffered broken ribs and torn fabric and had to be struck down for repair. Herculean efforts by the riggers had made it serviceable in a quarter of an hour and the crew, who had begged permission to be allowed to go on their own, took off only 24 minutes after us.

We gradually climbed to 8,000ft, passing through a layer of filmy cloud. Pat told me that we now seemed to be another aircraft short. We did not know at the time, but L5Q had lost its long range tank which, as she was one of the bombers, had been strapped to her torpedo rack. The engine had cut, but her pilot, Lieut. Morford, had managed to restart it. He now had insufficient fuel to fly to Taranto and back so had returned to *Illustrious* where, not expecting a friendly aircraft, they had opened fire on him, stopping, luckily, before he was hit.

So now we were only seven. It was a beautiful picture-postcard evening: there were only a few wisps of cloud below us, otherwise the sky was clear, and littered with a blaze of stars; to the south a three-quarter moon was throwing a golden pathway across the calm sea; the air was smooth giving hardly a judder. It would have been the most perfect evening to enjoy flying, had it not been for the reason for our flight.

It had become quite cold, but that could be expected in an open cockpit at 8,000ft. I asked Pat if he was comfortable and he replied: 'As far as might be expected'. I didn't envy him, jammed into the aftermost cockpit with vapour from the tank wafting around him while I was in reasonable comfort sitting on my parachute cushion, with an acceptable amount of space and surrounded by the familiar instruments.

Once again I found myself with very little to do. My old trick of putting my mind into cold storage didn't work this time. It seemed like riding through Egypt on a camel while pretending that the pyramids didn't exist.

After flying for more than an hour I noticed that the dark blue fabric of the horizon was torn by a patch of light. I pointed it out to Pat, who looked through his binoculars and said that it must be Taranto, but neither of us knew why it appeared to be flood-lit. As we closed the land, the light seemed to flicker and pulse until, when we were closer still, it began to look like a major firework display. With some horror I realised what it was: AA fire. What were they shooting at? The first strike should have been well clear long ago.

The loom of the land began to clarify and our two remaining flare droppers broke away to our right, heading for their zone over the oil storage depot. There was a healthy fire burning over there, so someone in the first strike must have achieved a hit.

We were still some ten miles from the harbour but, at a height of 8,000 feet, it was becoming clearly delineated in the bright moonlight and the glare from the tracer. Although partly obscured by smoke and gunfire it was a copy of the excellent photos given to us by the Marylands. Hale altered course slightly to port and we could see, to our right, the breakwater, Diga Di San Vito, and dead ahead the little Isoletto San Paolo with the larger island, San Pietro, just to the left of it. We began to open out the formation and slide into a well spaced line astern.

Our plan was to pass around and to seaward of the submerged breakwater, then to turn towards the east after passing Cape Rondinella, cross the land, then dive down behind the balloon barrage, turning south as we came over the harbour so that the battleships would be broadside on to us and actually overlap so that, if we missed our chosen targets, there would be a chance of hitting the next in line. It was a good plan on paper.

The northern shore of the basin was lit by flashes from gun batteries there and, now falling behind us, the island of San Pietro was spurting flames.

I could imagine Pat at that moment; he would be stowing his chart board and navigation gear in some relatively safe place, and checking that his 'G-string' was firmly secured and that his parachute pack was easily accessible, although there would not be much chance to use it during a torpedo attack. His usual calm voice came down the Gosport tubes: 'The course for *Illustrious*, incidentally, should be about 135°. You might like to set it now.'

He was being very thoughtful; if he were to be knocked out while I was still in one piece, I might be able to find my own way home.

We passed Cape Rondinella, starting to lose height and turn in over the land. The harbour was partly obscured by smoke from the guns and the burning oil depot, and also from another blaze to the north—the seaplane base or a crashed Swordfish? The ground was clear below in the moonlight. I could see streets of houses like a town plan with open spaces of parks or playing fields. I hoped that the residents were all in air raid shelters, as spent bullets and shrapnel must be raining down like lethal hail stones.

I followed the leader as he gradually lost height. Suddenly there was a burst of light to the eastward as the first flare ignited, followed by others until they hung in the sky like a necklace of sparkling diamonds. This seemed to drive the Italians to even greater fury—the flak doubled in intensity and the curtain of barrage below us now rose into a cone like a feathered head-dress: above us high-angle AA was bursting in crackling puffs of smoke. If the tracer was one in five there must be more metal than air. 'My God! No one can fly through that—shades of Balaclava!'; in the increasing chaos I lost sight of the other aircraft; no matter, a coordinated torpedo attack was not so important, as the targets were stationary; we must simply get amidst the battleships and do our own thing.

Ahead there seemed to be a partial hole in the flak just where I wanted to be—I aimed for it calling to Pat: 'Hang on, I'm going down.' 'OK. Do your worst. Good luck.'

I pushed the nose down, easing back the throttle to avoid over-revving the engine. The speed built up—140kts, 150kts, 155kts—I wanted to dive as steeply as possible, knowing that a gentle angle would only give me more time in the barrage. We were in it—the familiar red, green and yellow lines of tracer were crawling up towards us then hurtling past; ahead they appeared as a tangle of colour. The slip-stream was screaming through the struts and bracing-wires, and past my ears; my nose was filled with the stench of cordite; there was tracer above us, tracer below us and tracer seemingly passing between the wings. The dive was steepening and the speed building up—160kts, 170kts—we met a barrage balloon! No self-respecting balloon should have been at that height; its cable must have been shot away. I hauled the stick over to the left—I missed it. There was a tremendous jar, the whole aircraft juddered and the stick flew out of my hand. 'Christ! I've hit the balloon cable'—but the wings were still there. I grabbed the stick—it wouldn't move—we were complete out of control. It was no time for finesse. I applied brute force and ignorance. It moved most of its travel to the right but only partially to the left—was it working the ailerons?—No idea! I looked ahead. 'Bloody Hell!' We were diving almost vertically into the centre of the City of Taranto! I hauled the stick back into my stomach—were the elevators working? They were: an elephant seemed to be sitting in my lap but slowly we began to level out, but still curving round to the right. Were we going to make it? Buildings, cranes and factory chimneys were streaking past below us then we shot over the eastern shore of the harbour and were level over a black mirror speckled with the reflection of flames and bursting shells. I stirred the stick around and found that I had, at least, some sloppy lateral control. Air speed? Far too fast to ditch if we had to and too fast to drop a torpedo. I was determined to aim it at something after carrying the bloody thing all that way and having a rather hairy dive—I'd be damned if I didn't do something with it.

A quick glance around: to my right and slightly behind me was a massive black object covering most of the horizon and having a vast castle towering above it—a battleship. I heaved the stick over to the right putting us into a near vertical turn towards the target. I thought: 'That was a damned stupid thing to do; she might not go back.' She did. I levelled out after turning 180° and pointing towards the great black hulk of the ship. Height OK, judging from the level of her deck—air speed dropping nicely—angle of attack not ideal but the best that I could do—aircraft attitude for dropping a torpedo rotten. The only way that I could achieve a straight line was skidding with some left rudder and the right wing slightly down. Torpedoes don't like being dropped when not perfectly level. There was surprisingly little flak around us. I was forced to revise this opinion. She was awake and had seen us. Strings of lights prickled along her decks and multiple bridges and grew into long, coloured pencil lines drawn across the dark sky above us. She was giving us everything except her 15-inch guns but, thankfully, she seemed unable to depress her other guns low enough to hit us. Closer and closer we came, her decks ablaze with muzzle flashes, the superstructure towering above us. 'Look out! Don't get too close, these things have a safety range.' I pressed the button on my throttle lever, felt the torpedo release, held straight for a couple of seconds then threw the stick over into a vertical turn to starboard.

Inevitably, after dropping a nearly 2,000lb load an aircraft rises and E5H was no exception. We rose right into the ship's gunfire. I fought the sloppy controls to force her down, crying: 'Fly, you bitch'—poor thing, she was doing her best. There was another jar and shudder—we had been hit again—'Hell's teeth! Leave me something to take home.' Finally, I managed to push her down to skim the glassy surface, opened the throttle wide and knocked in the override. 'Careful; don't hit the water.' It's difficult to judge height over a smooth surface, particularly at night.

Ahead was the Diga Di Tarantola with its balloons. I must avoid that if I could but mustn't get too far to the right as there were more balloons over there. I edged over to pass slightly to the right of the Diga which, I hoped, would take me clear of balloons on each side. I could see San Pietro about three miles ahead; its batteries and those on the floating pontoons across the harbour entrance were still firing but were aiming towards the centre of the basin. Tracer from the battleship was passing over from astern of us and disappearing ahead. Scraping the surface of the sea we shot past the island into the wonderful, welcoming anonymity of the darkness. We were in clear air and still flying, although hardly in the approved flying school manner. Would she keep on flying long enough to get home?

I eased the nose gently upwards, took out the boost override and adjusted the throttle to give me normal climbing revs and boost. She started to climb. I moved the stick gently in all directions and applied a little rudder in both directions. Everything seemed to be working, at least to some extent but I could only keep a straight course with the starboard wing a few degrees down and by applying some left rudder. She was behaving much the same as she had when I was trying to attack the ship. No well behaved Swordfish should fly like that but, if she was prepared to fly at all, I would not criticise too much.

I swept my eyes over the dashboard. The engine instruments were showing no problems; revs steady; boost OK; oil pressure correct; oil temperature a little high but that was reasonable with the way that I had been treating the poor engine. Flight instruments were all over the place showing wing down, skidding and a general lack of keenness for the aircraft attitude. This, at least, showed that they were working.

It felt as though we had been over Taranto for hours, but in fact it could not have been more than a few minutes. Since Pat was always the perfect observer who never interrupted when I was involved with things that needed my full attention, I had not worried about his silence but I now felt a bit concerned, as I hadn't heard a word since we had left the coast. I lifted the Gosport tube and called rather tentatively:

'Are you alright, Pat?'

There was a sound like the heavy breathing at the beginning of an obscene telephone call, then:

'Yes—physically. What is your condition?'

I assured him that I seemed to be functioning normally and he said:

'Oh, that's good. I thought that you might be damaged as you have bent the aeroplane a bit'.

'It wasn't my fault. It was the bloody Eyeties.'

He assured me that he had not intended to imply any criticism of my flying which he felt must have been fairly competent, as I had got in and out and we were still flying, and he had had some doubts about that happening when we had entered the barrage. I explained what had happened during the dive and the attack. This caused silence for few moments: he had not appreciated that we had been completely out of control and nearly in a nasty mess in the centre of Taranto, nor that I had had only partial control during the attack. He had not noticed, either, that we were not able to fly straight even now. Eventually he said:

'I see. I suppose we should consider ourselves fortunate. Do you think that you can get what's left of this machine back to *Illustrious?*'

I replied that there was a fair chance and I was going to try, as I did not fancy eating spaghetti for the rest of the war. He quite agreed and said he left it entirely to me. I suggested that we would be more likely to find the fleet if he would give me a course to steer, and his reply was that I would have to wait a few minutes as all his navigation gear was somewhere in the bottom of the cockpit; meanwhile, I should steer roughly south east. I was already pointing more or less in that direction so left him to do some hunting around the floor.

After a remarkably short time he gave me a course. I set it on my compass but was not entirely happy about it, as magnetic compasses have a dislike of being skidded sideways. We compared it with the two observers' bearing compasses in the back, achieved a reasonable average and set it on my gyro. I had levelled out at 2,000 feet, which I felt was high enough since I had no wish to suffer from the cold as we had on the way in. The engine instruments were still reading healthy levels but the flight instruments were all over the place, which was not surprising in the attitude in which I was forced to fly. At a greater height I could have saved fuel and made more use of the mixture control but as, by my calculations, we had sufficient petrol for at least three hours, I felt that it should be enough to find the fleet or a piece of dry land.

Having done everything we could for our own preservation, we had time to discuss the attack. I had been rather too busy when over Taranto to think about things other than my own problems, so hoped that Pat could tell me something. He could not clearly tell what had happened to the battleships because with torpex torpedoes which explode underneath the target, hits are not immediately clear; he had, though, seen an explosion alongside one of the ships, the *Littorio* he thought, which had looked like a torpedo exploding on contact. He said that it had looked to be down by the bows. Another which he could not identify had seemed very low in the water. He had recognised our ship, which he said had been *Vittoria Veneto*, but I had to admit that I was doubtful of having hit it as our aircraft had been virtually out of control when I dropped. He was able to confirm that the seaplane hangars and the oil depot had been blazing. It seemed that we had done some damage.

I asked him about the health of our radio. He told me it was thriving and he could hear a great deal of chatter on the Italian naval channel, mostly in plain language; the word 'Taranto' kept cropping up but he couldn't understand the rest of it. He was able to get Radio Milan, which was playing gramophone records of Verdi operas. As my helmet was only fitted with Gosport tubes, I was not able to hear any of it. I just shut up and let him amuse himself.

My mind began to dwell on how I was to land this thing on deck—assuming that we found the ship—when there was no way that I could keep her level; furthermore, I had no idea whether we still had any wheels, or even undercarriage. I thought that I might be able to level out and kick her straight as I cut the engine then, even if she fell to pieces at that point I might, with any luck, manage to slither into the barrier.

When we had been flying for well over an hour since our hurried departure from Taranto, I began to feel that we should know something shortly. I was suffering from a very stiff and rather painful left leg from having to hold on left rudder all the time. I asked Pat if he had any idea of our ETA. In an incredibly calm and confident voice he replied: 'Oh Yes. About 25 minutes. I was just going to tell you that, I have picked up the ship's beacon.' My spirits rose about 200%. I asked for the change of course. He told me to carry on as I was. It took a few moments to accept this. We had been flying over the sea, at night, for well over 100 miles; there had been nothing with which to check dead reckoning navigation other than, perhaps, a back bearing of Taranto in the early stages, but he had brought us back to a moving fleet without a single alteration of course. I had always known that he was a good navigator but this was phenomenal. I didn't hurt his feelings by suggesting that there might have been an element of luck.

Fifteen minutes later I saw the foaming, phosphorescent swirl of a destroyer's wake as we passed over the screen then, directly ahead, the bulk of *Illustrious* loomed and became a familiar shape in the bright moonlight. I switched on the navigation lights and then, behind my left ear, came the flicker of Pat's Aldis lamp as he gave the recognition signal. A pinpoint of light from the ship's island acknowledged. I eased the throttle slightly back to lose height, then, as we passed over her flight deck, banked as far as possible into a left-hand turn to make a very wide circuit. Poor old E5H refused to turn to the left any more steeply. I pulled the little lever that should release the arrestor hook and hoped that it had worked. We were losing height at the desired rate and my gentle circuit seemed about right to bring me in line with the flight deck. As I turned on to the final approach the lines of lights delineating the landing area and the DLCO's illuminated bats became clear. Good. I eased back the speed to that approved for deck landing a Swordfish. The damned thing was immediately out of control. I banged the throttle open and at once achieved the old skidding but controllable attitude. The DLCO was giving me furious 'Too Fast' signals: hard luck. Fortunately, he was very experienced and good at his job, realised that there must be some reason for my wild progress and gave me a very early 'Cut' signal as I slid up to the round-down. We hurtled over the first arrestor wires, missing them all; the left wing started to drop—here we go—a terrific jerk—we had caught a wire while still airborne—a resounding thud as the wheels hit the deck—then we were stopped. I couldn't believe it: we were home with both of us unhurt. I sat like an idiot holding the brakes on, then suddenly woke up to the furious 'Come On' signals from the FDO. Releasing the brakes I taxied forward on to the lift, where the handlers instantly folded the wings while I cut the ignition switches. The faithful Pegasus gave a final splutter and cough, then subsided into glorious silence.

I said to Pat: 'Sorry about the landing.' His reply was: 'I thought it was quite good in the circumstances.' Praise indeed.

The lift dropped to hangar level and we were rapidly pushed into the brilliant lights. I was absolutely astonished at the scene. The hangar was full of Swordfish! Nearly everyone must have returned safely. I had expected that we would have been one of the few to survive. Pulling off my helmet, I heard cries from the fitters and riggers: 'Fookenell, mate; look at 'im!' 'Look at that ruddy wing!' See them bleedin' ailerons!' I followed their eyes: the rod connecting the ailerons on the port upper and lower wings was smashed with the jagged ends grinding together, resulting in one aileron being slightly up and the other slightly down—not surprising that I had suffered a loss of lateral control: the port, lower main plane had a hole about a yard long by half a yard wide. How on earth could any aircraft fly in that state? I did not think that anything but a Swordfish could have done it. At that point I would have happily subscribed towards a statue to the designer.

Japan Attacks Pearl Harbor

As it did during most of World War I, protected by huge oceans on its east and west coasts, and by equally huge land masses to its north and south, the United States looked on as the European nations committed themselves to another bloody conflict. It was plain, however, that for the most part, America's sympathies lay largely with the countries arrayed against the Axis powers of Germany, Japan, and Italy. When France surrendered in June 1940, Britain became the recipient of huge quantities of American aid and supplies, not to mention the less tangible but vital emotional support of the American population. After it was attacked by its erstwhile ally, Germany, in June 1941, the Soviet Union also helped itself to the cornucopia of America's good will.

However, the focus of world attention was clearly on the war in Europe. Feeling the pinch of restrictions and sanctions instituted by the Roosevelt administration in Washington, Japan found itself running out of metal and petroleum resources. Without these assets, of course, the government in Tokyo would be unable to continue its war in China and the rest of Asia. By 1941, Japan's situation had become desperate. There seemed little left but to drag the United States into the war and try to kick the Americans out of Asia once and for all. But how to strike the first blow to major advantage?

Then, the British Navy attacked Taranto. No one watched the events half a world away with more interest than the Japanese admirals, particularly Admiral Isoroku Yamamoto, arguably his country's greatest, most charismatic naval leader. Commander Minoru Genda, one of Japan's most respected young naval aviation commanders, had written a study on the possibility of an attack on Pearl Harbor for his superior, Rear Admiral Takijiro Onishi. Supported by Genda's proposal, Yamamoto created a plan to strike at the heart of the American Navy's Pacific fleet at Pearl Harbor, Hawaii.

Many American admirals had never wanted to move a major portion of the Pacific fleet so far west, to such an unconnected outpost. But Franklin Roosevelt would not be denied, and the fleet transferred to Pearl Harbor in the spring of 1940. By December 1941, most of the fleet's heavy warship contingent, including its aircraft carriers, along with many of the Army Air Forces' bomber and fighter units, was based in Hawaii and the Philippines. Thus, America's western defenses were often displayed, nested neatly at piers and anchorages, or lined up in ranks on flight ramps, ripe for a well-planned and well-executed first strike.

On November 25, 1941, after months of training, the Imperial Navy's Hawaii Operation task force, with six aircraft carriers, struck out from Japan and headed east. At dawn, December 7, it was only 220 miles north of Oahu, in perfect position for a two-wave attack on a quiet Sunday morning.

Walter Lord earned a well-deserved reputation as a historical reporter with A Night to Remember, *a description of the sinking of the ocean liner* Titanic. *His book on the attack on Pearl Harbor was the first such effort on the Japanese operation. Published in serial form in* Life Magazine, *complete with page-size illustrations by well-known illustrator Robert McCall,* Day of Infamy *became a best-seller, and until Gordon W. Prange's* At Dawn We Slept: The Untold Story of Pearl Harbor *(Penguin Books, USA, Inc., New York, 1982), it remained the best account of the events on December 7, 1941. Lord presented both sides of the operation, giving a humanity to the Japanese that had, understandably, been lacking so soon after the terrible fighting. Indeed, in 1957, barely 12 years after war's end, many of the participants were alive and active and had not yet begun to consider the Japanese side of things.*

With the recent interest in Pearl Harbor, largely fostered by the incredibly bad and often inaccurate, but nonetheless visually impressive, movie by the Walt Disney studios in late 2000, Day of Infamy *was reissued by a subsidiary of Henry Holt. Walter Lord's seminal work can be enjoyed by a new generation of readers.*

Walter Lord wrote several works of history, including A Night to Remember. *During World War II, he served with the Office of Strategic Services (OSS) in London. He died May 2002.*

From *Day of Infamy*, by Walter Lord

Lieutenant Harauo Takeda, 30-year-old flight officer on the cruiser *Tone*, was a disappointed, worried man as the Japanese striking force hurtled southward, now less than 250 miles from Oahu.

He was disappointed because last-minute orders kept him from piloting the *Tone*'s seaplane, which was to take off at 5:30 A.M., joining the *Chikuma*'s plane in a final reconnaissance of the U.S. fleet. And he was worried because—as the man in charge of launching these planes—he feared that they would somehow collide while taking off. True, the two ships were some eight miles apart, but it was still pitch black. Besides, when the stakes are so high, a man almost looks for things to worry about.

Nothing went wrong. The planes shot safely from their catapults and winged off into the dark—two small harbingers of the great armada that would follow. Admiral Nagumo planned to hit Pearl Harbor with 353 planes in two mighty waves. The first was to go at 6:00 A.M.—40 torpedo planes . . . 51 dive bombers . . . 49 horizontal bombers . . . 43 fighters to provide cover. The second at 7:15 A.M.—80 dive bombers . . . 54 high-level bombers . . . 36 more fighters. This would still leave 39 planes to guard the task force in case the Americans struck back.

By now the men on the carriers were making their final preparations. The deck crews—up an hour before the pilots—checked the planes in their hangars, then brought them up to the flight decks. Motors sputtered and roared as the mechanics tuned up the engines. On the *Hiryu*, Commander Amagai carefully removed the pieces of paper he had slipped into each plane's wireless transmitter to keep it from being set off by accident.

Down below, the pilots were pulling on their clean underwear and freshly pressed uniforms. Several wore the traditional *hashamaki* headbands. Little groups gathered around the portable Shinto shrines that were standard equipment on every Japanese warship. There they drank jiggers of *sake* and prayed for their success.

Assembling for breakfast, they found a special treat. Instead of the usual salted pike-mackerel and rice mixed with barley, today they ate *sekihan*. This Japanese dish of rice boiled with tiny red beans was reserved for only the most ceremonial occasions. Next, they picked up some simple rations for the trip—a sort of box lunch that included the usual rice balls and pickled plums, emergency rations of chocolate, hardtack, and special pills to keep them alert.

Now to the flight operations rooms for final briefing. On the *Akagi* Commander Mitsuo Fuchida, leader of the attacking planes, sought out Admiral Nagumo: "I am ready for the mission."

"I have every confidence in you," the admiral answered, grasping Fuchida's hand.

On every carrier the scene was the same: the dimly lit briefing room; the pilots crowding in and spilling out into the corridor; the blackboard revised to show ship positions at Pearl Harbor as of 10:30 A.M., December 6. Time for one last look at the enemy line-up; one last run-down on the charts and maps. Then the latest data on wind direction and velocity, some up-to-the-minute calculations on distance and flying time to Hawaii and back. Next a stern edict: no one except Commander Fuchida was to touch his radio until the attack began. Finally, brief pep talks by the flight officers, the skippers, and, on the *Akagi*, by Admiral Nagumo himself.

A bright dawn swept the sky as the men emerged, some wearing small

A6M2 Zeros aboard the carrier *Akagi* on December 6, 1941. The "padding" on the island is actually the crew's rolled hammocks, a practice derived from the days of the British Royal Navy's sailing days. The hammocks and their thin mattresses afforded a measure of protection against enemy fire. (Courtesy of Robert C. Mikesh)

briefing boards slung around their necks. One by one they climbed to the cockpits, waving good-by—27-year-old Ippei Goto of the *Kaga*, in his brand-new ensign's uniform . . . quiet Fusata Iida of the *Soryu*, who was so crazy about baseball . . . artistic Mimori Suzuki of the *Akagi*, whose Caucasian looks invited rough teasing about his "mixed blood." When it was Lieutenant Haita Matsumura's turn, he suddenly whipped off the gauze mask which had marked him as such a hypochondriac. All along, he had been secretly growing a beautiful mustache.

Commander Fuchida headed for the flight leader's plane, designated by a red and yellow stripe around the tail. As he swung aboard, the crew chief handed him a special *hashamaki* headband: "This is a present from the maintenance crews. May I ask that you take it along to Pearl Harbor?"

In the *Agaki*'s engine room, Commander Tanbo got permission and rushed topside for the great moment—the only time he left his post during the entire voyage. Along the flight decks the men gathered, shouting good luck and waving good-by. Lieutenant Ebina, the *Shokaku's* junior surgeon, trembled with excitement as he watched the motors race faster and the blue exhaust smoke pour out.

All eyes turned to the *Akagi*, which would give the signal. She flew a set of flags at half-mast, which meant to get ready. When they were hoisted to the top and swiftly lowered, the planes would go.

Slowly the six carriers swung into the wind. It was from the east, and perfect for take-off. But the southern seas were running high, and the carriers dipped 15 degrees, sending high waves crashing against the bow. Too rough for really safe launching, Admiral Kusaka thought, but there was no other choice now. The Pearl Harbor Striking Force was poised 230 miles north and slightly east of Oahu. The time was 6:00 A.M.

Up fluttered the signal flags, then down again. One by one the fighters roared down the flight decks, drowning the cheers and yells that erupted everywhere. Commander Hoichiro Tsukamoto forgot his worries as navigation officer of the *Shokaku*, decided this was the greatest moment of his life. The ship's doctors, Captain Endo and Lieutenant Ebina, abandoned their professional dignity and wildly waved the fliers on. Engineer Tanbo shouted like a schoolboy, then rushed back to the *Akagi*'s engine room to tell everybody else.

Now the torpedo planes and dive bombers thundered off, while the fighters circled above, giving protection. Plane after plane rose, flashing in the early-morning sun that peeked over the horizon. Soon all 183 were in the air, circling and wheeling into formation. Seaman Iki Kuramoti watched, on the verge of tears. Quietly he put his hands together and prayed.

For Admiral Kusaka it had been a terrible strain, getting the planes off in these high seas. Now they were on their way, and the sudden relief was simply too much. He trembled like a leaf—just couldn't control himself. And he was embarrassed, too, because he prided himself on his grasp of Buddhism, *bushido*, and *kendo* (a form of Japanese fencing)—all of which were meant to fortify a man against exactly this sort of thing. Finally he sat on the deck—or he thinks possibly in a chair—and meditated Buddha-fashion. Slowly he pulled himself together again as the planes winged off to the south.

It made no difference to Commander Fuchida. The Japanese leader didn't even try to cover his tracks on the flight back to the carriers. There just wasn't enough gas for deception. As fast as the bombers finished their work, they rendezvoused with the fighters 20 miles northwest of Kaena Point, then flew back in groups. The fighters had no homing device and depended on the larger planes to guide them to the carriers.

Fuchida himself hung around a little while. He wanted to snap a few pictures, drop by all the bases, and get some idea of what was accomplished. The smoke interfered a good deal, but he felt sure four battleships were sunk and three others badly damaged. It was harder to tell about the airfields, but there were no planes up, so perhaps that was his answer.

As he headed back alone around eleven o'clock, a fighter streaked toward him, banking from side to side. A moment of tension—then he saw the rising sun emblem. One of the *Zuikaku*'s fighters had been left behind. It occurred to Fuchida that there might be others too, so he went back to the rendezvous point for one last check. There he found a second fighter aimlessly circling about; it fell in behind, and the three planes wheeled off together toward the northwest—last of the visitors to depart.

At his end, Admiral Kusaka did his best to help. He moved the carriers to within 190 miles of Pearl Harbor. He wasn't meant to go closer than 200 miles, but he knew that even an extra five or ten miles might make a big difference to a plane short of gas or crippled by enemy gunfire. He wanted to give the fliers every possible break.

Now everything had been done, and Admiral Kusaka stood on the bridge of the *Akagi* anxiously scanning the southern horizon. It was just after 10:00 A.M. when he saw the first faint black dots—some flying in groups, some in pairs, some alone. On the *Shokaku*, the first plane Lieutenant Ebina saw was a single fighter skimming the sea like a swallow, as it headed for the carrier. It barely made the ship.

Gas was low . . . nerves were frayed . . . time was short. In the rush, normal landing procedures were scrapped. As fast as the planes came in, they were simply dragged aside to allow enough room for another to land. Yet there were few serious mishaps. As one fighter landed on the *Shokaku*, the carrier took a sudden dip and the plane toppled over. The pilot crawled out without a scratch. Lieutenant Yano ran out of gas and had to ditch beside the carrier—he and his crew were hauled aboard, none the worse for their swim.

Some familiar faces were missing. Twenty-seven-year-old Ippei Goto, who this morning had donned his ensign's uniform for the first time, failed to get back to the *Kaga*. Baseball-loving Lieutenant Fusata Iida didn't reach the *Soryu*. Artistic Lieutenant Mimori Suzuki never made the *Akagi*—he was the pilot who crashed into the *Curtiss*. In all, 29 planes with 55 men were lost.

But 324 planes came safely home, while the deck crews waved their forage caps. The men swarmed around the pilots as they climbed from their cockpits. Congratulations poured in from all sides. As Lieutenant Hashimoto wearily made his way to his quarters on the *Hiryu*, everyone seemed to be asking what was it like . . . what did he do . . . what did he see.

Now that it was all over, many of the pilots felt a curious letdown. Some begged for another chance because they missed their assigned targets. Others said they were dissatisfied because they had only "near-misses." Commander Amagai, flight deck officer of the *Hiryu*, tried to cheer them up. He assured them that a near-miss was often an effective blow. Then he had an even brighter idea for lifting their spirits: "We're not returning to Tokyo; now we're going to head for San Francisco."

At the very least, they expected another crack at Oahu. Even while Commander Amagai was cheering up the pilots, he was rearming and refueling the planes for a new attack. When Lieutenant Hashimoto told his men they would probably be going back, he thought he detected a few pale faces; but, on the whole, everyone was enthusiastic. On the *Akagi*, the planes were being lined up for another take-off as Commander Fuchida landed at 1:00 P.M.—the last plane in.

When Fuchida reported to the bridge, a heated discussion was going on. It turned out another attack wasn't so certain after all. For a moment they postponed any decision, to hear Fuchida's account. After he finished, Admiral Nagumo announced somewhat ponderously, "We may then conclude that anticipated results have been achieved."

The statement had a touch of finality that showed the way the admiral's mind was working. He had always been against the operation, but had been overruled. So he had given it his very best and accomplished everything they asked of him. He had gotten away from it, but he certainly wasn't going to stretch his luck.

Commander Fuchida argued hard: there were still many attractive targets; there was virtually no defense left. Best of all, another raid might draw the carriers in. Then, if the Japanese returned by way of the Marshalls instead of going north, they might catch the carriers from behind. Somebody pointed out that this was impossible—the tankers had been sent north to meet the fleet and couldn't be redirected south in time. Fuchida wasn't at all deterred; well, they ought to attack Oahu again anyhow.

It was Admiral Kusaka who ended the discussion. Just before 1:30 P.M. the chief of staff turned to Nagumo and announced what he planned to do, subject to the commander's approval: "The attack is terminated. We are withdrawing."

"Please do," Nagumo replied.

In the home port at Kure, Admiral Yamamoto sensed it would happen. He sat impassively in the *Nagato*'s operations room while the staff buzzed with anticipation. The first attack was such a success everyone agreed there should be a second. Only the admiral remained noncommittal. He knew all too well the man in charge. Suddenly he muttered in almost a whisper: "Admiral Nagumo is going to withdraw."

Minutes later the news came through just as Yamamoto predicted. Far out in the Pacific the signal flags ran up on the *Akagi*'s yardarm, ordering a change in course. At 1:30 P.M. the great fleet swung about and headed back home across the northern Pacific.

Payback Time: The Doolittle Raiders Bomb Tokyo

America was finally in the war! Japan's surprise or sneak attack, depending on just how liberal you were, on Pearl Harbor thrust Yamamoto's "sleeping giant" into the conflict with no way but forward toward the Home Islands. But for the first six months, the news for the home front was mostly bad. Japanese soldiers and sailors drove an unstoppable steamroller across the Pacific. Only logistical lines stretched rubber-band thin prevented her from mounting more attacks on Hawaii and farther east. Except for the ultimately unsuccessful campaign against the Aleutian Islands off Alaska, begun in June 1942 as a diversion to the main thrust at Midway, a nearly unnoticed attack by a submarine-launched floatplane in August 1942 was the closest Imperial forces got to the U.S. mainland. Something had to be done to raise home-front morale as well as to send a message to America's allies and enemies that she might be bruised and bloodied but was ready to carry the fight to Tokyo.

During a visit to Norfolk, Virginia, Navy Captain Francis S. Low first thought of putting Army bombers aboard a Navy carrier, which would transport them to a point off Japan. The bombers would launch and strike selected Japanese targets before flying on to recovery bases in China. President Roosevelt and his service chiefs approved the plan.

Four types of bombers were considered: the Douglas B-18 and B-23, the Martin B-26, and the North American B-25. The B-18, a derivative of the DC-2 commercial airliner, was too old and slow. The B-23's wingspan was too wide to operate from the narrow flight deck of a carrier. The B-26 was too hot and it still had many technical problems. The B-25 seemed to offer the best choice.

Navy Lieutenant (later Rear Admiral) Henry L. Miller taught the Army crews about flying from an aircraft carrier. He was as much a part of the mission as any Army member. After learning how to haul their twin-engine B-

25 Mitchell bombers off in less than 450 feet instead of the normal 1,200 feet to 1,500 feet of a standard runway takeoff, 16 all-volunteer crews—70 officers and 64 enlisted crewman—boarded the newly commissioned USS *Hornet* (CV-8) for the trip west. They were part of the Army's 17th Bombardment Group at Pendleton Field, Oregon, under the command of pioneer aviator Lieutenant Colonel James Doolittle.

Although the plan called to launch the B-25s 400 miles from Japan, an unexpected encounter with a Japanese picket boat nearly twice that distance from Japan required Vice Admiral William F. Halsey, the task force commander, to radio *Hornet*'s Captain Marc A. Mitscher to launch the B-25s. To the raiders, Halsey transmitted, "Launch planes. To Col [sic] Doolittle and his gallant command, good luck and God bless you!"

Incredibly, all the B-25s managed to claw their way into the air as the *Hornet*'s deck rose and fell in terrible swells. Pairing off in twos and threes, the Mitchells struck out for their targets, the first hitting Tokyo a little after noon on April 18, 1942. The tale of the ordeal of the Tokyo raiders is generally well known, aided by Ted Lawson's riveting account of the mission and his tortuous return to friendly hands.

The raid was a major strategic success, resulting in Japan's recalling several frontline fighter squadrons when they were badly needed elsewhere and speeding up the Midway operation at a time when Japan was not completely prepared for such a massive strike.

Jimmy Doolittle received the Medal of Honor and a brigadier general's star, bypassing colonel's eagles. His raiders continued their wartime service after returning, although two crews were captured and three men were executed by the Japanese, with one more dying in prison. For helping those crews that did make it to China, the Chinese paid a terrible price as the Japanese mounted a wide-ranging campaign to punish those who had assisted Doolittle's crews in any way. One crew that came down in eastern Russia was interned because the Soviet Union was not at war with Japan.

Ted Lawson lost his left leg because of injuries he suffered when his B-25 crashed in heavy rain off the Chinese coast. Returning to the United States, bearing other scars from his mission, he wrote his firsthand account of the raid, along with sports writer Robert Considine, and advised on the blockbuster movie made of his best-selling book. The movie starred Spencer Tracy as Doolittle and handsome Van Johnson as Lawson. It was good casting, and the movie played to packed theaters. Flight sequences were first-rate, and along with another movie from another popular book about carrier flying, "The Bridges at Toko-ri," "Thirty Seconds Over Tokyo" remains one of the best aviation war movies ever made. It received an Oscar for special effects in 1945, a nod to spectacular flying scenes featured in the movie. Promoted to major, Ted Lawson remained in uniform during the war, then worked in civilian industry. He wrote the flight manuals for North American's F-86 Sabre.

When he was well enough to travel, Lawson toured the country telling his story to appreciative audiences eager to hear his account. Like many returning war heroes, he occasionally found the grind of these appearances almost as fatiguing as combat. He died in 1992, survived by his widow, Ellen, and their children and grandchildren.

In this excerpt, Ted Lawson describes the frantic flight-deck preparations to launch the bombers after the task force's discovery by the Japanese picket. His crew is number 7, and they fly a B-25B (serial 40-2261) that they've named "The Ruptured Duck." It wouldn't be the only time Army crews flew from a Navy carrier, but it sure was the most exciting.

From *Thirty Seconds Over Tokyo,* by Captain Ted Lawson and Bob Considine

The flight deck of the *Hornet* was alive with activity, while the big voice of the looming island barked commands. The man I thought was responsible for our bad turret hurried by and I stopped him long enough to tell him what I thought of him. And was sorry, as soon as I did. Nothing was important now except getting off that wet, rolling deck.

Lieutenant Jack "Shorty" Manch, a Virginian who must be the tallest fellow in the Air Force, ran up to our plane, carrying a fruit-cake tin.

"Hey, Clever," he said to our bombardier, "will you-all do a fellow a big favor and carry my phonograph records under your seat? I'll take my record-player along in my plane and we'll meet in Chungking and have us some razz-ma-tazz," and Shorty practically trucked on away from us through the turmoil. Clever shrugged and put the can of hot records under his seat.

The Navy was now taking charge, and doing it with an efficiency which made our popped eyes pop some more. Blocks were whipped out from under wheels. The whirring little "donkey"—the same one that was supposed to have broken loose and smashed my plane—was pushing and pulling the B-25's into position.

In about half an hour the Navy had us criss-crossed along the back end of the flight deck, two abreast, the big double-rudder tail assemblies of the sixteen planes sticking out of the edges of the rear of the ship at an angle.

A formal MGM portrait shows newly promoted Maj. Ted Lawson while consulting on the production of the studio's movie of his book. (Courtesy of Ellen Lawson)

From the air, the *Hornet*, with its slim, clean foredeck, and its neatly cluttered rear deck, must have looked like an arrow with pin-feathers bounding along the surface of the water.

It was good enough flying weather, but the sea was tremendous. The *Hornet* bit into the rough-house waves, dipping and rising until the flat deck was a crazy see-saw. Some of the waves actually were breaking over the deck. The deck seemed to grow smaller by the minute, and I had a brief fear of being hit by a wave on the take-off and of crashing at the end of the deck and falling off into the path of the careening carrier.

The *Hornet's* speed rose until it was making its top speed, that hectic, hurried perfect morning of April 18th. The bombs now came up from below and rolled along the deck on their low-slung lorries to our planes. It was our first look at the 500-pound incendiary, but we didn't waste much time on it except to see that it was placed in the bomb bay so that it could be released fourth and last.

The Navy had fueled our planes previously, but now they topped the tanks. That was to take care of any evaporation that might have set in. When the gauges read full, groups of the Navy boys rocked our planes in the hope of breaking whatever bubbles had formed in the big wing tanks, for that might mean that we could take a few more quarts. The *Hornet's* control tower was now beginning to display large square cards, giving us compass readings, and the wind, which was of gale proportions.

I saw our take-off instructor, Lieutenant Miller, trot up to Doolittle's plane and climb in the bottom opening. For a time I thought that he was going along, but after a bit he came out and began visiting each of the other B-25s. We were in the Ruptured Duck now, all of us, and when Miller came up to the pilot's compartment he must have stood there a half minute with his hand stuck out at me before I came back to life and shook hands with him. I had so much on my mind. Miller wished all of us good luck, and he said, "I wish to hell I could go with you."

It was something of a relief when five additional five-gallon tins of gas were handed in to us. We lined them up in the fuselage beside the ten cans Doolittle had already allotted us. It was a sobering thought to realize that we were going to have to fly at least 400 miles farther than we had planned. But my concern over that, as I sat there in the plane waiting to taxi and edge up to the starting line, was erased by a sudden relief that now we wouldn't have to worry about running into barrage balloons at night. This, of course, was going to be a daylight raid. It was only a few minutes after eight in the morning.

Commander Jurika and "Nig" also came up to say good-bye and to shake hands. When they had gone I suddenly remembered that none of my crew had had breakfast and that all of us had lost sight of the fact that we could have taken coffee and water and sandwiches along. I was tempted to send Clever below to get some food, but I was afraid that there would not be time. Besides, Doolittle's ship was being pulled up to the starting line and his and other props were beginning to turn. The *Hornet's* deck wasn't a safe place. I found out later that one of the Navy boys had an arm clipped off by a propellor blade that morning.

Doolittle warmed and idled his engines, and now we got a vivid demonstration of one of our classroom lectures on how to get a 25,000-pound bomber off half the deck of a carrier.

A Navy man stood at the bow of the ship, and off to the left, with a checkered flag in his hand. He gave Doolittle, who was at the controls, the signal to begin racing his engines again. He did it by swinging the flag in a circle and making it go faster and faster. Doolittle gave his engines more and more throttle until I was afraid that he'd burn them up. A wave crashed heavily at the bow and sprayed the deck.

Then I saw that the man with the flag was waiting, timing the dipping of the ship so that Doolittle's plane would get the benefit of a rising deck for its take-off. Then the man gave a new signal. Navy boys pulled the blocks from under Doolittle's wheels. Another signal and Doolittle released his brakes and the bomber moved forward.

With full flaps, motors at full throttle and his left wing far out over the port side of the *Hornet*, Doolittle's plane waddled and then lunged slowly into the teeth of the gale that swept down the deck. His left wheel stuck on the white line as if it were a track. His right wing, which had barely cleared the wall of the island as he taxied and was guided up to the starting line, extended nearly to the edge of the starboard side.

We watched him like hawks, wondering what the wind would do to him, and whether we could get off in that little run toward the bow. If he couldn't, we couldn't.

Doolittle picked up more speed and held to his line, and, just as the *Hornet* lifted itself up on the top of a wave and cut through it at full speed, Doolittle's plane took off. He had yards to spare. He hung his ship almost straight up on its props, until we could see the whole top of his B-25. Then he leveled off and I watched him come around in a tight circle and shoot low over our heads—straight down the line painted on the deck.

The *Hornet* was giving him his bearings. Admiral Halsey had headed it for the heart of Tokyo.

The engines of three other ships were warming up, and the thump and hiss of the turbulent sea made additional noise. But loud and clear above those sounds I could hear the hoarse cheers of every Navy man on the ship. They made the *Hornet* fairly shudder with their yells—and I've never heard anything like it, before or since.

Travis Hoover went off second and nearly crashed. Brick Holstrom was third; Bob Gray, fourth; Davey Jones, fifth; Dean Hallmark, sixth, and I was seventh.

I was on the line now, my eyes glued on the man with the flag. He gave me the signal to put my flaps down. I reached down and drew the flap lever back and down. I checked the electrical instrument that indicates whether the flaps are working. They were. I could feel the plane quaking with the strain of having the flat surface of the flaps thrust against the gale and the blast from the props. I got a sudden fear that they might blow off and cripple us, so I pulled up the flaps again, and I guess the Navy man understood. He let it go and began giving me the signal to rev my engines.

I liked the way they sounded long before he did. There had been a moment, earlier, when I had an agonizing fear that something was wrong with the left engine. It wouldn't start, at first. But I had gotten it going, good. Now, after fifteen seconds of watching the man with the flag spinning his arm faster and faster, I began to worry again. He must know his stuff, I tried to tell myself, but when, for God's sake, would he let me go?

I thought of all the things that could go wrong at this last minute. Our instructions along these lines were simple and to the point. If a motor quit or caught fire, if a tire went flat, if the right wing badly scraped the island, if the left wheel went over the edge, we were to get out as quickly as we could and help the Navy shove our $150,000 plane overboard. It must not, under any circumstances, be permitted to block traffic. There would be no other way to clear the forward deck for the other planes to take off.

After thirty blood-sweating seconds the Navy man was satisfied with the sound of my engines. Our wheel blocks were jerked out, and when I released the brakes we quivered forward, the wind grabbing at the wings. We rambled dangerously close to the edge, but I braked in time, got the left wheel back on the white line and picked up speed. The *Hornet's* deck bucked wildly. A sheet of spray rushed back at us.

I never felt the take-off. One moment the end of the *Hornet's* flight deck was rushing at us alarmingly fast; the next split-second I glanced down hurriedly at what had been a white line, and it was water. There was no drop nor any surge into the air. I just went off at deck level and pulled out in front of the great ship that had done its best to plant us in Japan's front yard.

I banked now, gaining a little altitude, and instinctively reached down to pull up the flaps. With a start I realized that they were not down. I had taken off without using them.

Something for the Home Folks: Taking Liberties with the Facts

The saying goes that the first casualty of war is truth. When you're trying to sell the war to the people at home, to reassure them that their servicemen have the best equipment and are using it to destroy the enemy, facts are occasionally twisted or changed outright. In the first two years of America's involvement in World War II—the European capitals had been guilty of regular deception since 1939—Americans read that although the Axis was enjoying some success, they would never have total or eventual superiority.

The next selection is, frankly, one of my personal favorites, ever since I ran across it in an anthology in 1960. Although we did our best to find either the author or publisher, we came up dry. A friend reported that the author had been a screenwriter in Hollywood, which might account for the dramatics and untruths he uses to tell his entertaining story.

"Battle Descending" is pure fabrication. No American carrier took Brewster F2A Buffalos into combat in any European theater. The tubby little plane, the Navy's first monoplane carrier fighter, generated a less than stellar record and reputation in the fighting it did see in U.S. colors during the first six months of the Pacific war. Prewar, the F2A's weak, stalky landing gear contributed to more than a fair share of carrier landing mishaps.

The F2A participated in a few raids against the Japanese, but is mainly remembered for its dismal showing at Midway when a squadron of Marine aviators ran into four Imperial carrier squadrons of Mitsubishi Zeros. The Buffalo served with the Royal Air Force in Burma (alongside the Flying Tigers) and the Dutch East Indies Air Force against the Japanese, but like their Marine counterparts, these hard-pressed Allied air services saw little to recommend the Brewster. Only the Finns, somehow, had success with the Buffalo against Soviet bombers and fighters.

The "villain" of this story is the Focke-Wulf 187. In reality, the FW-187 was an abortive though attractive twin-engine, single-seat fighter. Only a few examples were made, and the type never saw squadron service in the Luftwaffe. The author's inclusion of a rear-seat crewman-gunner is perplexing, unless he was thinking of the FW-189, a clumsy-looking, twin-boom, twin-engine observation type that did see considerable service, especially in Russia, and did include a rear gunner. I doubt, however, that at this point immediately after the war a civilian writer would have had information available to distinguish between these two types.

MGM's highly successful 1944 film starred Van Johnson as Ted Lawson and Spencer Tracy as Jimmy Doolittle. In this scene Doolittle briefs Lawson's crew before the mission. The lineup includes: Don DeFore (l.), Robert Walker (third from left), Van Johnson (fifth from left), and Spencer Tracy (r). The film gave a more truthful depiction of the people and events than the addendum to the 2000 Disney film "Pearl Harbor," which suffered major lapses in accuracy. (Courtesy of Jim Farmer)

Whatever the truth, "Battle Descending" is an example of giving the young readers at home a "real" look at how American naval aviators were taking it to the enemy, even if they had to leave their aircraft.

From *Battle Descending,* by Richard Sale in *Teenage Aviation Stories,* edited by Don Samson

"This," Brad said, as he slid back the hatch of his greenhouse carefully, "is where I came in."

His Buffalo fighter was afire. Her right wing was ragged with flame. He had mushed the stick down into the left-hand corner, and had right-ruddered the flying beer barrel hard, so that she side-slipped left and swept the flames off right. This prevented them from engulfing the cockpit and toasting one Lieutenant (j.g.) Ames, U.S.N., to a nice, brown turn.

With the cockpit open, he fondled his parachute, checked the compressed-air cartridges of his life jacket, and then pulled out the sub-machine gun which his mechanic on the flattop had slid in behind his armor "just in case." He cradled the tommy gun in his arms, adjusted his oxygen input. Then he stood on the seat, inhaled, and followed through with a billowing belly-whopper into the icy zones, five miles from the island of Malta. Five miles—straight up.

"This," Brad thought as he dropped in a free fall, "is what happens to hot-shots who outgrow their britches."

It was No. 2 on his getting-hit-parade. He had been shot down once before. The first time, somewhere west of Sardinia, an Eyetie Macci had put lead in his mill, but he had been able to dead-stick a landing back to the flight deck of the U.S.S. *Staten Island*, which flat-top was home sweet home to him. That time, there had been a good excuse for his getting the hook. This time, it had a bad odor all the way.

When he was clear of the Buffalo, he yanked on the rip cord. The pilot chute blossomed first, then the main parachute. It filled its silk belly with cold air and functioned perfectly. That was a relief. He would not have to go back for a new one. The jounce it gave him when it cracked open, clicked his teeth, jerked his neck, twisted his back, but it was paradise. His free fall stopped. Thence he began a 15-foot-a-second descent.

The Buffalo screamed seaward, like a skyrocket. She was all afire now and somehow she looked beautiful—violently beautiful. It reminded him of one of those sunburn cream ads, with flames curling back from a beautiful girl's shoulders. Well, the beer barrel had been a beautiful gal while she lasted. He had grown very fond of the old tomato in the two weeks he had been flying her. And she would have been turning up her rump still if he had not tried to fight her at an altitude for which she had never been built. 26,500 feet! And against a Focke-Wulf!

Below, the Buffalo fighter, in her fiery robe, disappeared. She dived into the stratum of cotton clouds, and only the white trail of her smoke hung in the blue, marking the trail she had taken. He tried to remember where that layer of clouds was located, and it came back to him that he had climbed through them around 16,000 feet. The clouds were still far below him. He calculated roughly that he had bailed out at 26,500. He had fallen free for

The portly little Brewster Buffalo saw little combat in American colors, and then only in the Pacific. Although enjoying some technical firsts, the Buffalo was quickly outclassed by its land-based opposition. It fought with the RAF in Burma and the Dutch in the East Indies. But only the Finns used it with any success against the Soviets in the so-called Winter War of 1942. (Courtesy of the National Archives)

some time to clear the plane, say, 500 feet. Chute opened at 26,000 even. He had been falling, so said his watch, around five minutes. Fifteen feet a second was 900 a minute. He had dropped, then, some 4,500 feet by chute and was approximately at an altitude of 21,500 feet. He figured further, and learned that it would be 24 minutes before he came down on earth or sea in the world below where a man walked, breathed, and had being in a normal sort of way.

Brad clung to the sub-machine gun tightly and looked around infinity as he floated down. For the first time, he remembered the F-W which had shot him down. There was a kite which had been built to racket its machine guns at 40,000, up where the Fortresses alleyed the sky. The F-W had lost little altitude in the brief scramble, and surely Jerry was still around somewhere.

Brad stopped cradling the sub-machine gun and held it at his hip. His mechanic aboard the *Staten Island* had had strong feelings about the tommy gun. "Look, Lootenant," his mech had said; "you just been transferred to the *Staten Island* from the *Scarab*. Okay. The *Scarab* is a real carrier, and the *Staten Island* is just a converted hog boat. Okay. On the *Scarab* maybe you was fighting a gentleman's war. On a converted C-3 in the Mediterranean, you're fighting Nazis, and it ain't a gentleman's war. Jerries are ratzi, see, sir? If they give you the works, and you have to hit the silk, a Jerry'll shoot your guts out while you're hanging by your shrouds. An Eyetie won't do it—he'll give you a break. But Jerries and Japs, they like to machine-gun sitting ducks, up there. So you carry the tommy gun, Lootenant, and I'll breathe easier down here. It won't add much more weight, and I can slide it in behind the armor of your back-rest."

"Okay, kid," Brad said; "anything to keep up the morale of the boys in blue." . . .

Brad scanned the sky as his parachute swung him lazily to and fro. He saw nothing above or around him. Nevertheless, he was happy about that tommy gun. He had a full drum hugging the barrel, 500 rounds of copper-clad slugs, each sitting in its bright nickel hull. The stuff was only good for short fighting; it arced off at any real distance. But he did not feel like a clipped-wing crow, sitting there on the unimpressionable ozone. He still had a stinger.

Tallyho!

Brad tensed. Under his heels he saw aircraft. He could not make out the marking for a moment, but then he recognized the stubby plumpness of the ships. Buffaloes. They were in echelon, on a line of right bearing, and they had their noses up and had just broken through the clouds down there. They were climbing. That would be Stinky Thomas in the No. 1 spot, having taken over command of the squadron.

Next instant, Brad heard them. At least, he heard an engine. Even through his oxygen mask and tight helmet, he heard the roundhouse baritone of a chattering mill, and the slap of a prop against air. He was puzzled, for the American fighters were little more than black spots in front of his eyes when he leaned over and stared down at them. Marvelous how sound carried, really marvelous.

Without warning, he was swung in the cradle of the sky like a sprout on an apple-tree swing. He felt his parachute jerked by a blast of wind and hurled northward. He was jerked bodily up into the air for a second, and then he took up a pendulous motion that would have made Barnacle Bill sicker than the Afrika Korps. Even as he grabbed one of the front shroud straps to steady himself, still clinging to the tommy gun with his right hand, he saw the black bulk of the Focke-Wulf as it passed him in a whistling dive and banked to come around again.

The F-W came back. Brad set his teeth and stared at its air-screw arc as the Jerry put the plane on a dead-level beam with him and roared in. Brad made no move to use the tommy gun. Not yet. The bum from Berlin could have shot him the first time. Maybe the guy was just going to give him a couple of thrills. He saw the number on the Jerry's fuselage: 2040. That was the same palooka who had shot the wing of his Buffalo afire.

The Jerry eased off just before he reached the chute, and when he passed, Brad got a good look at the guy in the greenhouse and the rear gunner. The pilot had a hard, set face. As he flashed by, he made a motion of cutting his throat.

It was obvious, even to an optimist like Brad Ames, that the cat was merely toying with the mouse, and the thought of it sickened him. Personally, he would never have pot-shotted any poor lost soul on the seat of a chute shrouds. But to rub it in and play with the victim before you turned your Rheinmetall-Borsig machine gun on him was lower than a mole's navel. The pilot was not going to use the air cannon on him. Brad was the rear gunner's meat. That substitute for Dracula abaft the pilot was going to have some target practice on a live sleeve.

Maybe, just maybe, Stinky and the boys would make the grade in time. Brad looked down. He groaned. The Buffaloes had come up above the cloud stratum high now, but they were engaging or being engaged by a squadron of Zerstorers. That meant the Jerries were there to protect a bombing squadron, which meant Malta was still under attack. It also meant that Brad was on his own.

The F-W 187 began to circle him now as he dropped, and the rear gunner finally leveled the machine gun at him and opened fire. The range was ridiculous, no more than a hundred yards, and the red line of tracer showed plainly against the critical blue of the upper sky where the parachute hung suspended.

The smoky line of it was too high at first. Then it moved down, put a burst at the silk, lowered through the shrouds, and finally fell off the target, missing Brad completely.

It was the swing motion which saved him. The pendulum their slip stream had made of him gave him the only evasive tactics he could use, but they saved his neck. The F-W seemed to be closing the bank tighter, and when they got behind him he could not turn himself in his harness.

Brad was falling faster. He could sense it, for the plane had to nose down a bit to keep with him. When it came around within the visible range again, he raised his tommy gun and started firing. The muzzle tended to waltz off the target to the left, and there was no tracer in the magazine to show him how he was doing. He didn't try for a man, he just tried for the plane, sweeping the path it was taking in the bank and leading it just slightly. He wished that he could see the results. The gun banged raucously. He followed the plane as long as he could, but then it moved out of his arc of fire.

Brad haunched his shoulders and tightened all his muscles, as if their rigidity would keep out alien steel. He had a sense of breathlessness and was aware of the noisy thunder of his heart. He glanced above to see the tracer line from the F-W's rear gun, but there was none. Soon the plane hove into view on his starboard, and he opened fire in short bursts.

Glory be. The rear gunner had caught a packet. He had fallen with the inclination of the Zerstorer's bank, and was leaning limply against the double-chambered magazine of his gun. Nor did the pilot realize it yet, for he continued to bank, assuming his gunner would fire when ready.

Brad leveled the tommy gun and smoothed the trigger back. He fired four long bursts at the pilot and then at the port engine, which cut off the pilot as the plane swept around. Nothing happened.

As the F-W passed behind him, Brad twisted his head to watch. He could not bring the tommy to bear, but he could, by superhuman twisting, see the black plane. The pilot had finally become aware that his rear gunner was non compos mentis.

There wasn't much of the F-W to see. It had come around, so that he was facing it. The Jerry was seeing red. Brad stared grimly at the silhouette. A line of black wing, two propellor arms, and the center, between them, a black nacelle muzzle topped by a glistening greenhouse in which sat a guy with his thumb over the solenoid that fired the twin guns.

It was a duck shot, and as much of a setup as a guy in a parachute could ever have. The plane was head-on. No necessity of a leading shot, not even an over or under. Just a plain, good, old-fashioned head-on shot. Despite the rhythmic cavorting he was doing on the harness of the swinging chute, Brad hoisted the sub-machine gun to his shoulder and aimed full on the greenhouse between the Daimler-Benz nacelles.

He was outranged, however, and he became aware of the gray balls of smoke which broke from the twin guns on the F-W. Tracer showed below to his left. He gaped grotesquely, holding his breath as the parachute swung him down through the trail of fire. His mask was pulled from his face, and hurled into space, and an excruciating pain shot through his right arm just above the elbow. Desperately, he tried to hold the gun, still firing, but the paralysis which swept through the arm numbed his grasp and it fell from him.

The F-W, however, stopped firing. Out of the blue dived a Buffalo, 3 F-5, which meant Stinky Thomas. Stinky went in recklessly at the tail, disregarding those rear guns. He didn't know the gunner was dead, and he was seeing red. He got a wonderful burst along the F-W's spine and broke her. She deflected into a dive, passed under him, nosing into a dead vertical drop, and screamed down. She struck the cloud stratum, a bare 1,000 feet below, her exhaust leaving a solid trail behind, and did not reappear. There had been flames. Stinky dived down.

Within forty seconds Brad dropped into the clouds, himself. As he went, he cast an anxious eye at his chute, for he realized from the wind on his face that he was falling faster than he should have. There were a few holes in the silk above him, and only the grace of God, and thin air, had prevented the tracers which made the hole from actually firing the parachute itself. But he'd got half a scalp. Stinky would have to split the F-W with him. He'd got the gunner!

There was security within the clouds. They were thick and heavy. It was a welcome relief. He felt sick at his stomach, and his arm was giving him pain he had not believed possible. But he was alive.

He broke into the clear 2,000 feet later, feeling incredibly weak and ineffectual. Under him this time there was nothing to mar the view. There lay the blue Mediterranean sparkling gaily in the afternoon sun. Off there to the east lay Malta, its stony crags and promontories looking as flat as the sea. He was amazed, for he had begun his scramble in the sky on the far western side of the island, 'way offshore. The wind, in the interim, had carried him northwest, so that now, descending, he could look down into the harbor of Valetta with ease.

Except when airplanes got in the way, as they were getting in the way now. The roar of them was around him, and as he watched, Brad saw as rough a hair-pulling as could be managed with wing-born guns. They filled the sky from the seventeenth floor on down to the bargain basement. Starting at the top, there was a confused battle raging. He sat right in the nub of it. Stinky Thomas and Fighter Squadron One from the Staten Island were mixing with a nest of Zerstorers like the one that had attacked him upstairs. It was some fight. He squeezed into it, weak and faint, with no difficulty at all, although the merging cannon tracer lines crossed and cracked around him with a familiarity which was heart-racking. On the sixth floor, downstairs, a batch of humpbacked Hurricanes were blowing the fight out of a flight of Ju. 87 Sturzkampflugzengs which had come in under the F-W protection above, to bomb the harbor of Valetta, into which the vast convoy was now moving.

That was a sight. The ships covered the blue sea westward for 25 miles, bearing toward Malta's north shore. Some of them were entering amidst the bombing. Sometimes the sea was lined with white geysers where the Stukas jettisoned their eggs.

Far off in the center of the cluster of ships he could see the stubby little freighter with a flat-top, which was the converted aircraft carrier, *Staten Island*, where he hung his pajamas in a locker.

There were some Eyetie torpedo planes down there, trying for an attack on the convoy, but their heart wasn't in it, with Spitfires on their backs, blowing their wood fuselages to splinters.

Brad tried to spill air from his parachute with his good arm, but he didn't have the strength. He wanted to free-fall through the scramble as quickly as possible. He saw a Buffalo come roaring in at him and then bank, firing its six heavy Brownings savagely, and an F-W behind his back plummeted off in an abrupt peel away from him. The Jerry had tried to machine-gun him.

The Navy Buffalo stayed with Brad then, circling him and protecting him while the rest of the squadron did the dirty work. It was 3 F-5, which meant that Stinky was back. Stinky waved at him once and made a thumbs-up motion. Brad waved wanly.

Downstairs, the Stukas broke for home. The Italians had long before given up and jettisoned their torpedoes. The Zerstorers stayed the longest, but finally they, too, departed for Sicily, minus three, and the entire Buffalo Squadron cut into echelon and followed a circle all around him wide, at varying levels in their step-down.

They stayed with him all the way to the main floor. It was wonderful to see them and to have them there, like an overcoat against a chill wind.

At 4,000 feet, the world began to swim, and he felt both seasick and drunk at the same time. Brad closed his eyes to stop the giddiness, and his eyes stayed closed. The next thing he knew he was under water. He had fainted during those last 4,000 feet of drop. The cold water revived him instantly. He felt as if he were halfway down to the sea floor. He flailed at the sea with his good arm and felt himself going up. It was a longer ascent than he had reckoned with, and he barely made the surface before he inhaled, gasping, sucking in the fresh air.

A hook caught his chute harness, and he felt himself held above water without the necessity of treading. An alien voice cried, "Easy does it; easy does it. Take care o' the lad's wing; she's broken clean. . . . Now then, Yank . . ."

Beside him, Brad saw a sleek, short hull with a number 109 on it. She was only 68 feet long, a British motor torpedo boat. As they raised him from the sea and lifted him over the row of depth charges astern, he fainted again, but this time he revived quickly as they got under way.

"Thanks," he gasped.

"All right," said a crooked-toothed redhead; "he's conscious. Splice the main brace."

They poured rum down his throat. It tasted awful. More brightly, he sat up, favoring the bad arm. "Thanks," he said. He saw his white parachute floating on the sea astern. "How did you ever manage to be there so expeditiously?"

"Take it easy," the redhead grinned. "When you buzzed your ship you were going to ride the silk, they buzzed us in Valetta, and we came out to follow you down, sir. Nothing to it; just a little teamwork."

"Teamwork," Brad said. He looked rueful. "That is a very nice word."

"Ain't as good as 'rum,'" the redhead said. "Gor, sir, that was nervy fighting the bloody Jerry from a chute. . . . Have another nip."

"Thanks," Brad said. "But how could you see? We were up above the cloud level—"

"There was a big hole over there." The redhead pointed. "The skipper, he had the glasses on you all the time. 'Is number was 2040, weren't it? He splashed in cold. Quite a show. One of your Yanks potted him. . . . Another nip of rum, sir?"

"I got the rear gunner," Brad said.

"We wondered," said the sailor. "Other Yank went in at the F-W's tail so easy. Tailing a 187 with its rear guns is fine suicide when you have a live Jerry at the trigs. You ought to split that one."

The motor torpedo boat measured in alongside the U.S.S. *Staten Island* as she made for Valetta, her fighter brood having returned to nest, her scouts up now. Brad was taken aboard through a beam opening and rushed to the sick bay. The flight butcher hadn't even had a chance to work on that arm when Stinky Thomas broke into the bay.

"Golly!" said Stinky. "I thought you were a goner, Brad. I thought I'd never get up there!"

"You got there," Brad said. "Bless your heart, dodo. And thanks a lot. I thank you. Sally thanks you. We all thank you."

"Oh, nuts," Stinky deprecated. "But that makes my fourth." His eyes glowed. "Four scalps on the belt. I'm catching up to the Maestro. . . . You know, Brad, I got so mad at that son of a so and so, I hopped right into that rear gunner's field, and he should have mincemeated me. I was so mad, I never even thought that I was a setup for him. All I could see was them potting at you. But that fool rear gunner never opened up. There was I, a setup, and he never threw a packet at me. What do you suppose happened to him?"

Brad started to speak, and then stopped. He swallowed. Then he tried again. This time words came out. "I wouldn't be a bit surprised," he said, "if your dive hadn't scared him so spitless that he couldn't move. In any case, the scalp is your'n, sailor, so let's praise the Lawd and pass you the citation."

When Stinky had fled in exuberance to stand treat on sodas all around, Brad leaned back and closed his eyes. The flight surgeon said, "Lie still, Brad. This is going to hurt you more than it does me. Very foolish, getting shot down. Decreases your longevity chances. You ought to know better. Toast to the tattered Jay-Gee. May you live and learn."

"No," Brad said faintly, wincing at the touch of the probe. "Let's put it this way: May I learn and live."

"Stop being noble," said the butcher, "and go ahead and faint. I'll wake you up later."

The Battle of Midway

Much has been written about the pivotal four-day engagement between American and Japanese task forces centered on the two-island outpost of Midway. With all the bad news of the preceding six months, even with the relatively good reports coming from the Battle of the Coral Sea near Australia, stopping the huge Japanese fleet headed for Midway became the number-one priority in the Pacific.

Japanese naval aviation began in 1912 when Imperial naval officers went to the United States, Great Britain, and France for flight training. They returned home with two American Curtiss seaplanes and two French Farman seaplanes. The *Hosho* was the first Japanese carrier, and the first ship built as a carrier. Her keel was laid in 1919, and she was launched in 1921. She survived the war only to be scrapped in 1947. By December 1941, Japan had 10 operational carriers, while the United States had 8, with just 3 in the Pacific. Great Britain had only one carrier. Japan's carrier aircraft strength was more than twice that of the Allies. Frankly, no other power had embraced carrier aviation as zealously as pre-war Japan, and this enthusiasm bore great fruit in the first six months of the war, as well as during the last three years of the 1930s.

In early May 1942, two American carriers and their supporting ships had taken on a superior Japanese fleet bent on attacking the northwestern coast of Australia. The two-day engagement in the Coral Sea resulted in the loss of one U.S. carrier, the *Lexington* (CV-2); the serious damaging of the second, *Yorktown* (CV-5); the sinking of the Japanese light carrier *Shoho*; and serious damage to another Imperial carrier, *Shokaku*. Although the U.S. could ill afford the loss of one of its prime aircraft carriers at this critical early stage of the war, this battle was considered a major strategic victory because it stopped the hitherto unstoppable Japanese advance.

The Japanese quickly turned their attention to their upcoming operation against Midway, which included a diversionary attack on the desolate Aleutian Islands off the forbidding coast of Alaska. Added impetus came from the Doolittle Tokyo raid of April 18, which had shocked the Japanese out of their smug complacency. As they were to confirm in two years, their cities were wide open to aerial attack, and the 16-bomber strike against heretofore "safe" population centers pushed military planning for the Midway operation into high gear.Capturing Midway became essential, not just because it was an important American outpost and submarine base, but because it led directly to further attacks on Hawaii.

The Imperial Navy, led by Admiral Yamamoto, knew that it had to engage the American fleet, particularly its carriers, which the otherwise highly successful raid on Pearl Harbor had completely missed because the two main carriers, *Lexington* and *Yorktown*, had been at sea during the Japanese strike. It was one of the few happy turns of events for the Americans in December 1941.

Yamamoto sent out a two-pronged operation, made up of four task forces and including eight carriers, 80 percent of his available flattops. The plan called for the initial strikes in the Aleutians to focus American attention away from Midway, and then for the main force to hit the atoll on June 5. However, unknown to the Japanese, American naval intelligence had cracked their complicated message codes and knew what Yamamoto was planning. Thus, the three available large carriers in the Pacific—*Hornet*, *Enterprise* (CV-6), and *Yorktown*, with their individual fleets—were ready. Crammed with aircraft, the three carriers headed out to meet the enemy.

A PBY Catalina crew found the Japanese on June 3, 1942. After an ineffective initial strike by B-17s and PBYs from Midway, the Japanese responded with an all-out attack on Midway on June 4. The aerial battles were ferocious, with American squadrons suffering the most devastation. *Hornet*'s VT-8, flying 15 obsolete Douglas TBD Devastator torpedo bombers, was annihilated, with only one pilot surviving. *Enterprise*'s similarly equipped VT-6 fared little better, losing 10 of 14 aircraft and crews. *Yorktown*'s VT-3's 12 TBDs were all shot down. It was a terrible cycle of missed checkpoints and lost escorts. Not one torpedo launched by the three squadrons, totaling 41 aircraft and 82 crewmen, hit its target. The Japanese sailed on.

But the sacrifice of the lumbering TBDs had not been in vain. As the squadrons flying more effective SBD Dauntless dive bombers headed into the battle, they couldn't know that the Japanese would be caught on their flight decks changing the bomb load of their aircraft, and that the Zero fighter combat air patrol had been brought down to deal with the initial strikes by the Devastator squadrons. There was no protective umbrella over their precious carriers and the four ships' flight decks were crammed with aircraft and highly flammable aviation fuel and ordnance.

In short order, the SBDs dropped in screaming dives, and three of the Japanese carriers were in flames. The fourth carrier launched a strike that found the *Yorktown* and put three bombs into her. Barely able to make way under all the patchwork hurriedly accomplished after Coral Sea, *Yorktown* staggered, but somehow remained afloat. But revenge was swift as aircraft from *Yorktown* that had recovered aboard *Enterprise* sortied with the "Big E's" own bombers and hit the fourth Japanese carrier, sending her to the bottom along with her three sister carriers. Besides the loss of her four largest fleet carriers, Japan had also lost 250 aircraft and many experienced flight and maintenance crews, which were never replaced to prewar standards.

Midway was a tremendous defeat for the Japanese, a setback from which many argue they never recovered, in effect losing the war barely six months after beginning it. The next two-and-a-half years were merely what it took for Tokyo to realize they had lost.

In this three-part section, we examine one admiral's role at Midway, and then how one young man fought his war at Midway. Finally, we take a look at how the Japanese consider their role and effectiveness at the carrier engagement known as the Battle of Midway.

The Midway Admiral: Raymond A. Spruance

Ray Spruance was a quiet, distant man, who was also considered one of the Navy's main intellects of the time. He was a member of the prewar "gun club," which espoused the battleship over all other surface vessels, including the upstart aircraft carrier. Apparently destined to be relegated to subordinate roles, he got his chance when William Halsey, hospitalized with dermatitis in May 1942, personally selected Spruance as his replacement at that crucial time, just before Midway.

Although not an aviator, Spruance understood the value of aircraft and aircraft carriers in the modern Navy, and he was not afraid to use them as the main offensive forces against the enemy. After the American victory at Midway, Spruance, who garnered a large part of the credit for the Japanese defeat, served as Chief of Staff for Admiral Chester Nimitz. He then received other combat fleet assignments and eventually rose to full admiral. His supporters steadfastly contested that Spruance deserved the fifth star of Fleet Admiral; however, that exalted promotion eluded him. He eventually replaced Nimitz as Commander in Chief, Pacific, in late 1945, just after the war.

Vice Admiral Forrestel's rarely seen biography of Raymond Spruance was published three years before Spruance's death in 1969. The following selection describes how Spruance fought the battle at Midway, sending his planes and crews into harm's way and securing a major victory for the United States when it was most needed.

Vice Admiral Emmet P. Forrestel was a native of Buffalo, New York, and graduated with the Annapolis class of 1920. His career with the surface Navy included duty in several ships. He commanded the USS Finch *in the mid-1930s. During World War II, he served as Aide to the Assistant Secretary of the Navy until 1943, when he went to the Pacific. In August 1945 he assumed command of the battleship USS* South Dakota. *Following the war, he served in various staff and operational assignments until he retired in November 1959. He died in December 1979.*

From *Admiral Raymond A. Spruance,* by Vice Admiral E. P. Forrestel

Fletcher and Spruance were in a better condition of readiness. They had assumed that the enemy carriers would come from the northwest and that an all-out attack on Midway would be launched at dawn on the 4th. Task Force 16, operating 10 miles to the southward of Task Force 17, kept pilots and planes of *Enterprise* and *Hornet* in readiness as a striking force while *Yorktown* of Task Force 17 launched a dawn security search to the northward and maintained the combat air patrol. At 0545 the Midway search plane contact report previously mentioned was intercepted: "Many enemy planes heading Midway bearing 320 distance 150," and seven minutes later, at 0552 another intercept reported, "Two carriers and battleships bearing 320, distance 180, course 135, speed 24."

This was the report Spruance had been waiting for. As he recalls it, "As soon as I got the contact report of the attacking groups headed towards Midway from the northwest, I turned TF 16 towards where we estimated the Japanese carriers would be when our attack groups would reach them. This was based on the assumption that they would continue towards Midway at 25 knots. I speeded up to 25 knots and made preparations to launch aircraft when we got within striking distance. I felt that we must strike their carriers before they could launch a second attack on Midway."

Under instructions from CTF 17, to "attack enemy carriers when definitely located," he sped to the westward and came within estimated striking distance and began launching at 0700. In determining his launching time Spruance had to make one of the crucial decisions of the day. In carrier warfare, even more than usual in war, the side which strikes first reaps great advantage. The possibility of striking after the planes of the first wave had returned and were on deck, unarmed and without fuel, was tempting, but this consideration was secondary to the great importance of striking the enemy carriers before an attack could be launched against his own. He could not be sure that his task force had not been observed (actually they were reported to Admiral Nagumo by a seaplane scout at 0728); and only two Japanese carriers had been reported to Spruance, although intelligence led him to expect the presence of four or five. Furthermore, his torpedo bombers had a range of only 175 miles, and if there were errors in the reported enemy position, course or speed, the distance could be greater than expected. He also had to consider his directive to attack when the enemy carriers were "definitely located."

Steaming to the westward for about an hour after the first contact report and before launching, the courageous decision of a courageous commander was made and involved a calculated risk which was to be rewarded. His original intention had been to launch at 0900, when the estimated distance of the enemy would be about 100 miles, but after receiving reports of the strike on Midway, it appeared to him that by launching at 0700 there was a good chance of getting in the first strike while the first Japanese attack wave was on deck in the process of refueling. He felt fairly confident of the enemy's location and began launching when he estimated they bore 239°, distant 155 miles. This estimate was unfortunately inaccurate because the attacks by Midway-based aircraft had caused the Japanese carriers to maneuver so that they had not reached their expected position. The first attack groups therefore had to expend valuable time and fuel, as will be seen, searching for their targets.

The Japanese carrier *Hiryu* on fire on June 5, 1942. Hit by SBDs the previous day, it was photographed by the crew of a plane from the light carrier *Hosho.* (Courtesy of the U.S. Navy)

The order of launching was: (1) VF for combat air patrol, (2) dive bombers armed with 500 or 1000 lb. bombs, (3) torpedo planes, (4) VF for torpedo plane escort. Launching continued until 0806, but at 0745, before completion, estimating that the Japanese Midway attack planes should now be returning and landing, the 33 dive bombers of the *Enterprise* group which were in the air awaiting assembly of the group were ordered by Spruance to proceed to attack. This precluded the coordinated bombing and torpedo attack which had been planned, but Spruance chose to expedite his dive bombing attack rather than to wait for completion of the torpedo squadron launching. When *Enterprise* torpedo plane group departed at 0806, they were unaccompanied by fighters, since, through a mistake in identification, their intended fighter escort accompanied the *Hornet* torpedo group instead; *Hornet* torpedo, dive bomber and fighter squadrons having departed for the target at about the same time.

The ensuing carrier air action of 4 June will not be dealt with in detail. The 1948 Naval War College study of the Battle of Midway (NavPers 91067), gives a good analysis of the air action. Fifteen torpedo planes of *Hor-*

net's VT-8, having missed interception at the expected point of contact searched northward, made the first contact at 0920. Commander Torpedo Squadron 8 (VT-8) had made a long search for his target, and by the time contact was made, his planes were running low on fuel. In reporting the contact, he requested permission to withdraw and refuel before attacking. One of the several decisions of the day was now made. Spruance knew the great value of surprise against his enemy. He knew the Japanese attached real importance to achieving surprise and that when surprised themselves they reacted in confusion and error. His instant reply to the request to withdraw was, "Attack at once."

By this time Torpedo Squadron 8 had lost contact with its fighter cover, but without hesitation the squadron commander pressed home an heroic attack in which no hits were scored and all fifteen planes were shot down, principally by fighters before the planes reached their launching points. Only one survivor was later recovered. However, the expected Japanese confusion was achieved, and the attack had an important bearing on the ultimate victory. *Hornet*'s dive bombers, when their expected interception was not made, searched to the southwestward and eventually returned to the parent carrier without sighting the enemy.

Enterprise torpedo planes, without fighter support, made a sighting at 0930, but skillful maneuvering of the Japanese ships kept them from reaching an advantageous launching position, and all torpedoes missed, while 10 of the 14 planes were shot down by fighters.

Spruance, the humanitarian, had time to think of the aircraft crews who had been sacrificed. "I felt very badly about the loss of these brave young men," he recalled. But Spruance, the practical fighter, could view his material losses unemotionally, saying, "The torpedo planes themselves were obsolescent and about to be replaced by the TBF." The failure to score damaging torpedo hits caused him disappointment and concern, but did not discourage him.

The ineffectiveness of the American attacks up to this time, both by the Midway forces and the carrier torpedo planes, led the Japanese to conclude that they had little to fear from their enemy's offensive tactics. But although the torpedo attacks were unsuccessful, they contributed greatly to the ultimate success of the overall attack by causing radical maneuvering of the ships, thus preventing launchings, and by drawing the Japanese fighter planes down from a high to a low altitude just prior to the dive bombing attacks which followed.

Enterprise dive bombing squadron VB-6 also failed to make its expected interception at 0920, but then searched to the north, and guided by a Japanese destroyer which was headed for Nagumo's Force, sighted the Force at 1005. Until this time, neither side had achieved telling results, and the preponderance of Japanese striking power still weighted the odds in their favor. The tide of the battle began to turn toward the Americans, however, at 1024, when *Enterprise* dive bombers scored hits in two carriers, *Akagi* and *Kaga*, while they were attempting to launch planes, and left them ablaze. At this point Admiral Nagumo shifted his flag from *Akagi* to the cruiser *Nagara*, completing the transfer at 1130.

The *Yorktown* attack group arrived on the scene about the same time and attacked a third carrier, *Soryu*. Their torpedo attack was also fruitless, but the dive bombers made several hits and the carrier was soon enveloped in flames and smoke. Only two of *Yorktown*'s 12 torpedo planes returned to the carrier, but no dive bombers were lost or damaged.

Spruance and his staff were elated and relieved when they received reports that three enemy carriers had been struck hard before any of their own ships had been attacked. He knew then that the advantage was on his side at least temporarily, and wanted to press his advantage vigorously and promptly. "I was very pleased, but I wanted to get the fourth carrier, which had been separated from the other three and not attacked at the time we hit the other three."

Admiral Nagumo's fourth carrier had not yet been located. Admiral Spruance later felt that, had the *Hornet* dive bombers not failed to locate their target, the fourth carrier might then have been located and later attacks on *Yorktown* by this carrier prevented. The first of these came at 1205, when 8 of 18 Japanese dive bombers broke through the screen and scored three bomb hits in *Yorktown*. Damage to the flight deck was quickly repaired, but her speed was reduced to zero.

Admiral Spruance sent two cruisers and two destroyers of TF 16 to assist *Yorktown*, and at 1313 Admiral Fletcher, CTF 17, shifted his flag to *Astoria*, it being impossible to exercise command from *Yorktown* while her communications were inoperable. By 1437, with her flight deck again in operation, *Yorktown* was able to make 19 knots, but at 1443, while launching the last of her planes, she was attacked by a flight of ten torpedo planes from *Hiryu* which scored two torpedo hits, causing complete loss of power, a jammed left rudder and an immediate 17 degree hit, which increased in 17 minutes to 23 degrees.

At about the same time, 1445, a *Yorktown* search plane reported the hitherto unsighted fourth carrier, *Hiryu*, in company with two battleships, three heavy cruisers and four destroyers bearing 279°, 110 miles from *Yorktown*. Faced with the probability of a new air attack while his ship was completely disabled, the Commanding Officer, *Yorktown*, ordered "Abandon Ship," and this operation began at 1500 while cruisers and destroyers circled the ship and destroyers picked up survivors, and by 1639 all survivors had been recovered. *Yorktown* planes landed on other carriers and were absorbed into their depleted air groups.

At 1712, CTF 17 decided to depart the vicinity of *Yorktown* to follow the movements of TF 16 with his cruisers and destroyers, leaving the destroyer *Hughes* behind with instructions to permit no one to board *Yorktown* and to sink her if necessary to prevent capture. His intent was to return to *Yorktown* at daylight and commence salvage operations.

TF 16 planes meantime located *Hiryu* and at 1700 *Enterprise* air group scored four dive bombing hits in her, starting furious fires and explosions and she was soon dead in the water with a heavy list. In the early morning she was sunk by Japanese torpedoes, unknown to Spruance, who continued searching for her, having received an incorrect report that she was retiring at ten knots.

At 1816, Admiral Spruance advised Admiral Fletcher, "*Hornet* and *Enterprise* groups now attacking fourth carrier located by your search planes. *Hornet* about twenty miles east of me. Have you any instructions for future operations?" To this, CTF 17 replied, "Negative. Will conform to your movements."

"When I sent my signal," said Spruance later, "the *Yorktown* had been abandoned and her personnel taken off by TF 17 cruisers and destroyers. As a result of a search by *Yorktown* planes, we received the position of the fourth, and undamaged, Japanese carrier, so I ordered *Enterprise* and *Hornet* to attack her. Admiral Fletcher was standing to the eastward after the *Yorktown* was abandoned, when I sent my message. He was my senior, and I did not know what he wanted to do himself, or what he wanted me to do. His reply was a message which I have always appreciated."

A Lost Aviator Disappears

A lot of young men grew up quickly at Midway, and some died with their newfound maturity. Ensign C. Markland Kelly was a Wildcat pilot with VF-8, part of the *Hornet*'s air group. He failed to return after flying a fighter escort mission for *Hornet*'s SBDs on June 4, 1942. His father died in 1981, still without understanding why and how his young son had died.

Retired Marine Major Bowen P. Weisheit wrote a self-published monograph on Mark Kelly's last flight, examining the events surrounding his disappearance. Weisheit had been a navigator on PBJs, the Marine Corps' version of the B-25 Mitchell twin-engine bomber. He had also served in PB4Ys, Marine B-24s.

Ensign Kelly's misfortunes began when his squadron leader Lieutenant Commander S. G. Mitchell began the return portion of the flight, desperately trying to keep his young fighter pilots in line.

From *The Last Flight of Ensign C. Martin Kelly, Jr.*, by Bowen P. Weisheit

The turn VF-8 made to reverse course must have been very gradual. At least two of the wingman Ensigns were so preoccupied with following their section leaders that they never even realized the squadron had changed course. Mark Kelly, in fact, would have had trouble picking the precise moment his own section leader, Gray, had committed to returning to the *Hornet.* When McInerny flew from the rear of the formation up to Mitchell's plane, Kelly probably figured at first that Mac had engine trouble to report; but when it happened again, he knew the real reason.

Meanwhile, he was still so occupied with maintaining the strict flight formation that again he had failed to check on the dive bombers for some time. Now his attention was riveted on Mitchell's plane, as it suddenly took a nose-over attitude and dropped lower and lower in altitude. Looking down finally, Mark was shocked to observe that the dive bombers were missing entirely. Next, he noticed the bright morning sun streaming through the windshield into his eyes; for over an hour it had been behind the squadron. Now, he realized that they had left the group, were descending to an altitude dangerous for combat, and were heading eastward, instead of westward toward the enemy. This could mean only one thing: the squadron was heading back to the *Hornet.*

Routinely, the fighter pilot's oxygen-demand regulator was set in the "Normal" position, i.e., providing 2.2 liters per minute of the 8.2 liters of air the pilot needed at 20,000 feet. At this setting, the oxygen would last over six hours—much longer than the fuel supply. But, on this flight, at least two of the pilots—Kelly and leader Mitchell—had set their regulators to "100%" at take off, which meant that the 900 liters in their oxygen tanks would last no more than 1.9 hours. This explains why Mitchell immediately started to let down after reaching the point-of-no-return in 1.76 hours; and why Kelly had switched his regulator back to "Normal" when he observed that the pressure had dropped precipitously. (Pressure varied from 1,800 psi to 300 psi in approximately direct proportion to the amount of oxygen remaining in the tank.) Now he was drawing 2.2 liters per minute of oxygen from his tank and the rest of his respiratory requirement (about 14 liters per minute at 8,000 feet) from the polluted ambient air in the cockpit. Easing the straps of his mask and rubbing it around had left him feeling a little better, even though the scent of the cockpit air was not pleasant.

Kelly realized what McInerny, the squadron fuel-consumption guru, had done. Now he looked to check his own fuel gauge again. After a moment required for his eyes to focus properly, he realized that success in reaching the *Hornet* could well depend on what course the ship had taken until VF-8 found her again. The fuel was over half gone, so it would be a close call; but the return would all be "down hill" from their 20,000-foot altitude, so they should make it all right. If only they had been given a point option to shoot for, he thought. Unfortunately, not even that would have guaranteed the squadron success in getting back.

The *Hornet*'s YE-ZB radio homing system operated in the VHF band and, consequently, provided line-of-sight propagation. So the maximum reception distance at VF-8's 20,000-feet altitude at the time of its turn around was about 165 nautical miles—1.15 times the square root of the squadron's height (in feet) above the ship's transmitting antenna—perhaps a little more if reception from the ground wave were taken into account. The homing signal from the shipboard YE section of the system emanated from a "hayrake" antenna (a graphically accuracy term) rotating at a rate of two revolutions per minute. It emitted a morse-code signal in each of twelve 30° sectors or lobes around the compass. For 30 seconds, the radio emitted the two-letter ship-identifier signal in all twelve sectors; for the next 4½ minutes it emitted discrete azimuth identifiers in the form of a single letter in each of the twelve sectors; after that, the five-minute cycle was repeated. A returning pilot could head for the ship by alternately weaving gently to the left and to the right of the common boundary line between two sectors—where the two sector identifiers merged. A slight turn to the left would gradually intensify the audio level of the left-hand sector's signal; then a slight turn to the right would eventually bring back the faint tones of the right-hand sector's signal, whereupon a slight left correction would be made; and so on, all the way to the ship. However, the pilot did need to have *some* sense of the direction in which the ship lay, because the same technique could take him "down the beam" *away* from the ship without his knowing it—until he passed beyond the range of the transmitter.

As the VF-8 pilots turned and tuned in the *Hornet's* homing signal on their planes' ZB part of the YE-ZB system, Mark felt confident of reaching the ship. After all, he had been using the system for months and never had failed to get back safely. Still, he did not find its use very simple in either theory or operation, and he was flying much further from the ship than he had ever flown before. In any event, the squadron was easily within the theoretical line-of-sight reception distance from the *Hornet.* By now, the enclosed cockpit had heated up, and Mark was having trouble keeping the sweat out of his eyes under the mask. It was simply another diverting element as he knew his next sense to be truly tested would be that of hearing.

For Mark and most fighter pilots, use of the YE-ZB radio homing system began with removing the cotton from their ears. Then they would tune in the receiver to the proper frequency. Kelly fought to adjust the gain control so that the audio reception in his headset accentuated the morse-code signal over the rush of static in the radio, the roar of the engine, and the whistling of the cockpit cowl-type ventilator. The first letter he isolated was a repeated "dit-dit-dah" (the letter U), which continued for several minutes. He waited for the ship's two identification letters. Sure enough, there was a pause and than a "dah-dah-dah" (O). Before he heard the second letter of the identifier, he observed Gray, his section leader, make a sudden turn to the right. Mark missed the second letter in his exertion to follow the maneuver. He knew another 4½ minutes would elapse before the identifier signal would be repeated. Probably, he had jotted down on his lap plotting board the list of sector codes and the codes for *Hornet's* identification letters: ---, .--- (OJ). Now convinced that VF was indeed headed back toward the *Hornet*, Kelly's attention returned to flying the plane, watching for the enemy, adjusting the leaky oxygen mask, and endlessly checking the engine, fuel, and other gauges.

Meanwhile, Mitchell, unable to tune in the YE homing signal, had turned the lead over to LT S. E. Ruehlow, the leader of his second section. Gray had picked up the "dit-dit-dah" (U) homing signal, and now *he* took over the formation leadership for the return trip. The leadership hand-over, he explained later, came from Ruehlow by a pat on the head and a pointed finger. Mitchell, apparently, was never aware of this pass along. After riding the U-sector signal for some minutes, Gray again turned to the right until he began to pick up the "dit-dit-dit" (S) signal of the adjoining sector; with a slight adjustment to the left, he thus continued the eastward course toward the *Hornet.* Mitchell's wingman, Ens. J. A. Talbot, also had succeeded in tuning the homing signal and thought that *he* had taken over the lead. With so many leaders, how could things go wrong? The sometime fickleness of the radio signal would not prove to be the culprit this day. A pat on the head and a pointed finger may have been the seed of the fatal ending.

While this pass-the-buck procedure was understood, used, and accepted throughout naval aviation, it certainly was not the safest under the ultimate stress of a critical decision—a decision that demanded an action affecting all the planes and which could come only from the one person who was leading the squadron. When Mitchell, Gray, Ruehlow, Talbot, and Tallman eventually saw a ship's wake, ships, carriers, or fleet (the precise term used depended on who was telling the story), the situation unequivocally called for all the planes in the squadron to alter course to close the ship(s) instantly for recovery before their tanks went dry. As we shall see, whoever should have made the decision to turn to the sighted ship(s) did not do so. By not making the right decision, he made the wrong decision.

Ens. Kelly's profound relief upon sighting the ship soon was replaced by a growing concern, overlayed with the ingrained discipline of following orders. The flight inexplicably flew on without changing course or altitude. By the time he realized that the flight leader(s) had no intention of landing on the ship that he had sighted, the ship was gone. Still heading in a southeasterly direction, Kelly knew that the nearest land was tiny Laysan Island, a 35-foot-high sand spit about 500 miles away, which they had absolutely no hope of reaching.

Had he been mistaken? The long, 100-feet-wide wake had looked so typical. He had seen it many, many times before. Had everybody else somehow missed seeing it? Why weren't they at least heading toward Midway? Why couldn't he think straight about what to do? Why hadn't they been intercepted by the combat air patrol planes as "bogies" (enemy planes) when they showed up on the ship's radar? (Probably unknown to Kelly, at least two of the fighter planes—Mitchell's and Talbot's—did have IFF radios sending signals identifying the planes as "friendlies.")

But the time for analytical thinking was over; now was the time for fast action. First, the plane of Ens. C. R. Hill (section-leader Ruehlow's wingman), flying near Kelly, suddenly dropped away as the engine sputtered and the propellor started to windmill. The dead-stick flying qualities of an F4F-4 compare well to the gliding qualities of a brick (actually the plane had a best glide speed of 120 knots, providing a glide slope of 8:1 and a 1400 ft/min vertical velocity). Hill simply dropped out of view in an instant.

By now the flyers, exhausted from the emotional tension, mental anguish, and physical stress, could have been expected to respond with a typical gripping sense of panic. In truth though, even if they had had time to think about it, this would not have happened. To panic would have required them to abandon the fighter pilot's strongest, overriding belief: "It might happen to the other guy, but never to me." No, for Mark and the others, it was more like a game of acey-ducey that can go on for a long time before the final die is cast and someone really loses. And hadn't Mark himself been in the water three weeks earlier and survived unscathed?

The VF pilots had all been briefed on the proper procedures for ditching in the ocean. The fact that no one had actually done it before (deliberately, that is), simply offered the usual challenge: doing it better than the other guy. So what was new; they just had a dangerous landing job to do. Every landing of their little, spindle-legged monsters on a heaving carrier flight deck was dangerous. Yes, accidents happened, but always to the "other guy."

Some distance back at the rear of the formation, LTJG M. F. Jennings faced his moment of trauma and starting down to the ocean. Suddenly he was fascinated by what his wingman was doing. Ensign Humphrey L. Tallman, USNR, scion of a fine old New England family, after long hours of tense anticipation, finally was firing his guns in anger. On that warm summer morning of June 4, 1942, he decided to lighten up the three tons of stocky little fighter he was about to crash into the open ocean. In anger and frustration, he threw on the arming switch of his six .50-caliber machine guns, pressed the trigger, and set the big guns roaring to life. The half-inch slugs blasted from the gun barrels at three thousand feet per second, tracing fiery criss-cross patterns arching far out over the endless blue water. In seconds, the departure of some 1440 rounds of ammunition weighing a quarter pound apiece left the plane three hundred sixty pounds lighter. With mushy controls, Tallman slowed the plane to eighty knots with power still on. Next, he released the catches of the hood cover, slid it back on its rails, and locked it full open. He didn't want to risk having it jammed or locked shut by the violence of the landing impact. Reaching forward in the process, he noticed the gun sight and made a mental note to take care to avoid it in the crash. The howler (warning horn) went off with a complaining blast as the rpms dropped below 1200 with the landing gear not down. The plane now nosed up into the wind and slammed into the oncoming waves. In the next fifty seconds or less, Ensign Tallman unbuckled his safety belt (no shoulder harness), jumped out onto the wing, jerked the CO_2 cylinder lanyards to inflate his Mae West jacket, unfastened the cover of the life-raft compartment, pulled the raft out and inflated it, grab his lemonade-filled (?) canteen out of the cockpit, jumped into the raft, cut the lanyard attached to the plane, and floated free. The heavy engine pulled the plane's nose under the water, and it sank quickly in a long spiraling path to its resting place nearly three miles below the surface.

Ens. Tallman had made it. After paddling over to join Jennings nearby, he spent the next hour angrily writing on the side of the raft his version of the disaster. Forty years later, in recounting details of the story, he was very puzzled over his inability to recall disconnecting his oxygen mask or to recall *any* details regarding the oxygen system, critical as it had been to survival.

VF-8's actual track, which eventually I was to plot, revealed the fateful irony of the situation when the squadron turned to fly back. McInerny and Magda unknowingly were out in front of the squadron, but their course constantly widened to starboard and carried them beyond view of the U.S. fleet, which they surely would have recognized because they had already seen the Japanese fleet when they turned for home many miles back. By contrast, the leader(s) of the rest of VF-8, *not* having seen the enemy, mistook our fleet for the Japanese fleet and flew on to disaster.

It seems that big McInerny had about as close a shave in his ditching as anyone. All went well to the point of his sitting back with relief in his life raft. As the plane started to slide beneath the surface, Mac discovered that the stout line was still tying the raft to the plane. The raft started under following the sinking plane as Mac sprang forward to brace his feet against the fuselage in an attempt to pull back and break the line. But his feet slipped, and both Mac and the raft started for Davy Jones' locker. As the Mae West life jacket slid under the water, it pulled Mac entirely up and out of the raft, which by then was fully submerged. Suddenly, the line let go, and the raft and the Mae West with Mac in it bobbed back to the surface. Johnny Magda, who had landed nearby at the same time, paddled over, and they soon bailed out the raft. Mac was destined to be very uncomfortable in his salty clothes until they were rescued by a PBY search-and-rescue plane from Midway on June 8.

In another few minutes, Kelly's fuel gauge left no doubt that he would be next to take the plunge. Of all the frustrated VF pilots, Mark worked hardest to the very end to try to calculate his position accurately. From well before they had leveled off at 8,000 feet about 0930 until he noticed the carrier's wake under Gray's wing, his YE-ZB receiver had been singing the same homing song—"dit-dit-dah, dit-dit-dah" (U,U) along with the *Hornet's* identification letters for thirty seconds every 4½ minutes. As Mark watched the ship's wake slide by to port, the squadron was so close to the transmitter that the discreet sector signals began to merge—the planes had entered that close-in circle of confusion that is the nature and bane of YE-ZB homing.

Of course, when the fading of the familiar "dit-dit-dah" and its merging with other sector codes failed to prompt any change of course from the squadron leader(s), Kelly's concern that something was very wrong was only heightened. They should have headed for the carrier's wake at once. Either the flight leader(s) had not seen the ship or they had mistaken it for the enemy, he conjectured. In a post-war interview many years later, Ruehlow was quoted as saying that he had seen what he thought was the enemy shortly before ditching.

Leaving the circle of confusion and traversing the S (•••), R (• - •), N(---), and M(--) sectors, the flight continued on a straight southeasterly track away from the ship, eventually receiving continuously the "dit-dah-dit-dit" code of the L sector for what was to be the remaining 20–30 minutes of the flight. Presumably figuring that he might as well try to ditch as close to the *Hornet* as possible, Kelly decided to reverse his course when it came time to go down. In any event, he was not dissuaded from his decision by the danger inherent in making a down-wind ditching or by a frantic radio call from Talbot, breaking radio-silence orders, to warn him to turn around into the wind.

At about 1022, Kelly's engine coughed twice, sputtered, and quit. Instinctively, but futilely, he hit the wobble pump. Talbot, flying almost beside Kelly, saw him fall back and downward and decided to follow him down with power still on. Starting down, Talbot first transmitted a May Day distress call for help, which later was reported to have been picked up in Oahu about a thousand miles away. It was then that Talbot was shocked to see Kelly's deliberate turn to a down-wind heading for his landing with propellor windmilling. Talbot came on the radio and desperately tried to raise Kelly to remind him of the danger of his maneuver. It was to no avail, and they ditched a mile or so apart. As a tidy specific, Talbot's watch stopped at 1036, when it went underwater as he struggled with his life raft.

The ditchings were not quite over yet. The squadron commander and the two senior section leaders flew on for another fifteen or twenty minutes before they all ditched together. Mitchell failed to free his life raft and ended up floating in his Mae West. On impact, Ruehlow's head slammed forward into the gun sight or some other obstacle, receiving a nasty gash (no shoulder harness then, only seat belts); he barely managed to free his life raft. The leader of section three (Gray), possibly the final member of VF-8 to ditch, also tempted fate. He reported that he landed with his *wheels down*. Normally, this took a lot of doing, because the wheels in the F4F-4 were not lowered by the flip of a switch; they had to be cranked down with some twenty-nine revolutions of a handle in the cockpit. Gray reported that he did it the fast way by simply releasing the crank and letting the wheels crash down. Nevertheless, he made it, too, successfully. So by 1045 or so, all ten planes of VF-8 were in the water, and at least eight of the less than clear-headed pilots were safely floating around as the planes continued their travel to earth at the bottom of the abyss some 2,500 fathoms down.

From the Japanese Viewpoint

For several years, no one wanted to hear the enemy's side of the story. Intelligence debriefers had gathered a tremendous amount of classified material from hours of discussion with German and Japanese participants at all levels. But this information was never distributed to the general public. Finally, a decade after the war, a few Germans began writing about their experiences. Aviators were particularly active. In *Battling the Bombers*, Wilhelm Johnen wrote about flying nightfighter Messerschmitt 110s against British Lancaster and Halifax fleets. Hienz Knoke described engaging American P-51s with his Messershmitt 109 in his memoir *I Flew for the Fuhrer*.

For their part, the Japanese were a little more reluctant, and, although they were unashamed, even proud, of their part in the war, they characteristically did not want to share their memories with anyone but their own people. However, by the late 1950s, a few books had appeared, most notably *Samurai!* by the top surviving Japanese ace, Saburo Sakai, who died in 2001. Enjoying a publishing resiliency, this autobiography has seen several printings in the more than 40 years it has been in print.

Another offering from Japanese authors is the source of the next selection. First published by the Naval Institute in 1955, it remains one of that organization's most popular books, and features a forward by Admiral Raymond A. Spruance.

One of the primary planners of the attack on Pearl Harbor, Mitsuo Fuchida led the first wave of 183 planes, transmitting "Tora, Tora, Tora," to indicate that the surprise had been complete and the attack was underway. Fuchida was at Midway aboard the carrier *Akagi*. He was hurt when his ship was attacked and sunk by American SBDs, and he returned to Japan to write a detailed report on the Japanese defeat. He finished his war as the fleet air staff officer during the Battle of the Marianas, June 1944.

Surprisingly, his postwar activities included farming, a conversion to

Christianity, and becoming an American citizen in the 1960s. He died in 1976.

Like Fuchida, Masatake Okumiya graduated from the Imperial Japanese Naval Academy at Eta Jima and became a naval aviator. He participated in the Japanese attack on the USS *Panay*, an American gunboat serving in China in 1937. This incident first inflamed American opinion against the Japanese war in China. Okumiya served throughout the war, reaching the rank of commander. He joined the postwar self-defense force and rose to major general before retiring in 1964.

In the following selection, we see the Midway operation from Japanese eyes and perhaps realize that the enemy was not the monolithic entity most Americans believed it was during the war. Young men fought and died just as quickly under a rising sun as they did beneath stars and stripes.

From *Midway: The Battle That Doomed Japan,* by Mitsuo Fuchida and Masatake Okumiya

Enemy Carrier Planes Attack

As the Nagumo Force proceeded northward, our four carriers feverishly prepared to attack the enemy ships. The attack force was to include 36 dive bombers (18 "Vals" each from *Hiryu* and *Soryu*) and 54 torpedo bombers (18 "Kates" each from *Akagi* and *Kaga*, and nine each from *Hiryu* and *Soryu*). It proved impossible, however, to provide an adequate fighter escort because enemy air attacks began again shortly, and most of our Zeros had to be used to defend the Striking Force itself. As a result, only 12 Zeros (three from each carrier) could be assigned to protect the bomber groups. The 102-plane attack force was to be ready for take-off at 1030.

After *Tone*'s search plane reported the presence of a carrier in the enemy task force, we expected an attack momentarily and were puzzled that it took so long in coming. As we found out after the war, the enemy had long been awaiting our approach, was continuously informed of our movements by the flying boats from Midway, and was choosing the most advantageous time to pounce. Admiral Spruance, commanding the American force, planned to strike his first blow as our carriers were recovering and refueling their planes returned from Midway. His wait for the golden opportunity was rewarded at last. The quarry was at hand, and the patient hunter held every advantage.

Between 0702 and 0902 the enemy launched 131 dive bombers and torpedo planes. At about 0920 our screening ships began reporting enemy carrier planes approaching. We were in for a concentrated attack, and the Nagumo Force faced the gravest crisis of its experience. Was there any escape? An electric thrill ran throughout the fleet as our interceptors took off amid the cheers of all who had time and opportunity to see them.

Reports of approaching enemy planes increased until it was quite evident that they were not from a single carrier. When the Admiral and his staff realized this, their optimism abruptly vanished. The only way to stave off disaster was to launch planes at once. The order went out: "Speed preparations for immediate take off!" This command was almost superfluous. Aviation officers, maintenance crews, and pilots were all working frantically to complete launching preparations.

The first enemy carrier planes to attack were 15 torpedo bombers.* When first spotted by our screening ships and combat air patrol, they were still not visible from the carriers, but they soon appeared as tiny dark specks in the blue sky, a little above the horizon, on *Akagi*'s starboard bow. The distant wings flashed in the sun. Occasionally one of the specks burst into a spark of flame and trailed black smoke as it fell into the water. Our fighters were on the job, and the enemy again seemed to be without fighter protection.

Presently a report came in from a Zero group leader: "All 15 enemy torpedo bombers shot down." Nearly 50 Zeros had gone to intercept the unprotected enemy formation! Small wonder that it did not get through.

Again at 0930 a lookout atop the bridge yelled: "Enemy torpedo bombers, 30 degrees to starboard, coming in low!" This was followed by another cry from a port lookout forward: "Enemy torpedo planes approaching 40 degrees to port!"

The raiders closed in from both sides, barely skimming over the water. Flying in single columns, they were within five miles and seemed to be aiming straight for *Akagi*. I watched in breathless suspense, thinking how impossible it would be to dodge all their torpedoes. But these raiders, too, without protective escorts, were already being engaged by our fighters. On *Akagi*'s flight deck all attention was fixed on the dramatic scene unfolding before us, and there was wild cheering and whistling as the raiders went down one after another.

**Editors' Note:* These planes were of the valiant VT-8 (Lt. Cdr. J. C. Waldron) from *Hornet*. All 15 planes were shot down, and the sole survivor was Ens. G. H. Gay, who was rescued from the water by a Navy Catalina next day.

Of the 14 enemy torpedo bombers which came in from starboard, half were shot down, and only 5 remained of the original 12 planes to port. The survivors kept charging in as *Akagi* opened fire with antiaircraft machine guns.

Both enemy groups reached their release points, and we watched for the splash of torpedoes aimed at *Akagi*. But, to our surprise, no drops were made. At the last moment the planes appeared to forsake *Akagi*, zoomed overhead, and made for *Hiryu* to port and astern of us. As the enemy planes passed *Akagi*, her gunners regained their composure and opened a sweeping fire, in which *Hiryu* joined. Through all this deadly gunfire the Zeros kept after the Americans, continually reducing their number.

Seven enemy planes finally succeeded in launching their torpedoes at *Hiryu*, five from her starboard side and two from port. Our Zeros tenaciously pursued the retiring attackers as far as they could. *Hiryu* turned sharply to starboard to evade the torpedoes, and we watched anxiously to see if any would find their mark. A deep sigh of relief went up when no explosion occurred, and *Hiryu* soon turned her head to port and resumed her original course. A total of more than 40 enemy torpedo planes had been thrown against us in these attacks, but only seven American planes had survived long enough to release their missiles, and not a single hit had been scored. Nearly all of the raiding enemy planes were brought down.*

Most of the credit for this success belonged to the brilliant interception of our fighters, whose swift and daring action was watched closely from the flagship. No less impressive was the dauntless courage shown by the American fliers, who carried out the attack despite heavy losses. Shipboard spectators of this thrilling drama watched spellbound, blissfully unaware that the worst was yet to come.

As our fighters ran out of ammunition during the fierce battle, they returned to the carriers for replenishment, but few ran low on fuel. Service crews cheered the returning pilots, patted them on the shoulder, and shouted words of encouragement. As soon as a plane was ready again, the pilot nodded, pushed forward the throttle, and roared back into the sky. This scene was repeated time and time again as the desperate air struggle continued.

Five Fateful Minutes

Preparations for a counter-strike against the enemy had continued on board our four carriers throughout the enemy torpedo attacks. One after another,

**Editors' Note:* Of the total 41 torpedo planes from the three American carriers, only six actually returned from the attack.

planes were hoisted from the hangar and quickly arranged on the flight deck. There was no time to lose. At 1020 Admiral Nagumo gave the order to launch when ready. On *Akagi*'s flight deck all planes were in position with engines warming up. The big ship began turning into the wind. Within five minutes all her planes would be launched.

Five minutes! Who would have dreamed that the tide of battle would shift completely in that brief interval of time?

Visibility was good. Clouds were gathering at about 3,000 meters, however, and though there were occasional breaks, they afforded good concealment for approaching enemy planes. At 1024 the order to start launching came from the bridge by voice-tube. The Air Officer flapped a white flag, and the first Zero fighter gathered speed and whizzed off the deck. At that instant a lookout screamed: "Hell-divers!" I looked up to see three black enemy planes plummeting toward our ship. Some of our machine guns managed to fire a few frantic bursts at them, but it was too late. The plump silhouettes of the American "Dauntless" dive bombers quickly grew larger, and then a number of black objects suddenly floated eerily from their wings. Bombs! Down they came straight toward me! I fell intuitively to the deck and crawled behind a command post mantelet.

The terrifying scream of the dive bombers reached me first, followed by the crashing explosion of a direct hit. There was a blinding flash and then a second explosion, much louder than the first. I was shaken by a weird blast of warm air. There was still another shock, but less severe, apparently a near-miss. Then followed a startling quiet as the barking of guns suddenly ceased. I got up and looked at the sky. The enemy planes were already gone from sight.

The attackers had gotten in unimpeded because our fighters, which had engaged the preceding wave of torpedo planes only a few moments earlier, had not yet had time to regain altitude. Consequently, it may be said that the American dive bombers' success was made possible by the earlier martyrdom of their torpedo planes. Also, our carriers had no time to evade because clouds hid the enemy's approach until he dove down to the attack. We had been caught flatfooted in the most vulnerable condition possible—decks loaded with planes armed and fueled for an attack.

Looking about, I was horrified at the destruction that had been wrought in a matter of seconds. There was a huge hole in the flight deck just behind the amidship elevator. The elevator itself, twisted like molten glass, was drooping into the hangar. Deck plates reeled upward in grotesque configurations. Planes stood tail up, belching livid flame and jet-black smoke. Reluctant tears streamed down my cheeks as I watched the fires spread, and I was terrified at the prospect of induced explosions which would surely doom the ship. I heard Masuda yelling, "Inside! Get inside! Everybody who isn't working! Get inside!"

Unable to help, I staggered down a ladder and into the ready room. It was already jammed with badly burned victims from the hangar deck. A new explosion was followed quickly by several more, each causing the bridge structure to tremble. Smoke from the burning hangar gushed through passageways and into the bridge and ready room, forcing us to seek other refuge. Climbing back to the bridge I could see that *Kaga* and *Soryu* had also been hit and were giving off heavy columns of black smoke. The scene was horrible to behold.

Akagi had taken two direct hits, one on the after rim of the amidship elevator, the other on the rear guard on the portside of the flight deck. Normally, neither would have been fatal to the giant carrier, but induced explosions of fuel and munitions devastated whole sections of the ship, shaking the bridge and filling the air with deadly splinters. As fire spread among the planes lined up wing to wing on the after flight deck, their torpedoes began to explode, making it impossible to bring the fires under control. The entire hangar area was a blazing inferno, and the flames moved swiftly toward the bridge.

Because of the spreading fire, our general loss of combat efficiency, and especially the severance of external communication facilities, Nagumo's Chief of Staff, Rear Admiral Kusaka, urged that the Flag be transferred at once to light cruiser *Nagara*. Admiral Nagumo gave only a half-hearted nod, but Kusaka patiently continued his entreaty: "Sir, most of our ships are still intact. You must command them."

The situation demanded immediate action, but Admiral Nagumo was reluctant to leave his beloved flagship. Most of all he was loathe to leave behind the officers and men of *Akagi*, with whom he had shared every joy and sorrow of war. With tears in his eyes, Captain Aoki spoke up: "Admiral, I will take care of the ship. Please, we all implore you, shift your flag to *Nagara* and resume command of the Force."

At this moment Lieutenant Commander Nishibayashi, the Flag Secretary, came up and reported to Kusaka: "All passages below are afire, Sir. The only means of escape is by rope from the forward window of the bridge down to the deck, then by the outboard passage to the anchor deck. *Nagara*'s boat will come alongside the anchor deck port, and you can reach it by rope ladder."

Kusaka made a final plea to Admiral Nagumo to leave the doomed ship. At last convinced that there was no possibility of maintaining command from *Akagi*, Nagumo bade the Captain good-bye and climbed from the bridge window with the aid of Nishibayashi. The Chief of Staff and other staff and headquarters officers followed. The time was 1046.

On the bridge there remained only Captain Aoki, his Navigator, the Air Officer, a few enlisted men, and myself. Aoki was trying desperately to get in touch with the engine room. The Chief Navigator was struggling to see if anything could be done to regain rudder control. The others were gathered on the anchor deck fighting the raging fire as best they could. But the unchecked flames were already licking at the bridge. Hammock mantelets around the bridge structure were beginning to burn. The Air Officer looked back at me and said, "Fuchida, we won't be able to stay on the bridge much longer. You'd better get to the anchor deck before it is too late."

In my condition this was no easy task. Helped by some sailors, I managed to get out of the bridge window and slid down the already smoldering rope to the gun deck. There I was still ten feet above the flight deck. The connecting monkey ladder was red hot, as was the iron plate on which I stood. There was nothing to do but jump, which I did. At the same moment another explosion occurred in the hangar, and the resultant blast sent me sprawling. Luckily the deck on which I landed was not yet afire, for the force of the fall knocked me out momentarily. Returning to consciousness, I struggled to rise to my feet, but both of my ankles were broken.

Crewmen finally came to my assistance and took me to the anchor deck, which was already jammed. There I was strapped into a bamboo stretcher and lowered to a boat which carried me, along with other wounded, to light cruiser *Nagara*. The transfer of Nagumo's staff and of the wounded was completed at 1130. The cruiser got under way, flying Admiral Nagumo's flag at her mast.

Meanwhile, efforts to bring *Akagi*'s fires under control continued, but it became increasingly obvious that this was impossible. As the ship came to a halt, her bow was still pointed into the wind, and pilots and crew had retreated to the anchor deck to escape the flames, which were reaching down to the lower hangar deck. When the dynamos went out, the ship was deprived not only of illumination but of pumps for combatting the conflagration as well. The fireproof hangar doors had been destroyed, and in this dire emergency even the chemical fire extinguishers failed to work.

The valiant crew located several hand pumps, brought them to the anchor deck, and managed to force water through long hoses into the lower hangar and decks below. Firefighting parties, wearing gas masks, carried cumbersome pieces of equipment and fought the flames courageously. But every induced explosion overhead penetrated to the deck below, injuring men and interrupting their desperate efforts. Stepping over fallen comrades, another damage-control party would dash in to continue the struggle, only to be mowed down by the next explosion. Corpsmen and volunteers carried out dead and wounded from the lower first aid station, which was jammed with injured men. Doctors and surgeons worked like machines.

The engine rooms were still undamaged, but fires in the middle deck sections had cut off all communication between the bridge and the lower levels of the ship. Despite this the explosions, shocks, and crashes above, plus the telegraph indicator which had run up "Stop," told the engine-room crews in the bowels of the ship that something must be wrong. Still, as long as the engines were undamaged and full propulsive power was available, they had no choice but to stay at General Quarters. Repeated efforts were made to communicate with the bridge, but every channel of contact, including the numerous auxiliary ones, had been knocked out.

The intensity of the spreading fires increased until the heat-laden air invaded the ship's lowest sections through the intakes, and men working there began falling from suffocation. In a desperate effort to save his men, the Chief Engineer, Commander K. Tampo, made his way up through the flaming decks until he was able to get a message to the Captain, reporting conditions below. An order was promptly given for all men in the engine spaces to come up on deck. But it was too late. The orderly who tried to carry the order down through the blazing hell never returned, and not a man escaped from the engine rooms.

Fighting for the Solomons: The Carrier Navy Joins the Marines

Coming barely two months after Midway, the Allied assault on the Japanese airfield on Guadalcanal on August 7, 1942, began a bloody six-month campaign on the ground, in the air, and on the sea. The Marines fought the Japanese defenders, who scurried into the surrounding hills and trees to harass the Leathernecks day and night. Marine fighter and dive squadrons arrived to protect the newly won airfield. The Navy, with three carriers, *Enterprise*, *Hornet*, and *Wasp* (CV-7), tried supporting the Marine aircrews, but soon found itself under attack by Japanese bombers and fighters staging from Lae, New Guinea. Eventually, by February 1943, the Japanese had given up, had recovered their remaining troops from Guadalcanal, and had left the area. It was another important victory for the Allies, and another major defeat for the Japanese.

While the Marine Corps squadrons (along with a few Army and New Zealand units) were the backbone of the defense of Henderson Field on Guadalcanal, Navy squadrons augmented the Allied effort when *Enterprise* was hit by Japanese bombs and forced to retire for repairs. Its fighter squadron, VF-10, was sent ashore to help bolster the two Marine squadrons, VMF-223 and VMF-224. (The *Wasp* was sunk by a Japanese submarine on September 15, 1942, while protecting a convoy. The *Hornet* followed on October 26 during the Battle of Santa Cruz.) Having survived several intense air engagements, the young carrier aircrews were willing to help out.

Then-Ensign E. L. Feightner was a junior member of VF-10, but he had already seen considerable combat. He would eventually gain nine aerial kills in the Pacific, four in Wildcats and five in Hellcats. But now, he was still learning his trade as a fighter pilot. I once pressed him for photographs for a book we were working on. I have never known any naval aviator who didn't have more than a few pictures showing himself in flight gear, by his plane.

Feightner, known as "Whitey" since his early days in the Navy because of his fair complexion, simply smiled and said, "We were just busy trying to stay alive. Who had time to take pictures?" Fortunately, later in the war, things were apparently more "relaxed," and then-Lieutenant Feightner posed in a VF-8 Hellcat.

After the war, Whitey enjoyed a successful career—eventually rising to rear admiral—as a test pilot and early member of the Blue Angels flight demonstration team. He is, in fact, one of only two aviators to have flown the Chance Vought F7U-1 Cutlass with the Blue Angels. Quiet and unassuming, with an easygoing manner and ready smile that perhaps belie his fighter-pilot's tenacity, Whitey Feightner describes those difficult days as a temporary member of the Cactus Air Force fight to keep the Marines on Guadalcanal.

This book was actually the first of three collections Captain Wooldridge edited using interviews mainly from the Naval Institute's extensive oral history collection. It con-

tains many fascinating entries that tell the story of the rise of the aircraft carrier as the Navy's primary capital ship, replacing the traditional battleship.

From *Carrier Warfare in the Pacific: An Oral History*, edited by E. T. Woolridge

In October 1942, while assigned to VF-3, Butch O'Hare's squadron, I received unexpected orders to join VF-10 on the *Enterprise* at Pearl Harbor. The commanding officer of VF-10 was Lt. Comdr. Jimmy Flatley, who was one of the most amazing people I had ever encountered. Talk about a true leader. Jimmy Flatley just sort of epitomized what you think of as a squadron commander. He got along well with everybody in the squadron, was a good pilot, and had the respect of all the aviators in the squadron. He was a very

An SBD rear gunner mans his two .30-caliber machine guns. Enlisted crewmen were important factors in the success of the Dauntless, considered obsolete before Pearl Harbor. (Courtesy of the National Archives)

religious and patriotic man, and felt very deeply that anything we did as a squadron should be done with an eye toward helping not only the squadron and the ship, but also the country. He was concerned about such things as safety to the point that he invented the shoulder harness, although it was really a chest harness. We actually modified all our airplanes; we had a strap around the chest, which we religiously put on, at his insistence, of course.

He was a great person for sitting down with people in a small group to talk tactics. He would talk about the philosophy that he thought the Japanese lived by, what we should expect from them. He was great for writing letters to the squadron. It worried him when we lost somebody. I can remember distinctly the first day in combat we had with the squadron. Ten days out of Pearl Harbor, we thought we had lost 11 people on the first day of combat. It turned out we hadn't really lost that many; some of them were sitting on other ships out there in the task force, but we thought we'd lost them. That really hurt him, the fact that we had lost that many people.

He was extremely innovative. For instance, the day the *Chicago* was sunk, 30 January 1943, we were on combat air patrol. There were 12 Bettys that came after the *Enterprise*, and we managed to turn them away and they then headed for the *Chicago*. We shot down some of them before they got there, but the rest of them bunched up together and since they could outrun us in a dive, they were out of gun range when they reached the *Chicago*. They dropped four torpedoes into the *Chicago*, which was under tow, and they sank it. But, once they were out of the dive and down on the deck, we were able to slowly overtake them. I was flying wing on Jimmy Flatley, and we were very slowly overhauling one which was out of range in front of us. Jimmy Flatley pulled his nose up, fired, and put tracers in front of the Betty, which panicked him. He turned right and headed for a cloud layer at about 2,500 feet. Being a dumb ensign, I went after him in the cloud. Apparently he turned upon entering the cloud layer, and the next thing I saw, in the cloud, was this big shape directly in front of me. I fired and managed to avoid him somehow and broke out on top of the cloud layer, but he didn't come out on top. I immediately went back down below, joined up on Jimmy Flatley, who gave me a thumbs up and said, "You got him." That's the kind of man he was. When we got back aboard, he would take no credit for that airplane at all; he gave me full credit for it. He said, "No, all I did was turn him. You shot him down."

Shortly after I reported to VF-10, we left Pearl Harbor, and 10 days later the *Enterprise* and the *Hornet* were in the battle of Santa Cruz, in October 1942. I'll never forget, on the day before the battle actually started, 25 Octo-

Early in the war, crewmen ready an SBD for action aboard *Enterprise*. The two white lines on the tail were a lineup aid for the LSO guiding the pilot toward the flight deck. (Courtesy of the National Archives)

ber, we had reports that the *Zuikaku* and the *Shokaku* were up north of us and coming our way, so we organized a search-and-attack group. The group commander, Comdr. Dick Gaines, was flying a TBF, and we had about five TBFs with torpedoes. I think there were 6 SBDs, all carrying 1,000-pound bombs, and we had 8 fighters. This was late afternoon, along about 3:30, when we launched. We got out to 175 miles, and it was absolutely one of these days you could see for 100 miles. We still didn't have a task force in sight, so we flew another 75 miles northeast of there. When we got to that point, according to our original mission, we should have been landing aboard ship. I was completely confused. I thought I was a navigator, but by this time we'd made those two unscheduled turns and we fighters were continuously weaving over the bombers. There was no radio transmission at all; we were under complete radio silence, so nobody touched his radio. By now I didn't have any idea what was going on. If somebody had said, "Go home," I would have had a hard time finding my way back to the ship.

We finally headed back for the carrier. About 20 minutes out on that leg, headed back toward the ship, Lt. F. D. Miller, one of the lieutenants in "Killer" Kane's division, dropped back, and I turned to look at him because I was tail-end Charlie in the other group. This pilot bailed out of the airplane. I saw his chute open, and he went drifting back. I didn't know what to do, but everybody kept going, so I kept going with them. He obviously had run out of fuel. We were all down to about 42 gallons, and we were still a long ways from home at this point. Right after that, it got dark. All at once the group commander started to circle; this was where the carrier should have been. It was now dark, and there was nothing visible down there. Still no radio transmission. We had the YE in those days, but I don't think the ship had it on. There was a concern that it might be a beacon for the Japanese and bring them in. We knew they had the code, and they also had some captured equipment they could home in with. It had been decided that we wouldn't break radio silence because we knew the Japanese task force was somewhere around, but we didn't know where. The ship wasn't there because some submarines had fired torpedoes at the force, and so they headed off out of the rendezvous area. We were expendable, I guess.

We started to spiral down through a ceiling of broken clouds with bases at 600 to 800 feet. The moon wasn't up yet; it was really black. The next thing I know, I saw clouds going by me and we're down under this cloud deck. We've got 25 or so airplanes all milling around under the cloud layer. No carrier was visible, and about this time, unbeknownst to us, the bomber leader decided to drop the bombs that they had been carrying around all this time. When they dropped them, there was the biggest flash you've ever seen. One of those bombs didn't safe, and when it hit the water it blew. When I saw that flash, I thought one of our people had inadvertently flown in the water, but it was actually one of the bombs going off. About that time, I noticed I could see my wing lights reflecting off the water. I was flying stepped down on Swede Vejtasa. He had let down, and, boy, I immediately pulled up above him. He was down low, making small turns, and all of a sudden he just straightened out and headed off. He later said he had found an oil slick from the task force with his wing lights, down on the water. We were literally only 10 or 15 feet off the water at this point, and everybody else was following us. Forty-five miles away, we found the task force. Swede did it by just following that leak. We wouldn't be here today if it hadn't been for somebody as innovative as Swede.

I'll tell you, talk about a really competent aviator—now, there's one. Swede Vejtasa was about as cool and laid back as anyone you'd ever want to meet. Early in the war, he and his rear-seat man shot down several airplanes in the SBD. He switched over and became a fighter pilot, and I'll tell you, he was just superb as a fighter pilot. He never wanted to be anything else. I remember the day he shot down seven airplanes in one flight. He got in the midst of this group of torpedo planes and just stayed there picking them off one by one, from the back to the front. Swede Vejtasa was single-minded; he was an airplane driver whose mission was to find and kill the enemy.

We found the task force, and they decided that, since the fighters were the lowest on fuel, we should get aboard first. Swede went around, and I was right behind him. That was my first night carrier landing, and it didn't bother me in the least. I'd done it on the field before, but I'd never landed aboard the ship at night. That's how short we were on training in those days. We got aboard and the group commander was still circling in his TBF, but the other

Lt. Whitey Feightner in a Hellcat, which displays 9 Japanese flags to indicate his tally of kills. (Courtesy of E.L. Feightner)

TBFs started dropping in the water. In fact, the executive officer of the squadron came around and was in the groove when he ran out of fuel, and went under the fantail. Most of them made good water landings, and they survived. Then the group commander came around and landed last. He had not been carrying a torpedo. We got all the fighters aboard except for Miller, who had bailed out on the return flight. This was the night before the big battle, so, of course, we were minus a few airplanes.

The next morning, 26 October, was the day of the battle of Santa Cruz. We put together a strike group, and we launched about 8 A.M. I was a "spare" or "standby" for this mission, but I ended up going anyway. I remember I was flying with Flatley that morning, in the second section. We were about 20 miles from the carrier and had just formed up and were headed out, when Flatley signaled us to check our guns. Everybody was checking to make sure that their guns were charged. We looked up, somebody yelled, and Zeros were all over us. They had gotten up about two hours before we did, I guess, and they were already that close to the task force. They took out three fighters on the other side of the formation, plus several other planes out of the attack group. Those early Japanese pilots were good. Flatley didn't let us turn away a bit. He just said, "Keep going." The Zeros made that one pass and went on to our task force, and we headed off on to their task force. Apparently, unbeknownst to us, there were about three waves of attackers that actually hit the *Hornet* and the *Enterprise* while we were gone. We found the enemy task force and made an attack. We headed back home, and I'll never forget the sight. I looked up on the horizon, and the first thing I saw was our happy home, listing and smoking heavily. We got there, and it turned out to be the *Hornet*, not the *Enterprise;* the *Enterprise* was about another 30 miles beyond the *Hornet* group.

We got back to the *Enterprise* really low on fuel, got in the traffic pattern, looked down, and saw torpedo wakes. The ship was zigzagging, S-turning like mad, but continued taking us aboard. I landed aboard, and there were airplanes stacked up clear back across the barricade; there weren't any barricades up at all. The landing signal officer was standing out there giving us the cut. The ship was in a terrific right turn when I came aboard. Here I am in a left-hand pattern, and they're making a right turn. I came around and got to the stern, he gave me a cut, and I landed. I'll never forget the shock; they taxied me forward, and I got out of the airplane and there was no no. 2 elevator, it was stuck in the down position. I got down to the hangar deck and was wading around in water halfway up to my knees, fuel, and dead bodies. Something had burned on the ship; you could smell this all over the place.

They had taken bomb hits forward and amidships. In the meantime, not only are these torpedoes going by, but there's a dive-bombing attack going on at the same time.

At some time during these attacks, Robin Lindsey, the landing signal officer, got in the backseat of an SBD on the flight deck and started shooting those twin .30-calibers at an airplane coming across the fantail of the ship. Robin Lindsey was almost legendary among LSOs. Here was a person that everybody respected as much as a pilot as they did as an LSO.

Things were pretty hectic. It turned out, when we got organized, that we had a bunch of airplanes airborne with no room to land them. What was left of the *Hornet*'s air group was now aboard the *Enterprise*, or else was still airborne. We had so many airplanes that we had to send 13 planes to Espíritu Santo. By 30 October, the ship was in Nouméa, New Caledonia, where they were going to do major repairs, while the air group camped ashore at the grass field of Tontouta.

We were based ashore in Nouméa until 11 November, when they pulled the ship out long enough to take us up close, and launch us for Guadalcanal. The whole concept was that we were to search and find a bunch of enemy transports coming down to reinforce Guadalcanal, and that's about all the information that we had. Red Carmody, flying search in an SBD, ran across the ships on 14 November. That afternoon 8 SBDs with 1,000-pound bombs and 12 F4Fs led by Jimmy Flatley launched to attack the transports. We were cruising at about 25,000 feet, when we saw the Japanese task force, about 25 miles away. All of a sudden, we looked up and about 40 Zeros were coming right at us, maybe 5,000 feet above us. I've never heard anybody so calm in my life. Jimmy Flatley got on the radio and said, "Don't anybody flash any wings or canopy. Just sit still. Our mission is to attack the ships down there." With these airplanes going overhead, he assigned every bomber a separate target. Fortunately, the Zeros didn't see us, and we pressed right on in. There were still 7 of the original 11 transports coming down the slot; the rest had been taken care of by repeated strikes by planes from Henderson Field. We went in and strafed, and there were people all over those decks. I have vivid recollections of seeing people leaping over the sides of the ships. Eight airplanes strafing down the deck of a transport, with people all over the topside, must have been devastating. By the time we had finished strafing we had Zeros all over us, but we managed to get the bomber group together, and we covered them all the way back in to Guadalcanal. That attack was pretty successful, but there were still five transports heading for Guadalcanal. By

A typical deck scene involves deck crews unfolding the wings of an F4F-4 Wildcat. The little Grumman gave a good account of itself in the first three years of the war in the Pacific and in the Battle of the Atlantic. (Courtesy of the National Archives)

nightfall, after further attacks from the *Enterprise* and Henderson Field, there remained only four damaged transports still headed down the Slot.

We flew to Guadalcanal to land. They put the bombers on the bomber strip and sent the fighters to another field called a fighter strip. This was a grass field with some Marston matting along the sides of it, where they were parking the airplanes. I landed and off to my left a big geyser of dirt went up. I thought the place was being bombed. It turned out the Japanese had some howitzers up on the side of the hill, and they were shelling the airfield. I looked around, and there was a Marine standing over there, beckoning me to come over under the trees, so I taxied over there. They grabbed the tail and pulled it around underneath the trees. I jumped out of the airplane and had the shock of my life. There was a hole about two-and-a-half feet across, in a big winding spiral down into the ground. I was told that it was a 14-inch shell hole, where the battleships had been shelling the place. I almost fell in that thing.

They put me in a Jeep, went around collecting pilots, and took us over to a tent area back under the trees, while these guns are still shelling the field. Right in the middle of this, a Marine with a campaign hat on and smoking a big cigar comes strolling across the middle of the strip. It was Joe Foss; he's strolling along across to the area, saying "Hello" to all these people as if nothing was happening, and he welcomed us to Guadalcanal. At the time we landed, the Marines had three fighter airplanes that were operational, and that night, that guy up in the hill managed to bag a couple more airplanes; both of them were ones that we had brought in. I guess we ended up with about 20 fighters left out of that whole melee.

The next morning, when it got to be daylight, we were about to get up, and a Marine came by and said, "Stay in your tent." The next thing, we hear all this automatic fire, and the Marines are out there machine-gunning the tops of the trees. The Japs had infiltrated during the night, and were sitting around up in the tops of the coconut trees, waiting for us to come out. The Marines got several of them out of those trees, and then we were allowed to come out.

I remember, in the tent at night, we were all lying there, and it was muggy and hot. Until they'd start shelling from the ships out in Savo Bay, we'd try to get a little bit of sleep. One night we could hear the Japanese walking around outside and trying to climb up into the trees. Of course, every bird sounded like a Japanese, so people were really on edge. There were maybe 20 of us in the tent when all of a sudden one of them let out a blood-curdling scream, and I heard a .45-caliber being cocked. Somebody in that tent said, "I'm going to shoot the first guy who moves." Well, I didn't even breathe. A little bit later, some guy said, "I'm all right." He had been lying there barechested, and a lizard had crawled across his chest. He thought it was a Japanese that was after him.

While we were there in November, we were part of the First Marine Division, Reinforced. Anybody who was available flew together. We had such a tenuous hold on that place, and it was just touch and go whether we were going to survive and stay there. The P-39s were flying coffins in those days; in air-to-air combat, they didn't have a chance against a Zero. They were hanging bombs on them, and when the Japanese would set up these guns up in the hills, they would go up and dive bomb them from low altitude; practically skip bombing them is what they were doing, and they were pretty effective. Some P-38s moved in while we were there. The pilots had very little time in the airplanes at all. I talked to one pilot who had never fired the guns in it; somebody stuck him in a P-38 and said, "Go."

Those days on Guadalcanal I wouldn't care to repeat. Thank goodness, on 17 November, we left to begin the return trip to Nouméa. We turned all our airplanes over to the Marines and got a new set of airplanes that came out on a "jeep" carrier. We came back later in February 1943 for a short stay. I tell you, you can have my share of being a ground soldier. I want no part of that.

There Were Heroes on Both Sides

By 1944, the once-omnipotent Zero had been overpowered by new Allied types like the Grumman Hellcat and Vought Corsair. The portly F6F eventually tallied more than twice as many kills (5,216 in the Pacific as well as limited engagements in the European theaters) as its nearest competitor, the F4U (2,140), producing more aces than any other naval fighter. Yet the Zero was still a dangerous adversary when flown by one of the surviving skilled Imperial Navy aces, many of whom had been killed by 1945.

The air war remained a dangerous venue, and none of the young ensigns and jaygees who saw their first combat in June 1944 would have believed any newsman's account of how the Pacific had become a "Navy lake." Not when so many friends and experienced flight leaders were still failing to return to their carriers, and seats in the wardroom were empty.

Much of the Japanese carrier fleet had been sunk by late 1944, its last appearance in strength coming in the epic Marianas Turkey Shoot of June 1944, and in a greatly reduced amount in the Battle of Leyte Gulf in the Philippines that October. In contrast, the American carrier fleet had grown to more than 100 ships. Only the end of the war would curtail the staggering production schedule back home. As Japanese leaders feared in 1941, American industrial strength was smothering any offensive ambitions, snuffing out the flame of Japanese military might. Yet there were often flashes of the Samurai aviators who ruled the Pacific skies for 18 months after Pearl Harbor.

Richard Newhafer wrote *The Last Tallyho* using his experience as an F6F Hellcat pilot. Although the novel reached the bestseller lists when it was published in 1964, it disappeared from most book stores quickly, which is unfortunate. Newhafer's paean to the Pacific fighter-pilot's war is one of the finest fictional treatments of the subject, ranking with F-86 pilot James Salter's 1958 tale of Sabre pilots in Korea, *The Hunters*. But where Salter's book was recently reissued, Newhafer's novel has yet to reappear after nearly 40 years.

As a note of interest, the stalwart squadron commander in this excerpt, Lieutenant Commander Harry Hill, had a true-life counterpart, at least in name. Lieutenant Harry E. Hill shot down seven Japanese aircraft as a member of VF-5 (one of the squadrons in which Richard Newhafer served) aboard USS *Yorktown* (CV-10). However, unlike the fictional Harry Hill, he survived the war and retired as a commander in 1968.

Born in Chicago in 1922, Richard L. Newhafer received his gold wings and ensign bars in May 1943. He flew Hellcats with VF-5 and VF-6. He was credited with 2.5 aerial kills while with VF-5. He enjoyed a career as a successful writer before his premature death in 1974. He was an early member of the Blue Angels in 1949 and 1954, serving as the team's public affairs officer. Hellcat ace Zeke Cormier, who led the Blue Angels in 1955, wrote, "A better choice could not be imagined for team public affairs officer than the irrepressible, free-spirited Lieutenant Commander Dick 'The Old Newf' Newhafer." (Wildcats to Tomcats: The Tailhook Navy, *Cormier, Schirra and Wood, with Tillman, Phalanx, 1995)*

Newhafer proved a little too *free-spirited, and after several incidents involving his ground conduct in and out of uniform, he and the Navy parted ways. Cormier wrote, "His [Newhafer] decision saved me further embarrassment, but I would miss a good friend, and the team would miss considerable color."*

Newhafer had received the Navy Cross for a daring attack on the Japanese battleship Ise *in July 1945, as well as three Distinguished Flying Cross awards for other late-war missions.*

Although he left active duty in October 1950, he was recalled later that year. As a member of VF-23, flying F9F Panthers, he flew combat in Korea from the USS Essex. *He later wrote for television, including the series "Combat" and the detective show "Cannon." He authored more than six novels, but none achieved the success of his first,* The Last Tallyho.

From *The Last Tallyho,* by Richard Newhafer

Crowley had come into the fight with trepidation. After Hill had led the formation into the first encounter Crowley had swung wide and entered the fight from the starboard side. He came in fast, leaving his wingman, Anders, fifty yards behind. This suited Anders perfectly because he had no inclination to fight by Crowley's side in such an engagement. Anders watched Crowley barrel into the melee, then he swung over and up, trying to climb two or three thousand feet so he would meet the enemy with an altitude advantage. It took Anders one minute to climb twenty-five hundred feet, and when he had, his airspeed had dropped off to one hundred knots. He came out of his climb just as a Zero opened fire at him. Anders grinned in his cockpit because his speed disadvantage would win this one for him. Easily now, his slow speed allowing him to turn quickly inside his opponent, he brought his guns to bear and opened fire, and the Zero disappeared in a mass of debris. Anders continued hard in his turn, rolled over onto his back and screamed down into the fight.

Desperate to avoid his opponent's fire, Harry Hill put his plane into too tight a turn. Instantly his damaged wing faltered under the strain and he tumbled off on one wing in a semi-spin, a half-stalled condition. Yamota was on him like a hungry tiger. Hill managed to kick hard rudder and shove the stick all the way to the corner of the right side of his cockpit, but now he knew he couldn't beat this man, not with his ruined wing. His eyes swept the sky, looking for a way out of his dilemma.

In desperation he headed for the cloud bank that hovered over Dublon Island.

As Yamota turned to follow him, Marriner and Winston sped to the scene from the east, late for the fight but not too late to get in their licks.

Yamota closed on Hill from the rear, his throttle all the way forward, his eyes peering through his gunsight as he measured Hill for the kill.

This is not the way to kill, he thought. Not when your opponent is wounded, his wing shattered and his plane unable to defend itself. But it goes this way sometimes. I would let this man go, but what then would I have to go on, in other fights? You can't let all your beaten enemies go. This man is good, but his plane is beaten, and so he can at least die knowing it was his plane that faltered and not himself. Yamota closed within fifty yards of Hill, just as Hill slid into the cloud bank over Dublon Island.

Yamota smiled because he knew his enemy would have to come out. A man can't fly in a cloud bank forever. Not when his carrier is two hundred and fifty miles away. He circled around the cloud bank and took up a vigil at its base.

One moment Hill was flying in bright sunlight, and the next he was immersed in the milky nothingness of the cloud. He flew straight ahead, and

Borrowing heavily on his own experiences as a Hellcat pilot, energetic Richard Newhafer wrote a classic novel of the late Pacific air war. He stands on the far left of this lineup of the Blue Angels 1954 team. Cdr. Zeke Cormier, the officer in charge (second from right) was, himself, an ace. At this point, the Blues were flying F9F-5 Panthers. (Courtesy of the NMNA)

soon he was in murky blackness, flying on instruments as the air grew rough and shook his plane relentlessly.

Marriner wasted no time. He put his sights on a Zero, scored a hit with a two-second burst and watched the Zero tumble burning through the sky. Winston had sheared off and taken on another Zero, and they were wheeling in tight circles, each trying to get inside the other's radius of turn, when Marriner, coming to Winston's aid, saw Yamota against the cloud base where Hill had taken refuge. He swung in that direction.

Harry Hill cursed. He was not in the habit of losing. He circled carefully in the cloud, flying blind, knowing his fuel would not allow him to stay in there forever.

Marriner saw the speck that was Yamota against the cloud base and threw a quick glance at Winston. Winston was holding his own, so Marriner defied the rule book that forbade breaking up the two-plane teams and went for Yamota.

Crowley had done nothing. He had skirted the fringes of the violent fighting until now he saw a single Japanese plane flying along the base of a cloud. He had no way of knowing it was Isoku Yamota who was more than a match for two of him. So he went after the lone plane, at the same time that Dick Marriner was closing from the opposite direction.

And as Crowley and Marriner headed for his opponent, Harry Hill decided the time had come to take his chances and flew his plane out from the pillar of cloud into the sunlight. He had only a moment to adjust his eyes to the light, for Yamota gave him no more time. He pounced from above. He might have preferred a more even fight, but he was a fighter pilot and he did what he had to do.

So he wheeled in on Hill from the rear, narrowing the distance between the two planes. Hill's plane loomed ever larger in Yamota's gunsight rings. In a few seconds his shells would be hitting Hill's plane, violently buffeting it in the sky.

Marriner arrived just in time to see Crowley swing his plane into position behind Yamota. Good, he thought. He's got that Jap bastard cold. Then as he flashed by on a reciprocal course Marriner glimpsed Hill's number on the side of his battered plane.

Crowley was within extreme firing distance of Yamota now and directly behind him, but Crowley did not fire. He watched as the Zero filled his gunsight, and still he didn't fire. His eyes were squeezed almost shut, and his finger was slippery on the trigger. Dimly, far back in his mind, there lurked a fear that if he fired now the enemy would turn on him in anger. It was ridiculous and illogical, but Crowley was past lucid thought. So he closed on Yamota without firing a round. Crowley was letting Harry Hill die.

Marriner flashed past on a reciprocal heading.

"God damn it," Marriner's voice yelled. "Open fire. Open fire."

Marriner swept up in a tight reversal and started back down again, but he was far out of range. Only Crowley could save Hill, and Crowley did not do it.

Crowley's finger was loose now on the trigger. He felt like a spectator sitting in a hundred-dollar seat at ringside. He was dimly aware that what he was doing was wrong, but he could do nothing as his mind and muscles refused to function.

Harry Hill knew he could do nothing more. Through his rear-view mirror he had seen Crowley's plane swing in behind the Zero and for a second had thought he was saved. But he wasn't, and he knew it now. Yamota's slugs slammed through his fuselage, and he banged the stick over to the right, twisting the plane around. He saw the Zero turn with him, and then Yamota's shells tore into the side of the cockpit and ruined Hill's body between the neck and the waist. Hill doubled over in the cockpit, and blood sprayed from his mouth onto the instrument panel. His body pushed against the stick, and the plane nosed over in a steep dive.

Yamota knew the kill had been made, and he turned away to take on the Hellcat behind him. Crowley continued to fly on blindly, not making even a token pass at Yamota.

But now Marriner was sweeping down on Yamota's tail, all guns firing. Yamota heard the sounds of the shells before he saw the Hellcat and instinctively rolled over and pulled into a split-S. Marriner was with him all the way, rage welling in him.

Hill knew he was going to die. In his acceptance of the fact resignation mingled with regret. Well, I always expected this only now that it's come I realize I am not quite ready to go. I would like to kill that bastard in the Hellcat who was behind the Jap and never did a damn thing. I saw him there just before my body fell apart. Hill made an effort to focus his eyes, and saw his altimeter needle unwind past three thousand feet.

Three thousand feet to live. Over the roaring of the wounded engine Hill's mind groped at the memories that crowded his dying brain.

The pain was filling him now, rising up to his throat and into his mouth. Well, Christ, it has been a good go, and I've had all the best of it. The fighting was not quite finished yet, but I suppose others will carry on for me. I hope they get that sonofabitch who shot me.

The altimeter needle swept past one thousand feet, and Harry Hill knew he had only seconds left.

With a superhuman effort he pushed himself upright and looked down the nose of his plane at the onrushing ocean.

A half mile away from the place where Hill died, Marriner and Yamota went at it a second time. Some time during the fight the knowledge came to Yamota that he had again found the man he had fought over Tarawa. Perhaps it was his excellence in avoiding the clever traps Yamota set for him, traps that had worked successfully in a hundred other skies. Whatever it was, Yamota knew he had found his man again. Marriner had recognized his foe when, after his first firing burst, the Japanese executed a half-roll and a snap

One of the enduring personalities of the Pacific air war, VAdm. Marc Mitscher wears his long-billed ball cap. An early aviator, Mitscher commanded Task Force 38 up to Japan's doorstep. (Courtesy of the Naval Historical Center)

from the inverted position. Then too, no two Japanese pilots could possibly have had that enormous number of flags decorating their planes.

This time Marriner felt no inclination to spare his opponent. He had seen Yamota kill Harry Hill, and was grimly determined to kill him in turn. So they fought again as they had at Tarawa, all over the sky in the greatest exhibition of fighting and flying that the other fliers who saw them had ever witnessed. There were opportunities for other pilots, one side or the other, to slide in and try for a kill on one of the two men, but none tried it. They left the two alone, and that is as it had to be.

First Yamota took the advantage, then Marriner wrested it away from him. There are only so many maneuvers that an airplane can perform and the two men employed every one known. Neither could best the other, and so they fought doggedly, waiting for the first mistake, the slightest error in judgment.

But there was to be no victory this time. For the second time in as many encounters, the two pilots were destined to fly away from each other with the question still unresolved as to which of them was the master.

As in their first encounter, it was a matter of ammunition on Yamota's part but to this was added the question of Marriner's fuel supply. Yamota had expended most of his ammunition in his kill of Hill. Marriner had two hundred and fifty miles to fly back to his ship, and as his needle reached below the quarter tank mark, he knew he would have to postpone a decision in this fight a second time.

Marriner was behind Yamota, striving to get on his tail and within firing range, when Yamota waggled his wings and banked his Zero sharply and headed for the airstrip on Dublon Island. Marriner started to follow him, but then he looked at his gas gauge again and considered the foolhardiness of following his opponent low over the deck across an enemy airstrip. Every gun on the island would have him boresighted. Anger flashed through him that this second fight had to end in a stalemate.

Yamota, watching through his rearview mirror, saw the Hellcat peel away and reverse course. Another time, he thought. Another time, and we will settle this. I have no way of knowing for certain that you are the man I met at Tarawa and who spared my life, and yet I am certain you are. Perhaps it is the way your plane refused to be decoyed or the manner in which you hold your own with me in a tight turn, which the engineers have told me is an aerodynamic impossibility. They said the Hellcat could not possibly stay with the Zero in a high-g turn, yet you have done it and confounded me every time I thought I had you. You are the best I have encountered, and so I know you are the man from Tarawa. Well, for the second time I am out of ammunition, and you must be short of fuel, so both of us must say 'Until next time.' And there will be a next time because I am sure that the two of us cannot go on living in contested skies. Sayonara.

Marriner climbed toward the rendezvous sector east of Eten Island, and heard Crowley's voice.

"All Strike Able planes. Rendezvous in sector X-ray immediately. Angels ten."

Three minutes later Winston slid alongside Marriner and held up two fingers to signify two kills. Marriner nodded his head. He looked around the sky and saw Bates climbing from below to join them.

It took Strike Able forty-six minutes to fly back to Task Force 58, and for half that time Marriner tried to convince himself that Crowley had had some excuse for not opening fire on the Jap that had killed Hill. But he could not convince himself, and with the formation almost within sight of the task force he seriously entertained the idea of shooting Crowley down in cold blood.

He gave his plane a bit more throttle and pulled his division forward alongside of Crowley and Anders.

"Red Leader," he called, "fire your guns."

Crowley turned his head. "What the hell are you talking about?"

Marriner jockeyed his plane in closer to Crowley's until the two planes were flying almost wing tip to wing tip.

"I said fire your guns. I want to see if they work."

"What goes, Dick?" Anders called Marriner.

"Hill's dead," Marriner replied. "Crowley was behind the Jap that got him. I want to know if his guns are jammed or not."

There was a silence, and then Anders said, "Better fire your guns, Red Leader."

Crowley flew along straight and level.

Marriner dropped back and slid in behind Crowley, "You've only got a few seconds to fire those guns," he called.

"I can get him from this side," Anders' voice said.

"Don't leave me out of this." It was Winston.

"Me either," Bates cut in as he pulled out to the side from Marriner.

"If we all fire together they'll have to hang us all," Anders said, and there was no laughter in his voice.

Crowley realized that he was face to face with death. "God no," he cried into his mike, "my guns are jammed. They jammed."

"All right," Marriner called. "Give him the doubt. For the time being, anyway." Crowley might be telling the truth. Guns jammed easily at high altitude. But Marriner shook his head because he still was not convinced.

On the bridge of the *Concord*, Delacrois and Balta heard the radio transmissions of the inbound strike.

"It seems your pilots are raising hell with each other again. Christ, Sam, this has to stop." Then he realized that Balta had a drawn look around his eyes.

"Hill?" Delacrois asked softly.

Balta nodded slowly. "I don't think Hill is coming back. I think he missed, finally." Balta's voice broke.

"Well," the Admiral said and stopped. He wanted to say more, but there was nothing more to say.

Balta turned to Delacrois, and now there was scorn in his voice. "From the conversation up there I take it that Crowley's guns jammed at a time when he might have saved Hill. That's par for the course with Crowley. He hasn't been on time yet." Balta turned away so the Admiral would not see the tears of anger in his eyes.

"I'll want to speak to the men on that hop. Crowley and the division leaders. As soon as they land." The Admiral spoke flatly, carefully keeping emotion out of his voice.

Balta and Delacrois moved to the aft end of the bridge. The incoming strike was overhead now, and the first divisions were breaking off and diving for the landing circle. Crowley's section came around first and swept into the wires. Behind them Marriner and his division came, answering the landing signals with precise application of rudder and stick, sailing up the groove nose-high and steady at eight knots. Over the fantail and, receiving the cut, into the arresting gear. One by one they taxied forward and shut down their engines on the port elevator.

So Young to Command

Britain's small force of aircraft carriers was among the first to see action in the European war that began in September 1939. Although these ships were on the whole fairly old, and certainly smaller than other carriers in other navies, their crews sailed them with skill and élan that contributed to the few victories the Allies enjoyed in the first three years of the war. By 1945, along with their larger American counterparts, the British fleet ranged from the Atlantic to the Pacific, playing an important role in the long march toward Tokyo. In Europe, however, except for the U.S.-led invasions of North Africa and southern France, the Royal Navy's flattops fought alone against the combined might of the German and Italian air fleets staging out of bases in northern Europe and southern Italy.

We've included two excerpts that describe the experiences of Fairey Swordfish crews, and perhaps a brief word of explanation is required. Sometimes overlooked by historians, especially American writers, the old "Stringbag," a nickname more often used than its formal one, was one of the few aircraft to fly and fight throughout the war. It participated in several early, pivotal engagements, including Taranto, the attack on the two German cruisers *Scharnhorst* and *Gneisenau* in February 1942, the hunt for the *Bismarck* in May 1941, and the early battles in North Africa. Its biplane design was obsolete by the early 1930s, and certainly by the war's start, the once-plentiful type had begun leaving fleet service. But the Swordfish found gainful employment and, indeed, was much loved by its crews, who flew the big, single-engine aircraft at speeds approaching those at which many other, more advanced types stalled. The Stringbag was one of the unsung heroes of the naval air war.

One of the more unusual aspects of the air war in World War II was the quick rise of many junior aviators to senior positions. This was particularly true in Europe, where squadron commanders were majors and lieutenant colonels at only 23 or 24. There were examples of wing commanders and group commanders, full colonels, at 25 or 26. Royal Navy aviator John Godley enjoyed such a rise in responsibility and discusses it in the following excerpt.

Godley flew 67 missions. Here, he describes the surprise of his promotion to lieutenant commander and squadron command at only 24. His book has become a minor classic, although there are nearly a half-dozen well-written memoirs and histories on the Stringbag and her crews. It is well worth searching out.

When his father died in 1950, John Godley became Lord Kilbracken of Killigar, a small village in the remote northwest of Ireland. He is an accomplished author and has produced other books of prose and poetry.

From *Bring Back My Stringbag,* by John Kilbracken

I returned to Maydown on January 6th, 1945. Expecting to rejoin *Adula* within a week, then back to the familiar MAC-ship routine. But I was in for a surprise, one hell of a surprise. Summoned on the 13th by the Captain.

—I'm pleased to tell you that you have been appointed to take over command of 835 Squadron. You are promoted to Lieutenant-Commander. Congratulations.

Completely speechless apart from a muttered thank-you sir. In peacetime it normally takes eight years as a two-striper to reach this dizzy height. In my case sixteen months. Promotion to equivalent rank when barely twenty-four was not unusual in the Army, still less in the RAF. But it would make me one of the youngest two-and-a-halfers in the Wavy Navy.

—The squadron operates from HMS *Nairana.* It is at present at Machrihanish, due to re-embark next week. You will take over command the day after tomorrow.

So soon. Oh Jesus. Jesus, would I be able for it?

—Aye, aye, sir. Is it Stringbags, sir?

—A mixed squadron. Let's see, fourteen Swordfish IIIs and six Wildcat VIs. They've been having a rough time, it seems. Convoys to Murmansk and a shipping strike off Norway. Morale, I gather, is not particularly high.

My apprehension and exultation growing simultaneously. Two-and-a-half stripes. Commanding twenty aircraft. Forty officers and nearly 200 men. But would I be able, would I be able? This, after so long, my very first appointment to one of His Majesty's ships. No problem, I thought, in the air leading the Stringbags. (The Wildcats, which were fighters, would fly and fight independently.) But knowing nothing at all, after the total informality of Coastal Command and MAC-ships, of all the RN mysteries. Such a weight of responsibility. And morale 'not particularly high'.

—Has the previous CO bought it sir, I enquire.

—No, he's due for his rest period. Overdue, I think. Along with a few others. They really seem to have earned it. His name is Jones, he's an observer. Lieutenant-Commander Valentine Jones, RNVR. You know him? All the better.

Dazed to my cabin. Many thoughts confused. The power and the glory, yet such feelings of inadequacy. For the first time in my life. But now I'd be the boss. Somehow or other I've just got to pull it off. On Russian convoys, Christopher twelve days old. To say farewell to Charlie after fifteen months of flying with him. Put on a bold face, for Christ's sake. But all the White Ensign rigmarole: requestmen, defaulters, up spirits, darken ship. What the procedures? The paper work, the flying programmers?

Was I even fit for the flying? Leading thirteen Stringbags. And today the thirteenth. But thirteen lucky for some. Telegram to Penny. My new address HMS *Nairana*, c/o GPO, London. Though floating in Arctic waters. Get my uniforms altered, find a tailor in Derry. A half stripe, you understand? By tomorrow. Leaving the next day for an unknown destination and destiny.

How had I come to be chosen? There were many two-stripers senior to me. But Val's successor was needed at once, so had to be a pilot or observer serving in home waters, or ashore in the UK, preferably in a Swordfish squadron. And by now, at last, these were rare: apart from 835 and 836, there was I think only 813 in *Campania*. So the field was greatly narrowed. I wasn't the most senior of the two-stripers available but must have been thought the best qualified. I'd never have commanded a squadron and reached this lofty rank if Val's rest period hadn't come due at this moment. In another three months, both 835 and 813 had been disbanded. So this my last and only chance.

Two days of relentless ribbing by my 'P' Flight shipmates. Yes sir, no sir, three bags full sir. Drinks with Charlie in Derry. And meeting Douglas Graham, my classical tutor at Eton, the only master I'd respected, now a Navy padre. 'You seem to have caught up with me.' Appearing sheepish with my halfstripe in the Mess, not really believing it.

Taking off next morning in abominable weather for whatever fate awaited me. Without an observer, just another pilot as passenger. But I could find my way blindfold. Ceiling less than a hundred feet, visibility half a mile. Picking up the Mull of Kintyre, following its steep crags northwards to the low-lying golf links, well remembered. Sneaking in, over its recognized fairways and bunkers, as a driving rainstorm swept me. The only aircraft to take off or land that day.

To find things far worse than I'd expected. I'm hardly out of the Stringbag when an unknown young officer approaches.

—Are you the type who's taking over from Val? Well there's one or two things you'd fucking well better know. To begin with, the fucking Captain's crazy . . .

What to do? Pull my rank? Demand he call me 'Sir'? No question of it.

—Pipe down about the Captain. If I'm to hear anything, I'll hear it from the C.O.

—I'll take you to him. He'll tell you the fucking same. Thank Jesus I'm leaving with him tomorrow.

That's something anyway. Finding Val in his office. Extreme fatigue in his eyes. He does his best to fill me in. The squadron seems near breaking point. Incessant flying in foulest Arctic weather. He's afraid my job not easy. Several of the boys had been far too long in ops, and seven of these, six pilots and an observer, would depart with him next day, when replacement aircrews arriving. The main trouble, yes, was the Captain. They call him 'Fly-off' Surtees. A wan smile. Also known as 'Strawberry'. From his complexion.

—*I'd* tell him conditions weren't fit for flying. Wings would tell him the same. But Strawberry—he isn't a pilot—had always the same reply. 'Get the engines started, the crews aboard. We'll turn into wind and then we'll see.' Once he'd said that, we already knew what would follow. Invariably. As soon as we were headed into wind: 'Fly off.'

—Have you lost many crews?

—That's the extraordinary part of it. He must have the luck of the devil. We've never lost a life through flying in bad weather.

How could I replace Val? Everyone loved him. From the aircrews down through the chiefs and petty officers to the lowliest air mechanic. He departed next morning with the others leaving the squadron. I felt out of my depth. Shit-scared and alone.

Mustn't show it. A week at Machrihanish before embarking. Literally sick with apprehension, not sleeping, couldn't eat. *But mustn't show it.* If I collapse the whole squadron will collapse. Telling the boys we'll be rejoining the ship on Monday. Greeted with a groan. *Now that's enough of that.* We've a job to do and we're damn well going to do it. 'But Strawberry, sir . . .' You mean Captain Surtees? Well pipe down about Captain Surtees. Did he ever lose a crew to the weather? Grudging silence. Well what's the bloody beef then? Trying somehow or other to pull the squadron together. When I need pulling together. Hostile eyes facing me. I'll do my best to stand up for you but we do what the Captain orders. 'The boss always said . . .' But I'm the boss for Christ's sake. I don't want to know what the boss said. What Lieutenant-Commander Jones said. We fly on board on Monday. Have some pride in your squadron.

Everyone pissed as hell round the piano every night. But that's normal. At least I know all the songs. Who's that knocking on my door. Cats on the rooftops. The harlot of Jerusalem. The ball of Kirriemuir. Can even teach them some new ones. *But that's not enough* I remember telling myself with unhappy certainty. Knowing the songs is not enough.

Ah Jesus. And it'll be worse, far worse, on board. How the hell can I pull it off?

I chose to fly with the Senior Observer, named appropriately Strong. Due soon for his second stripe, with long operational experience. But not *too* long, no sign of the twitch. Quiet but wholly reliable, George Strong at once impressed me. Steady as Ailsa Craig. So let's get in the air then. Just have a look-see together.

Our Swordfish IIIs the very latest, known to all as pregnant Stringbags from the great bulge in their bellies to house their ASV Mark X, the best airborne radar yet. They have to be two-seaters, just a pilot and observer, to make room for all the ancillary equipment. No guns of course; what would we want with guns? A further refinement, Rocket Assisted Take-Off Gear, known by its acronym RATOG, making maximum warload possible however light the wind. A rocket under each wingstub. Press the appropriate tit when abeam of the bridge, off go the rockets and you are forthwith propelled skyward. Good as an in-built catapult. And the Pegasus XXX engine, last of the long line of Peggys, better than ever. So here they were in 1945 taking what was basically the same old Stringbag, which had been flying for nearly a decade and was generally believed to have passed long since from front-line duties, still finding it worthwhile to work on her and modify her, giving her such of the latest trappings as she could carry and still sending her forth on the most demanding missions. As I would soon discover.

Flying as much as I could. Whilst getting to know the officers and men I now commanded. And trying to instil some sense of cohesion and pride, some flash of *esprit de corps*, a lifting by its own boot-strings of morale. Leaning heavily on my chiefs and petty-officers who knew everything. Identifying the stalwarts among my aircrews, those who were the trouble-makers and those (the majority) who were somewhere in between, just ordinary honest

Lt. John Godley stands between his two Swordfish crewmen. Godley led 836 Squadron's "P" flight. Lt. Jake Bennett is on the left, while Leading Airman Charlie Simpson is on the right. (Courtesy of John Godley)

flyers. Working on all of them as best I could but feeling my best not good enough to have an outfit ready for action the next week.

No trouble with the fighter-boys. They came under my command but operated as an almost autonomous unit, led in the air by Allen Burgham, an already decorated New Zealander. Their Wildcats, under lease-lend from the States, were relatively modern though not a patch on the latest Spitfires even.

Nairana was one of no fewer than forty-three escort carriers that had now been completed, of which only *Audacity* and *Dasher* had been lost (and *Archer* retired). All but six had been converted in the States. There were also the two new fleet carriers, *Indefatigable* and *Implacable*, commissioned in 1944, and three light fleet carriers, *Colossus*, *Vengeance* and *Venerable* (the names they chose!) though the last three were only starting their trials. *Nairana*, one of the six British conversions, namesake of a packet-boat converted to launch seaplanes in 1917, was attached to the First Cruiser Squadron in the Home Fleet, based at Scapa. But now lying at Greenock. My squadron her total establishment. Her last operation set out to be a strike against shipping off Norway but proved totally abortive: it was the night of the full moon, a week after Christmas, but the weather was so unfavourable that, despite Strawberry's reputation, the order to fly was never given. Much moaning because Hogmanay spent at sea on this fruitless mission. Before that a very tough run to Murmansk and back.

I dreaded the time we would join her. Believed all my shortcomings would at once be laid bare. The disembarked maintenance crews had gone overland to Glasgow two days previously. The Wildcats already on board. We took off from Machrihanish on January 22nd, two vics of five Stringbags, one vic of four Stringbags, flights in line astern. Make it good and close boys. Forming up over the sea to westward, then in very tight formation back across the airfield. Ceiling 800 feet. Skirting Arran, skirting Bute. To locate our mother-ship precisely on time in the firth, already turning to windward for us. I signal echelon starboard, we break formation one by one at half-minute intervals to land.

And as soon as we landed an immediate change for the better. Contrary to everything I'd expected. I seemed to come alive and all the squadron with me. Such a hustle and bustle. Six times as many aircraft as *Adula* and everyone six times busier. The Wildcats already below-decks in the hangar. The first to land, I am signalled right forward to the port side of the flight deck, my wings already folded before I cut the engine. The second, my number two, wings folded and alongside me within twenty seconds. Thirteen Stringbags to be crammed ahead of the barrier before the fourteenth lands. Ground crews at the double to shepherd all in place. The whole operation completed in well under ten minutes.

So now I must face the Captain. Face Strawberry, face Fly-Off, face Captain Villiers Nicholas Surtees, DSO, RN. I make my way to the bridge. On the uncomplex bridge of a MAC-ship, no one but the mate whose watch it is, probably in civvies. Perhaps the ASO. Everything so informal. Nothing informal here. Three or four officers, all in full uniform. Caps, collars and ties! A couple of signalmen. A chief and petty-officer or two. The Captain hunched in front of them all, binoculars raised to eyes, gold-peaked cap pulled down to the binoculars, smaller than the rest, his bright scarlet complexion if nothing else proclaiming his identity.

—Lieutenant-Commander Godley, sir. Come on board to join.

Strawberry hears me. Doesn't lower his glasses.

—Ah Godley. Ah yes.

In his own time turns to see me. Round red face with twisted half-smile. Summing me up. Me bare-headed in flying gear, red silk scarf, helmet and gloves in hand, fur-lined jacket over battledress, Mae West over jacket. I must have looked a schoolboy. Extends his hand to me, welcomes me. Whatever he's thinking he doesn't show.

Is this such a devil incarnate?

I leave the bridge, set out with a petty officer, one of my right-hand men, Landcastle, to begin learning the layout of the ship. The hangar deck, to which one by one my Stringbags are being lowered on the lift, then hauled forward by a tractor to their allotted positions, hardly an inch of space between them. Chocked up, securely lashed to the deck, driptrays under each engine. My cabin very far aft, on the deck below the hangar, my suitcase and kitbag waiting to be unpacked. I think a safe haven. 'You'll not sleep here often where we're at sea, sir.' More movement, explains Landcastle, than anywhere else on board. Because so far aft. As it was to happen, I had little time to sleep *anywhere* when at sea. Visiting the messdecks, first the men's, then the petty officers', then the chiefs'. The hierarchy on the lower deck no less evident than in the wardroom. *Nairana* seems ten times bigger than friendly *Adula* though this an illusion. Every square foot utilized, not the wide open spaces above the well decks fore and aft. I hear two bells, make an effort to remember. Yes it must be 1300. So to the wardroom for ginning up. Thank you, Landcastle, as he salutes. I find I've a real appetite for the first time in a week.

We dropped anchor two hours later. Summoned by the Captain. He and I alone together in the austerity of his cabin. As though he wishes to forge a personal bond. He already knows all my background. Makes no reference to my age, my inexperience, my total lack of knowledge of Royal Navy procedures. Takes all that for granted, my petty officers will look after me, what matters is the flying. This is a fighting ship, Godley. And it's your aircraft do the fighting. We exist for one purpose only, to keep them flying. We sail tomorrow night for Scapa, arriving the next evening, the 24th. Now this is for your ears only. The 28th is the full moon. We're going back to Norway. The last full moon, as you know already, we never got off the deck. We could have taken off and landed, but not fly the length of a fjord or attack shipping effectively if sighted. This time we may be luckier.

—What the squadron needs, sir, is one successful op.

—The squadron will be all right, Godley. I rely on you for that.

No mention whatever of shattered morale, of the essential need to rebuild confidence. This too taken for granted. He knows already I must be aware of it. Or if I'm not, nothing he says can make the smallest difference.

We are joined by Commander (Flying), as he is known though not yet a three-striper. He plays the same role in an escort carrier as the ASO played in MAC-ships. The only flyer on board senior to me (though of the same rank), with longer experience in the air, or older. The only intermediary between me and the Captain in flying matters. He will lay down the broad flying requirements, leave me to fill in the details. John Ball has been many months with *Nairana.* Experience going back to Sharks and Seals. Lined and beaten face, tough as leather, someone to help me when I need it.

—Come in Wings. I've been telling Godley our programme. He thinks we need just one success to set the squadron on its feet.

—Maybe next week, says Wings.

—Let's hope so. Well that's all, Godley. Good luck.

Getting the boys together. No, getting my officers together. Telling them we sail for Scapa tomorrow night, no boats ashore this evening. Waiting for the groan. It doesn't come. Now we've got to maintain 100 per cent serviceability just as long as we can. Each of you to work personally on his aircraft tomorrow. Don't just leave it to the men. There'll be night deck landings in the firth after sailing for the new boys. That includes me. A ripple of laughter but friendly. To the open sea by midnight, picking up our escort. Then a Stringbag on defensive patrol, two standing by as back-ups, till after we round Cape Wrath.

The planned programme goes through without a hitch. We sail in perfect weather, a calm sea, frosty air, brightly moonlit sky. The night deck landings are prangless. Our nonpareil batsman, Bob Mathé, soon to become perhaps the only 'bats' to be awarded a DSC—in fact, two DSCs. The luxury of having all of fourteen Stringbags so that continuous single-aircraft cover can easily be provided, if necessary, for over twenty-four hours with no one flying twice. A daylight Cobra with George between Lewis and the mainland. Dropping our hook soon after dark in Scapa.

Someone has written a new song. To the tune of Stand up, stand up for Jesus. We bawl it out round the piano fortissimo, all but the second line. Which is rendered *ppp*, to reach our own charmed circle only.

Fly off, fly off for Christ's sake,
For the Captain wants a gong
Fly off, fly off for Christ's sake,
For the Captain can't be wrong . . .

Was I wrong in feeling somehow that the last line was sung with less irony than I'd have expected a week earlier?

The hard-pressed enemy was bringing large supplies of vital war material from northern Norway at this time, especially iron ore. Shipped in convoy or independent vessels close inshore. These ships had been attacked by the RAF in daylight using Mosquitoes and had taken to night sailings after suffering heavy losses. It seemed there was nothing, though to us it appeared incredible, that the RAF—whose job it should have been—could send against them at night in the limited airspace available for a low-level attack, the only way of making certain of results. So the Stringbag had once again been called upon. With her absurdly low speed and extremely tight turning circle, she was still, apparently, even in 1945, the only kite for the job.

But the operation ahead of us seems in retrospect so foolhardy, so unnecessary. Our target area was well within range of RAF bases, from which hundreds of modern bombers were being sent out every night, putting only themselves at risk. How could it be that none of these was able for the job in conditions of brilliant moonlight? And to bring two squadrons of obsolete biplanes—835 in *Nairana*, 813 in *Campania*—to within range of the shipping lane, no fewer than nine warships would sail to within sixty miles of the enemy-occupied coast. In waters where U-boats might be lurking, within easy range of shore-based bombers, even fighters. A cruiser (*Berwick*), two destroyers (*Algonquin* and *Cavendish*) and four smaller warships besides the two carriers. All at risk, very probably for nothing, as four weeks earlier. To fly just one operation against a possible absent enemy. And in any case no target of real importance expected. Then scuttling back to Scapa.

But ours not to reason why. We sailed at 1900 on January 27th and would take twenty-five hours to reach our takeoff position, which was 140 miles north-west of Bergen and less than half that distance from the coast. The weather perfect, not much swell, maximum visibility. From the small hours of the 28th, we and *Campania* between us kept at least one Stringbag airborne to sweep the sea ahead of us for U-boats. During the few daylight hours—our most northerly latitude would be sixty-two degrees—we also had a pair of Wildcats ranged permanently on the flight deck in case a shadower should find us, but never needed. One of my Stringbags and two of *Campania's* were unfortunately damaged on landing, which reduced our combined striking force to twenty-five.

Briefing at 1830 by Wings, already within 100 miles of Norway. We are to fly off in two waves. The first strike to scramble at 2000, seven Stringbags from *Nairana* and six from *Campania*. Each of these flights to act independently, with its own specified search area. We in 835 to head for the entrance to Rovde Fjord. This isn't a fjord in the usual sense, but an open-ended channel—the route followed by all shipping—between the mainland and a group of offshore islands, a mile or two wide, with mountains rising precipitously up to 3000 feet on either side. We were to follow it for forty miles, attack any target seen. *Campania's* six aircraft had a similar search area some forty miles to northward.

Wings tells me I can best establish my position by heading first for Ristø, a small offshore island, easily picked up by ASV. Inhabited, he assures me, by nothing more hostile than farmers and cows. Unexplained why, after coming so far, we will be covering only forty miles of the enemy shipping lane. Which will take only half-an-hour. Allowing for ninety minutes to get there and back from the ship, we could have spent at least two hours on patrol, covering four or five times that distance. Even more if on reaching Ristø we split up into subflights, two or three aircraft in each, each with its own search area, whistling up the others if a target sighted.

The second strike, six Stringbags from each carrier, to scramble at 2040. Would orbit safely offshore, in their respective back-up positions, awaiting orders from their leaders who would then be completing their patrols.

I give orders that one aircraft in each strike is to be armed with six bombs, the rest with eight armour-piercing rockets. By 1915, in the most brilliant moonlight, the seven Stringbags of the first strike have been ranged on the flight deck. My own, A for Able, waits in front, facing fore-and-aft amidships. The others, wings already spread, angled inwards in pairs astern of it. We board at 1940, check equipment, start up engines. At 1955 the nine ships turn together into wind so that both flights may scramble simultaneously. *Campania* close on our starboard beam.

At 1959 I get my green light from the bridge, chocks away, open up full throttle, fire my RATOG after eighty yards, am propelled at once skywards into the brilliance of the night. The freezing slipstream blowing in my face. I keep my airspeed below seventy, make a long sweeping turn to port, so that the others can quickly come up with me. (The CO of 813, to avoid any risk of collision, is doing the same to starboard.) In less than five minutes we are in extended echelon formation. I settle down on a course of 042 at a comfortable ninety knots.

I feel perfectly calm and confident. No therapy to compare with commanding an operational squadron! The silvered full moon in a sky without a cloud, so bright that it's nearly like daytime. We can see the great snow-covered mountains of the enemy coast at a range of thirty miles. And Ristø soon visible ahead, not many minutes after George has located it by radar and given me a new course to bring us directly overhead. Where the first excitement awaits us. For reasons I've now forgotten if ever I knew them, my orders are to overfly the little island at cherubs twelve (1200 feet), then use my own discretion. Instead of safely at sea level, remaining so much longer undetected by enemy radar. So I'm leading my six aircraft in open formation, straight and level at ninety knots, between it and the moon at this most vulnerable altitude.

Whereupon two batteries of Bofors guns change all our previous notions of farmers and cows. Streams of multi-coloured tracer come streaking up towards us. At first so slowly, as always, then with lethal acceleration.

We should have been dead ducks. George and I anyway. For most of the flak was aimed at our leading aircraft. And we presented the simplest possible target. But once again our slow speed saved us. Accustomed to Mosquitoes at least three times as fast, the gunners simply couldn't believe that we were stooging in at ninety. For a couple of seconds the tracer came streaming up some thirty yards ahead of us. That was all their chance. Before they could correct their aim, I've put A for Able into a steep diving turn to port, the rest of the boys without delay behind me. In a series of twisting turns to sea level, alternating port and starboard, which the gunners cannot follow. To the safety of the wavetops beyond their range.

But they will have alerted the gun positions ashore. Already before diving I've seen the entrance to our fjord. Now though still several miles distant I see streams of tracer crisscrossing it, from batteries on the mainland and the nearest offshore island. What did they think they were firing at? But knowing for sure that I must lead my Stringbags through this deadly curtain. No way I can escape this. My thoughts at this moment fly to Christopher, four weeks old. Surely not now of all times.

To make matters worse, I see that one of my junior pilots, in the excitement of the moment, has accidentally switched on his navigation lights. He's lit up like a Christmas tree. George to him on the radio: HOPSCOTCH X-RAY, FOR CHRIST'S SAKE DOWSE YOUR LIGHTS. Message not even acknowledged. HOPSCOTCH X-RAY, THIS IS HOPSCOTCH ABLE. HOW DO YOU HEAR ME? Still no response. We pride ourselves on our radio techniques, but at this critical moment his not working. We send the message by Aldis lamp. The penny drops as we approach the mainland, but our presence now certainly known to every gun position.

Perhaps I'm wrong but I think he'd done us a favour. To the well-guarded fjord entrance, prepared for a hot welcome. In very open formation, flying as low as we dare to avoid being silhouetted. And for some of the time, anyway, the gunners would be unable to fire without shooting at one another. But they never open up. And I believe they must have reached the conclusion, accustomed as they were to ultra-modern aircraft, that this little clutch of single-engined biplanes, dawdling past at under 100 knots, with wheels lowered (of course we couldn't raise them, but how were they to know this?) and lights so boldly flashing, just couldn't be the enemy. Whatever else they might be, perhaps trainees who had lost their way. How else to account for the absence of opposition from these known batteries, which minutes earlier had been firing away at nothing?

Now flying at zero feet up the narrow waters of the fjord. Lights coming on ahead of us in homesteads on both shores as we are heard approaching and men come to their doors, seeming to show us the way. The whole country deep in snow, the white mountains bathed in moonlight rising high above us on either side. Till at last we sight a target. A merchantman sailing towards us on her own. I lead the flight over her, hesitate, decide to fly the last ten miles of our patrol in search of a larger vessel. But the rest of the fjord is empty. Turning back, climbing to 1500 feet, the perfect height for rockets.

I can hardly believe it now, but I had no thoughts whatever for the men sailing in this coaster. Which we were now about to sink. It was nothing compared with the nightly devastation caused by the air forces on both sides. And it was a specific target of strategic importance, not a city where those who suffered most were the civilians. But it amazes me that I never thought for a moment, as I still clearly remember, of the men on board that ship. It was wholly impersonal.

My aim as I banked to dive, absolutely calmly, at the correct angle of twenty degrees towards this doomed vessel, was to strike her with my rockets just below the water level. Flak may have been coming up but I simply wasn't aware of it, I was concentrating solely on my aim. I'd had so much practice and the conditions were so perfect that I expected to be accurate within a very few feet. And I was. Firing my eight rockets in a ripple of four pairs at a range closing from 1000 to 600 yards. Seeing them all striking the water as planned, a yard or two short of the merchantman amidships.

A steep turn to starboard to watch the others. The next two pilots, Roffey and Payne, also score several hits. My own rockets alone would have been enough. Now I know she has no chance of surviving. She is stopped and on fire. I call on the four others to break off the attack, to seek targets ashore. Two of them, Gough and Supple, choose the flak positions at the entrance to the fjord and must have silenced them. In any event they never fired again.

The six aircraft in my flight are soon heading independently back to the ship. I return at low altitude to where the stricken vessel is now low in the water, lifeboats being lowered. And I make one more sweep of the search area. To find two more merchantmen on the point of entering Syvde Fjord, which runs steeply inland to southward. At once calling my second flight, arrived recently offshore. Led by my Senior Pilot, Geoffrey Summers. Three to attack each ship. Waiting for them to approach, then leading them to their targets.

All around me the silent mountains. I can clearly see each cottage on the rock-strewn mainland shore. The steady hum of my Peggy. Summers, Paine and Cridland dive to attack one ship, scoring hits with at least six rockets. Two of my most junior pilots are badly off aim as they swoop down on the other but Provis with his bombs scores a direct hit and two near misses. The first target is on fire but the crew manage to beach her before she can sink. The second is settling rapidly. But I can't stay to observe her. I've been airborne over three hours. Our normal maximum endurance is four-and-a-quarter hours with a war-load and we still have to locate *Nairana* and land on.

We approach her shortly before midnight. I check on my petrol, find I still have nearly forty gallons, enough for over an hour. Well there's a damn good engine for you. The six other aircraft of the first strike have been back for over an hour. Now some of the pilots in the second strike, who think they've been hit by flak or developed engine trouble, are requesting emergency landings. I assure Wings by radio that I've plenty of gas though airborne forty minutes before them, so they are allowed to pancake ahead of me. In the end I'm almost the last to land, at 0017, after four hours and eighteen minutes.

We are already heading home. And quite expecting a counter attack. We had been using radio and the position of warships that included at least one carrier must have been well known to the enemy. But none develops. After a long debriefing, I go below to the hangar to see what damage done. Not a single scratch. And the Stringbag damaged the previous day has been repaired so that all my aircraft are again serviceable.

A couple of hours' sleep. At first light we range the Wildcats in case of an attack by bombers, a couple of Stringbags for defensive patrols if ordered. Their services not needed.

Back to the hangar to check the aircraft, speak to the men. All spirits high. Great show last night, sir. A piece of cake. But my fitter looks down to me from working on the engine with something less than a grin.

—Do you know how much petrol you had left when you landed, sir?

—Yes, Titchmarsh. That's one hell of a good kite. Over thirty gallons after four-and-a-quarter hours.

—Well sir, I don't know whether to tell you. But I didn't see how it was possible. So I took a reading with a dipstick. Sir, your fuel gauge was unserviceable. You had so little gas it didn't register on the dipstick. Three or four gallons at most.

Three or four gallons. Enough for five or six minutes. A moment or two of silence, then gales of laughter.

—Well, Titchmarsh, for Christ's sake fix that fuel gauge.

—Sir, I've already done so.

—Well for Christ's sake keep it fixed.

Campania's twelve aircraft had flown an inconclusive mission. Apparently they'd carried out an attack but without known results 'in the absence of a flare-dropping aircraft'. It seems strange, it was nearly bright as day, but that's what they reported.

We dropped our hook in Scapa Flow a few minutes after 2000. Everyone in high spirits. By luck we'd had that one success to set the squadron up. Just when we most needed it.*

We spent six days in Scapa. During this time I began to realize what was at the root of all the tensions that had made *Nairana* so much less than a happy ship and were still frequently evident. They had little, I came to see, to do with Strawberry, though as Captain the general antagonism felt by the squadron had become focused on him. In fact he was a great and gallant officer, as air-minded as any of us, who drove his aircrews to the limit but, as the record showed, not beyond it. Or not all that far beyond it anyway. He wasn't a flyer but this was true of almost all captains. The real trouble was the completely unacceptable length of time that some members of the squadron had been kept flying on operations in such perilous, exhausting conditions. Bob Selley, for example, had been in 835 for almost three years without a break. This could never have happened in the RAF, where crews were rotated on a regular, far less demanding basis. The eight who had departed for their 'rest' at the time I joined the squadron, I discovered, had all just been diagnosed as medically unfit to fly—in our language, twitched to bits. And so had come to regard as unreasonable what they would once have seen as a challenge and accepted without question.

There was also the unhappy fact that a majority of the non-flying officers on board, including the Commander, belonged to the old school and still declined to recognize that aircraft had a vital part, let alone a supreme part, to play in naval warfare, even though they were serving in a carrier. Or perhaps *because* they were serving in a carrier. Hence the resentment they felt towards the aircrews, the ship's sole *raison d'être*, especially when all of us were without exception Wavy Navy.

Captain Surtees had none of this. He'd been at pains to make clear to me that he believe what should have been self-evident, that the sole function of *Nairana* was to provide a floating platform from which my squadron could operate. But where did this leave his non-flying executive officers? It left them right in the background, as mechanicals whose main responsibility was to make it possible for us to fly. This they didn't like at all.

My aircrews must have come from every kind of background, I never asked them. They'd won their commissions neither because they'd been to the right schools nor because they knew the right people. But on the sole grounds of being expert flyers. Well reasonably expert. They possessed none

*Long after the war I was able to discover from enemy records that all three ships were in fact sunk—and, happily, not a single crewman killed or injured.

of the Dartmouth attitudes of so many straight-ringed officers. And thank the Lord for that. They were just a bunch of kids whose flying for the moment was their life and quite likely their death, knowing nothing whatever of so-called upper-class pretensions and caring even less. That they could successfully operate their out-moded aircraft in nearly impossible weather meant absolutely nothing to their RN fellow-officers (or many of them). Did they have watch-keeping certificates? Could they take charge of a whaler? Did they speak the King's English?

It was ridiculous and in retrospect seems very nearly unbelievable. But it was true. The straight-stripers would so much rather have been serving in a real ship (as they saw it), a cruiser or a destroyer, where a Wavy Navy officer would know his place. Their sentiments overflowed when a senior member of this group described us (and was overheard) as 'all these little buggers masquerading as gentlemen'.

I couldn't take it in silence. Knowing their prejudice was unalterable, I thought the best course was to encourage the squadron to laugh at them. Thus, I hoped, further lifting morale, which such attitudes had done so much to damage, though never before expressed so explicitly. And I wrote a new song, which began with the lines:

We're eight-three-five, masqueraders are we,
Trying to hide our i-dent-i-tee.

And this was the last verse:

So though we may not be quite upper crust,
We'll go on flying as long as we must,
And raise this epitaph over my dust:
He was in eight-three-five.

How we loved bawling this doggerel within all-too-easy hearing of the scowling officer, crimson with rage, who'd made the offensive remark! He knew well we were referring to it but was powerless to stop us, which would have meant admitting it. It was specially good when, as once happened, Strawberry joined us at the piano and we taught him the words, which he thoroughly enjoyed, without knowing their esoteric significance.

Surtees did have shortcomings. For instance his eyesight was so bad that he couldn't make out an attacking torpedo-bomber at 800 yards nor read without spectacles. Moreover he would very seldom wear his spectacles outside the privacy of his cabin, which would have meant admitting to his disability, but ask for signals to be read to him.

Mad however he wasn't. I have to say this because a wild rumour had swept through the squadron some time previously that he might be declared insane—a measure of the resentment he had attracted from some officers in the squadron. The gist of it was that his second-in command, the much disliked F. J. ('Fat Jack') Cartwright, and the ship's senior medical officer, 'Doc' Waterman, had got together when at anchor in Russian waters to consider such a course, *à la* Caine Mutiny, to relieve him of his command, confine him to his cabin (I never heard any actual mention of irons) and set up Cartwright in his place.

I've never been able to determine how much truth resides in this extraordinary tale. But the evidence of insanity put forward is so flimsy that I totally discredit it.

As will be seen, I was able to establish a good working relationship with Surtees. I didn't even find this very difficult. He was a hard man—akin perhaps to Ahab in Moby Dick—but far from impossibly so.

Two days ashore at Hatston for flying practice, with bashes around Kirkwall in the evenings, but *Nairana* never idle long. The day after our return on board, February 4th, Wings and I are summoned to Strawberry's cabin. For the news we are expecting. We are sailing after dark next day with the First Cruiser Squadron to take a convoy to Murmansk. We have to be ready for anything. How's your serviceability, he asks me.

—All aircraft serviceable, sir, except one Wildcat.

—Try and have her ready by noon tomorrow. We're expecting opposition, every aircraft needed.

I go below to my cabin, write a brief note to Penny. I may be away a while, give my love to Christopher. Later singing all the same old songs. But turning in early for my last real sleep till we reached Kola Inlet nine days later.

CHAPTER 3

The Early Cold War, 1946–1953

The Navy Wants a Piece of the Action: Nuclear Bombers Aboard Carriers

One of the most important new developments of World War II was the atom bomb. It had stopped the war, pure and simple, and canceled the terrible prospect of a million Allied casualties during the planned invasion of Japan. At war's end, the United States enjoyed a complete nuclear monopoly, but soon other countries, including the new nemesis, the Soviet Union, had gained the terrible new weapon. Delivery of nuclear bombs became a high priority for Navy and Army interests.

Although the American carrier fleet stood at more than 100 in 1945, it had predictably reduced in strength after the Japanese surrender. But new opportunities, as well as new enemies, required a fairly high number of operational ships and deployments.

After the highly publicized success of two Army B-29s that dropped the two nuclear bombs on Japan, that service, and after 1948, the independent U.S. Air Force, wanted a clear monopoly of its own in who would be the primary carrier of America's growing nuclear arsenal. There was some justification for this selfishness, but the Navy wasn't about to give up without a fight, nor a demonstration of its own capability to bring the bomb to a potential enemy's doorstep. New aircraft and several new carriers were already coming.

Given a carrier's limited landing space, it was always interesting when someone tried to bring larger, heavier aircraft aboard. There was also the inevitability of entering the jet age. Britain flew jets aboard one of her carriers in December 1945—then-Lieutenant Commander Eric Brown piloting a De Havilland Vampire aboard HMS *Ocean*—and a U.S. Navy XFD-1 Phantom, flown by Lieutenant Commander James Davidson did it on July 21, 1946, aboard USS *Franklin D. Roosevelt* (CV-42). But besides these developments, Navy leaders were concerned about carving a slice of the pie of nuclear bombers for themselves. Thus, when a senior naval aviator proposed bringing aboard a truly large aircraft to demonstrate its ability to fly from carriers with a nuclear bomb load, he got his chance.

Norman Polmar wrote this book when he was only thirty, fairly young for the author of such a large compendium of information. He has completely revised the book and the new edition should be published within the next two years. A native of the Washington, D.C., area, Polmar is a well-known military analyst, specializing in naval subjects. He is occasionally seen on various popular cable programs such as the Discovery Channel, and maintains a busy schedule as a consultant and author. This book on carriers remains a familiar reference, and it is seldom omitted from other authors' bibliographies.

From *Aircraft Carriers: A Graphic History*, by Norman Polmar

The earliest known Navy discussion concerning a carrier-based nuclear strike capability took place late in 1945. The subject was discussed in detail by Vice Admiral Mitscher, then Deputy Chief of Naval Operations (Air), Captain William Parsons, and Commander John T. Hayward. Parsons had participated in the development of the atomic bomb and had armed the A-bomb in the B-29 "Enola Gay" en route to Hiroshima; Hayward had become involved in the atomic bomb project while working in rocket development. At the time of the Mitscher-Parsons-Hayward talk in 1945 the Navy was well along with blueprints for a class of 60,000-ton "super" carriers; at the same time the Navy was holding design competition for a carrier-based aircraft capable of carrying a 10,000-pound payload. By coincidence, the "Fat Man" bomb exploded over Nagasaki had weighed about 10,000 pounds.

North American Aviation won the design competition. The new aircraft would evolve as the AJ Savage, a high-wing plane with two piston engines under its wings and a jet mounted in its after fuselage. Commander Frederick Ashworth, who had armed the atomic bomb dropped on Nagasaki, inspected a mock-up of the AJ at the North American plant. The plane was originally intended as a limited research project to investigate the problems of operating large aircraft from carriers. Ashworth and Captain J. N. Murphy, both assigned to the Bureau of Aeronautics, decided that the aircraft design could be modified to carry the new Mark VI atomic bomb, an improved version of the "Fat Man."

Commander Ashworth drafted a letter for the Secretary of the Navy to send to the President requesting permission for the Navy to proceed with the aircraft. When Forrestal received the draft (he was still Secretary of the Navy) he decided presidential action was not required and told the Navy to start work on the project. On June 24, 1946, North American was awarded a contract for production of the AJ Savage.

But the Savage would not be ready for delivery to fleet squadrons until the fall of 1949 and the Navy was not willing to lose time nor opportunity in the development of a carrier-based nuclear strike capability. The only naval aircraft which could carry a 10,000-pound bomb load and stood any chance of getting off a carrier deck was the P2V Neptune, a new twin-engine, land-based patrol bomber. With a wing span of 100 feet, a length of 78 feet, a height of 28 feet, and a take-off weight of some 60,000 pounds, the Neptune was considerably larger than the B-25s which had flown from the *Hornet*, and the R4Ds which had flown from the *Philippine Sea*.

Extensive tests with the Neptunes were conducted ashore. One of the lead pilots in this project was Commander Thomas D. Davies who, in September 1946, flew an XP2V-1 Neptune named "Truculent Turtle" from Perth, Australia, to Columbus, Ohio, in 55 hours and 17 minutes. The flight shattered the world's record for distance without refueling by setting a mark of 11,235.6 miles. The Neptune's distance record stood for almost two decades.

After hundreds of practice take-offs from airfields ashore, on April 27, 1948, barges brought two P2V-2 Neptunes alongside the large carrier *Coral Sea*, moored at Norfolk, Virginia. The planes were lifted to the carrier's deck by crane, looking like giant, pre-historic monsters to the carrier's sailors who were used to seeing "bombers" with only half the wing span of the Neptunes. During the night the *Coral Sea* steamed out into the Atlantic while mechanics tuned up and fueled the aircraft.

At 7:16 the next morning the throttles of the first Neptune were pushed to full power by Commander Thomas Davies; he released the brakes and as the plane started forward the auxiliary rockets fired, pushing the plane into the air. Minutes later the second Neptune, piloted by Commander J. P. Wheatley, started down the *Coral Sea*'s deck, fired its JATO bottles, and lifted into the air.

The Navy now began a program to develop this test into a operational capability. As an interim heavy attack aircraft the Navy ordered twelve Neptunes to be produced with special features for carrier take-offs, these being designated P2V-3C. The -3C was equipped with special high-altitude engines and carried 4400 gallons of fuel, almost double that of the standard P2V-3. And, the plane was equipped to carry a Hiroshima-type atomic bomb (Mark VIII). This weapon weighed about 9000 pounds and produced an explosion equivalent to approximately 14,000 tons of high explosives. (The AJ Savage would be able to carry the larger, Nagasaki-type Mark VI weapon;

With the assistance of JATO, a Lockheed P2V Neptune launches from the carrier *Coral Sea* in 1948. (Courtesy of the U.S. Navy)

the Nagasaki bomb itself had produced a blast equivalent to about 20,000 tons of high explosive—a 20 kiloton yield.)

During this period the production of atomic bombs from the relatively limited amounts of available uranium 235 and plutonium was a major interservice controversy. The Navy wanted a portion of the fissionable material allocated to the development of weapons which could be carried by carrier-based aircraft; the Air Force wanted the available material to go into the production of proven bomb designs—which could best be carried by the larger B-29s. Approval for the weapons the Navy wanted was won from the Joint Chiefs of Staff and the crucial Military Liaison Committee of the newly established Atomic Energy Commission through the efforts of Admirals Parsons, Radford, and Forrest Sherman. Parsons, promoted to rear admiral, was in charge of the Navy's efforts in the development of atomic weapons; Vice Admiral Arthur W. Radford had succeeded Admiral Mitscher as Deputy Chief of Naval Operations (Air), holding that post from January 1946 to February 1947, and was then Vice Chief of Naval Operations from January 1948 to April 1949. Rear Admiral Forrest P. Sherman was Deputy Chief of Naval Operations (Operations) from December 1945 to January 1948, and became Chief of Naval Operations in November 1949. The Navy's cause was helped further by Commander Hayward serving as Director of Plans and Operations for the Armed Forces Special Warfare Project and Commander Ashworth being Executive Secretary of the A.E.C. Military Liaison Committee during the critical years 1947 and 1948.

The Navy commissioned Composite Squadron Five in September of 1948 to evaluate heavy attack aircraft and to develop the doctrine and tactics for delivering nuclear weapons from carriers. Captain Hayward was commanding officer of the squadron and Commander Ashworth was his executive officer. Additional "heavy attack" squadrons would be formed from this unit as more men were trained and the new AJ Savages became available. Ashworth would command the first of the later squadrons, Composite Squadron Six formed early in 1950.

All twelve of the specially configured P2V-3C Neptunes were delivered to Composite Squadron Five by January 1949. After extensive practice take-offs from land bases, three of the squadron's Neptunes were hoisted aboard the *Coral Sea* on March 4, 1949. Three days later, off the Virginia Capes, the Neptunes took off with the aid of JATO boosters. Captain Hayward was at the controls of the first plane, which weighed 74,100 pounds including a 10,000-pound dummy nuclear weapon. The other Neptunes weighed 65,000 and 55,000 pounds, respectively.

While the second and third bombers flew to a nearby airfield, Captain Hayward's Neptune flew across the country, dropped its "bomb" on the West Coast, and then returned across the United States to land at an East Coast airfield. The plane was airborne for almost twenty-three hours and flew almost 4500 miles.

The flight indicated that aircraft with a gross weight of 74,000 pounds could be flown from the large *Midway*-class carriers, deliver an atomic bomb on targets as far as 2200 miles away, and either return to the carrier and ditch nearby or fly off in a different direction after dropping its bomb and rendezvous with a rescue ship or submarine. However, the Navy was still a far way from having a practical carrier-based nuclear strike capability.

The Neptunes made another twenty take-offs from the *Midway*-class carriers during March, April, May, and June. All three of these ships had their flight decks strengthened to operate the loaded Neptunes and were modified to handle nuclear weapons. In addition to secure storage spaces for the nuclear weapons, spaces were needed to assemble the bombs, a job which required the efforts of up to forty highly trained technicians for forty-eight hours. (The bomb had to be assembled immediately before loading in the aircraft.)

In September of 1949 the Navy gave a sea power demonstration off the Atlantic coast for several senior officials of the Army, Air Force, and the Department of Defense. The climax to the demonstration came when Secretary of Defense Louis Johnson, Secretary of the Air Force Stuart W. Symington, and General Omar N. Bradley, Chairman of the Joint Chiefs of Staff, climbed aboard a P2V-3C on the deck of the *Midway*. Captain Hayward took the pilot's seat, Secretary Johnson sat in the co-pilot's seat, and Secretary Symington and General Bradley took positions aft. Johnson, as will be related, had just ordered the Navy to halt construction of a flush-deck aircraft carrier. Hayward turned to Johnson and said: "If anything happens on this take off, we will have a flush-decked carrier, with your approval or not!" Then Hayward took off—with the Neptune's wingtip missing the *Midway*'s island structure by a few feet—but without incident and returned his distinguished passengers to Washington.

A more meaningful Neptune flight from the *Midway* came on October 5, 1949, when Commander Ashworth took off as the carrier steamed in the Western Atlantic, flew his plane to the Canal Zone, then north to Corpus Christi, Texas, and finally landed at San Diego, California. The flight had spanned 4880 miles in just under twenty-six hours.

One P2V-3C was fitted with an arresting hook and Captain Hayward made 182 arrested landings ashore. He recommended that arrested landings be made aboard ship, but permission for the test was not forthcoming and the proposal was soon overtaken by the availability of the AJ-1 Savage.

Deliveries of AJ-1 Savages began in September of 1949. This was the first true carrier-based heavy attack aircraft. With a gross weight of 47,630 pounds the AJ-1 was designed to carry an atomic bomb or 10,000 pounds of conventional weapons to targets a thousand miles from the launching carrier. The plane had no guns, relying on fighter escort and speed for protection. To facilitate handling aboard ship the plane's wings (71½ feet across) and tail (20½ feet high) folded.

By the end of 1949 there were six AJ-1 Savages in addition to P2V-3C Neptunes in Composite Squadron Five. Although the Savages had not yet operated aboard carriers, on January 6, 1950, the squadron was declared ready to launch an atomic strike from *Midway*-class carriers. But in 1950 the practicality of operating heavy attack aircraft from carriers was still questionable. The carriers had to be "pre-loaded" with aircraft in friendly ports because the Neptunes could not land aboard and the flattops would then be unable to operate their regular aircraft until the Neptunes had been flown off, either back to their staging bases or to deliver a nuclear attack. In view of the number of Air Force bombers in service at the time, the Navy's nuclear strike capability seems to have been minor and of questionable potency.

Marines on Carriers in Korea

The subject of Marine aviators flying from Navy carriers has always been a touchy one. By the early 1930s, carrier qualification periods for Marines were regular affairs, and certainly, by the American entry into World War II, Marines had become at least occasional parts of the regular Navy air groups of the time. During a period of perhaps 18 months, most Marine Corps flight crews operated from shore bases, offering long-range escort services for bombers of all the other services.

Marine Corps squadrons were the first to operate the F4U Corsair, but from airfields in the Solomons. Vought's big, blue gullwing was deemed too hot, too treacherous to fly aboard ship. It wasn't until the Marines began flying their "U birds" from smaller, less accommodating escort-class carriers that both the Corsair and the Marines found their new home afloat.

During the last 12 months of the war in the Pacific, Marine Corps Corsair squadrons formed major portions of the late-war air groups that carried the conflict to Japan's many doorsteps, as well as protecting the vulnerable task forces from the kamikaze suicide pilots who attacked ships off Okinawa.

With the introduction of jet aircraft to carrier air groups of the late 1940s and early 1950s, it appeared that the days of the propeller-driven fighter of the earlier war were numbered. Then, during the Korean conflict, the Navy found that it lacked suitable aircraft to fly from carriers and strike at Communist facilities and troop concentrations, or protect the beleaguered United Nations ground forces far inland, under constant threat from an implacable enemy aided by the Soviet Union and the People's Republic of China.

Hurriedly organizing, the 1st Provisional Marine Brigade departed the West Coast in July 1950 and arrived in Korea on August 2. On August 3,

Corsairs of VMF-214 flew the Corps' first of many sorties in the new Asian war. For several months at a time, Leatherneck Corsairs operated from the few Navy carriers assembled off the Korean peninsula at any one time. Often, the incoming Marine pilots traded their F4U-4Bs for F4U-4s. The Bs were armed with four 20mm cannon, and the ammunition required more storage consideration than the less touchy .50-caliber ammunition of the dash-4. But the Marines coped as they usually do, and along with their shore-based compatriots, the various VMF and VMA squadrons served well aboard ship. (In June 1952, several squadrons changed their fighter designation to attack.)

One of the more eloquent proponents of Marine carrier squadrons was then-Captain John S. Thach, one of the Navy's premier fighter aces and tacticians of World War II, who would rise to four-star rank. As the following excerpt proves, Thach couldn't say enough about his jarhead aviators. As captain of the small carrier USS *Sicily* (CVE-118), Thach and his crews were in the thick of heavy fighting in the last part of 1950.

This book is the second volume in Tim Wooldridge's trilogy of first-person accounts on naval aviation. It offers an enjoyable mix of accounts of the 25 years following World War II.

From "The Black Sheep in Korea" by Admiral John S. Thach in *Into the Jet Age: Conflict and Change in Naval Aviation 1945–1975,* edited by E. T. Woolridge

I had one squadron of twenty-four airplanes aboard—VMF-214, the Black Sheep squadron. That's about the number that you could operate most efficiently, where you had plenty of room to move them around, and that was the size of a marine close air support squadron. Later, another squadron, VMF-323, relieved them. They were also real pros, but that was much later in the Korean War.

These marine pilots of VMF-214 were all quite experienced; they weren't young kids. Most of them were married and had children, and they took their work seriously. They really were the top pros in the business, I think, in the whole world. Many of them had been in World War II. They were heavily decorated and they knew the business of close air support. Every one of those pilots had infantry training. From time to time later in the Korean War, pilots from the *Sicily* squadron would rotate into the front lines as air controllers for a while and then back to the *Sicily*. The marine troops knew them by their first names; they knew who they were and what they could do.

The close air support problem as it existed at the time had a bearing in a way on why things went the way they did. Of course, the U.S. Air Force wasn't too interested in close air support. The people in high command in the air force were primarily focused on the big bomber idea, that you really didn't even need troops to win a war—just fly over and bomb them, and then wait for a telegram saying that they surrender. So the air force was utterly unprepared to do close air support the way it had to be done if you were going to help the troops at the front lines.

It wasn't a matter of just being a communication problem; it was a matter of education over a long period and experience and doctrine built up. I just couldn't believe it could be so bad, and neither could Rear Adm. J. M. Hoskins, who commanded Task Force 77. He'd send these planes down and the pilots would come back and say, "We couldn't help. We wanted to. We were there and we couldn't get in communication with people." Those that did would get in touch with the Joint Communications Center and they either wouldn't have any target or by the time they got it it was old information.

The air force went on the principle that any aircraft committed had to go through a pretty high echelon of command before it did anything. They wanted to keep tight control in high places. Never would they let anybody in the trenches control one of their airplanes; they were going to do it themselves from higher up. Well, they had to get some sort of a working arrangement with the carrier-based aircraft who were used to being controlled from the ground in the front lines. They did it on every amphibious landing, and their tactical air control squadrons had a great deal of history and experience, and the navy-marine teams were real pros in this business.

Another thing that made it difficult was that the Fifth Air Force wouldn't accept a navy tactical control party. Help would be offered by navy air control and they didn't want to take it. They were more interested in getting control of all aircraft than they were in helping the army! When General MacArthur and Vice Adm. A. D. Struble, commander, Seventh Fleet, went to Formosa, Gen. George Stratemeyer, commander, Far East Air Force, quickly held a conference. The official record of it says that they deployed four generals and a colonel to face one captain and two commanders and two lieutenant commanders to make sure they controlled the decisions that were to be made. Nevertheless, something good could have come out of this, but it didn't.

They wanted to be sure that they would get operational control of all aircraft in the Korean theater.

As a result of the conference, they put out a memorandum saying that first priority for carrier operations would be in close support, second was interdiction south of the 38th parallel, and the third priority was to strike Bomber Command targets beyond that line, if requested by the Bomber Command. Coordination for the attack south of the 38th parallel was to lie with the Fifth Air Force. Of course, you'd have to get permission from the Far East Air Force to attack any targets otherwise, which were supposedly bomber targets. They interpreted this as operational control and this disturbed the navy quite a bit. It had implications that they would be telling the carriers where to go—right rudder, left rudder, and so forth.

We couldn't stand that and Admiral Struble, since he was not even consulted before this memorandum was put out, didn't want to go along with it at all. He tried to get away from it but there was not much he could do—he didn't get help from anybody. So the Joint Operations Center (JOC) was formed by the air force at Taegu, and the navy fast carriers tried very hard to cooperate because of these urgent and emergency calls for help by the army. The army must have been giving the air force hell at that time, because the air force even started repeating their screams for help to the fast carriers. So the carriers would send everything that they asked for.

I have listened on the close air support radio many times. By this time there really wasn't any Fifth Air Force, Korea. The Fifth Air Force was back in Japan because, as somebody said, the best way to defeat an air force is to walk into the airfield and keep the airplanes from flying, and that's what the North Korean army did. The enemy got so close to Taegu they were within the landing circle, so that field had to be given up. By this time the field at Pohang, we called it Pohangdong, was being exchanged sometimes every night.

I know one time we sent some planes over there because the army had a patrol that really wanted help and our pilots were willing to work at it. I heard this on the radio and also the pilots informed us; the army controller was so enthused about the *Sicily* Corsairs, the marine pilots, that before releasing the planes after they'd expended their ammunition, they begged them to come back the next day. The words were sometimes, "Please, please, come back tomorrow. We'll take that airfield back again. If you'll just come back tomorrow we can do it together." It would almost bring tears to your eyes to realize how much these army troops over there wanted some real good close air support. They hadn't ever had it before. One of them said, "We had close air support like I've never heard of before. This is something I didn't realize could happen."

The F-80s would come over from Fukuoka, Japan, and the front line was just near the end of their range. They'd call the controller and say, "Give me a target, give me a target. I've only got five minutes more. Got to go back." They would be asked, "What is your ordnance?" "I've got two 100-pound bombs. Hurry up." I heard that so many times and finally I heard, I think it was an air force controller, say, "Well, take your two little firecrackers and drop them up the road somewhere because I've got something coming in that has a load."

On 7 August the First Provisional Marine Brigade, which had been landed at Pusan and got in position under Brig. Edward A. Craig, attacked westward from Masan toward Chinju. This was the first time that the escort

Armed with rockets, a Black Sheep F4U-4B launches from *Sicily* in 1950. Never intended as a ground attack fighter, the Corsair flew the lion's share of such Navy and Marine missions in Korea. (Courtesy of the NMNA, Thach Collection)

carriers with the Corsairs and the marine ground forces got into action with everything there for coordination. It was a beautiful thing to listen to. I couldn't see it but I knew what was going on. It was just like going from confusing darkness into bright daylight. The coordination was perfect and everything clicked just the way it should. You should have seen those pilots when they came back. Each one would heave a big sigh of relief and say, "Now we're doing what we're supposed to do in the right way."

Unless we were working with the marines, the pilots still had to go through that Joint Communications Center. Working with the marines, we didn't have to bother with the JOC, because we had our own tactical air control group. The forward air controller, the ground controller, often an aviator, would sometimes be out in front of the front lines. Sometimes they'd be in a little jeep or a tank or just crawling along and dragging the communications equipment, in the bushes.

One time the controller said, "I want just one plane of the four to come down and make a dummy run. Don't drop anything. I'm going to coach you on to a piece of artillery that's giving us a lot of trouble. I'm very close to it but I can't do anything about it. It's just over a little knoll." He described the terrain, just where it was and so forth. So the leader came down in his Corsair and he was coached all the way down, the air controller practically flying the airplane for him.

The controller said, "Now, do you see it?"

And the pilot said, "Yes, I see it."

"Okay, then, go on back up and come down and put a 500-pound bomb on it. But be very careful." So the pilot came down and he released his bomb and hit it with a big explosion.

The controller said, "Right on the button, that's it. That's all. Don't need you anymore."

The pilot said, "Just a minute. While I was on the way down, on the right hand side of my gunsight, I saw a big tank. It was under a bush, but how about that target?"

And the controller said, "I told you I was close. Let it alone. That tank is me."

A Whitehat's Viewpoint

Of necessity, military organizations have always been divided into two distinct societies, enlisted personnel and commissioned officers. From its earliest days, military aviation has usually included both types of members, with pilots also coming from both groups. However, while other countries such as Britain, Germany, Italy, and Japan in World War II took full advantage of enlisted aviators, other nations relegated non-pilot duties to the enlisted ranks. Thus, many memoirs and histories come from or describe the officer experience, leaving the enlisted cadre to float in the peripheral atmosphere of the overall narrative.

By the beginning of the Korean War, most of the Navy's enlisted flight crewman were serving as radar operators in multiplace aircraft that required more than one man to fly the plane and operate its mission equipment. This situation was not the same in the Marine Corps, which used a larger number of enlisted pilots, some of whom had been officers in World War II but had mustered out and then returned to active duty as noncommissioned-officer aviators. It was somewhat rare, however, to find enlisted crewmen in carrier squadrons, and one of the rarer types was the AD Skyraider crew.

In the new but rapidly developing area of electronic defense and snooping, the Navy had modified the big Skyraider to accommodate one or two crewmen in fuselage stations below and behind the pilot. These AD-4Ws and their derivatives were part of the carrier fleet until 1969, flying the Skyraider's last combat mission during the Vietnam War.

Jack Sauter served with composite squadron VC-12 as an electronics technician and eventually flew 21 missions from USS Champlain *(CV-39) during the last two months of the Korean War, earning his combat aircrewman wings, a highly prized piece of jewelry amongst enlisted ranks. This excerpt from Sauter's excellent book about his experiences describes how these men flew their huge, single-engine ADs in the waning weeks of the war.*

From *Sailors in the Sky: Memoirs of a Navy Aircrewman in the Korean War,* by Jack Sauter

July 25 was a rare clear day for this time of the year, and again the task force broke its record with 746 flights. Our team flew four ASW sorties, and all the crewmen got air time. Calley and I were up with Lt. Williams, as the four carriers pounded out close to 30 knots to bring their squadrons back on board. We still had another two hours left in our mission, so we were treated to a rare view of the whole force steaming below us.

The carriers were headed into the wind on a course taking them directly into the coast of North Korea. We asked the pilot why the ships hadn't turned. He surmised that some of the jets probably didn't have enough fuel to wait for the carriers to come around again. When only five planes remained, Williams said, "Watch closely and you'll see some fancy maneuvering." Admiral Johnson was quickly running out of sea room, and he'd shortly be in range of the North Korean shore batteries.

Finally, the last aircraft touched down, and before the wire was released from the hook, the entire task force made a 180-degree turn. A dozen ships turned as if guided by one hand, their long wakes cutting white circles in the blue sea. *Life* magazine later ran a spread on the *Champ*, and what we saw that day was captured on film.

Those on board had a different viewpoint, but an equally interesting story to tell. The coastline was clearly visible and getting closer by the minute. As the last plane was in her final approach, the bullhorn blared, "Stand by for an emergency turn to port." All hands hung on. When the order "execute" was given, the ship heeled over and everything not secured rolled across the deck. More than one tool box and some loose pieces of cowling went over the side.

During a later launch, a plane handler was sucked into a jet engine and critically injured. This increased tempo of operations was taking its toll.

That night after chow, all the technicians went to work. Every one of our planes had electronic troubles and had to be ready by the following morning. Three of us manhandled a heavy radar unit down three ladders and into the shop, where we spent five hours locating and repairing the problem. On top of this, the APN-1 electronic altimeters were out of alignment. Their sensitive circuits just weren't suited for the rough and tumble of carrier operations. It was akin to repairing a sensitive timepiece next to a railroad trestle. Later, the APN-1 became my ticket to a brief moment in the spotlight, but that was still in the future.

Well after midnight, the radar in 703 checked out and we all went below to the mess deck for some cool drinks. Tired as we were, we still needed a drink before sacking out. The cooks recognized us and wanted to make something special, but we didn't even have the energy to eat. In spite of the heat and humidity, we collapsed in our bunks with everything but our shoes on.

It was two hours, but it felt more like two minutes, when we awakened to the bong, bong, bong of General Quarters and the urgent voice on the P.A. system shouting: "General Quarters! General Quarters! All hands man your battle stations!"

Previously all G.Q. drills had been announced beforehand. Our compartment lights were turned on, and everyone started running. I heard someone shout, "This is no drill!" (Where had I heard that before?) needing no further motivation, all hands on the *Champ* responded with something bordering on pandemonium. We just made it to the Ready Room before all the hatches were dogged down behind us.

To save our night vision, the Ready Room was bathed in an eerie blue light as we changed into our flight suits and Mae Wests. The teletype revealed many bogies closing the task force. The usual banter and horseplay were missing as we received our assignments. All three of VC-12's planes would be launched.

I could hear the jets being "catted" above me as I climbed into the backseat of 703 on the deck-edge elevator. With Lt. Williams as our pilot, we would operate as the backup AEW plane. Oddly, it was 703 that had kept us up half the night. The *Lake Champlain* was pounding out many knots as I propped open the hatch to get some air. From our perch out over the hull, I could watch the boiling sea race by. A few yards from the foaming bow wave, the ocean was black beyond description. I wouldn't want to be down tonight.

On previous flight ops, when a launch was delayed, we usually passed the time telling jokes or exchanging the latest scuttlebutt. Tonight was different. Each of us sat silently with our own thoughts. All of us, I'm sure, made impossible promises to God, and I was one of them. My gut was wound so tight, it was hard to breathe, no less talk. For the umpteenth time, I tightened the harness of my chute. I remember praying, "Whatever else happens, don't make me bail out of this thing!"

In spite of the tremendous anxiety, between the heat, the darkness, and our overall fatigue, we both dozed off. The next thing I remember was Williams shaking me. We'd been told to stand down and secure from General Quarters, he said. Whatever it was out there had disappeared as our jets closed for interception. We should all get some shut-eye, as there was a full card on for tomorrow, and all of us were scheduled for a 0930 launch in the same aircraft. It was now 0455.

It was a dog-tired crew that shuffled back to their living spaces, carrying steel helmets and fire-fighting equipment. George and I fell asleep immediately in the big leather chairs in our Ready Room, still wearing our useless

Aviation Electronics Technician Second Class Jack Salter poses by his Skyraider. (Courtesy of Jack Salter)

.38s. It was easier, and far more practical, than dragging ourselves back to our compartment. Besides, the Ready Room was gloriously air-conditioned.

That day, CAG-4 flew 166 sorties, a *Champ* and TF-77 record. Our jets flew deep into North Korea, cratering airfields and attacking communications centers. The props operated around Wonsan. Flak was intense, and many planes returned with holes in their wings and fuselages.

During a deck launch, an AD4B from Attack Squadron 45 went into the sea after an engine failure. The pilot, Lt. Brumbach, was recovered by our helicopter uninjured. While recovering aircraft from our early afternoon attack, Airman L. Woods, a plane handler, was seriously injured when he was blown into a tractor by a jet blast. More serious was the news that Ensign Broyles of VF-22 failed to rendezvous after a bombing run over the North. An extensive search came up empty. He was last seen in a 40-degree dive at 6,000 feet. No ejection was observed, and he was listed as missing in action.

When flight operations ceased that day, crewmen were sprawled everywhere. They were in passageways, on mess tables, in catwalks, and on wings. It was as if the *Champ* had been struck by a gas attack. Some poor bastards were later detailed to take part in a rearming from USS *Vesuvius* (AE-15), and I saw about a hundred of them straining under the floodlights as I headed below decks about 2130. Except for the activity on the hangar deck (which ended at 2300), the ship was a tomb.

On 27 July, another hot, humid day, I flew my twenty-first mission with TF-77. (George and I ended up tied for the most flights in the war zone.) In the final approach, while I was taking a photo of the *Champ*, word was received that truce terms had been accepted by both sides. We became the last plane of the final combat strike to touch down. All offensive operations ceased. After 37 months and 53,629 combat deaths, the Korean War had ended.

Fiction As True As Fact

Although there have been many works of aviation fiction, beginning as far back as the Greek myth of Icarus and progressing through Jules Verne and H. G. Wells by the early 20th century, most are works of technical hyperbole and emotional exaggeration. Even today, at the beginning of the 21st century, most authors still struggle to convey the peaks and valleys of flying, especially for the military. During the 1930s and 1940s, the most popular author of the genre was French aviator Antoine de St. Exupery. He disappeared in an F-5, the reconnaissance version of the famed Lockheed P-38, while on a mission in 1944 at the advanced age—for a military pilot—of 44. Although I've given him a good try, I've found St. Exupery a crashing bore, too long-winded, and too wrapped up in his own philosophies of life to make his books entertaining or at least worthwhile.

The current crop of writers is led by people like Tom Clancy, who resuscitated the military novel with a hefty infusion of high-tech facts and imagination. Steven Coonts, a former naval aviator who flew A-6 attack bombers in Vietnam, brought the genre one step further. Recently, Commander Ward Carroll, an F-14 radar intercept officer, entered his novel, *Punk's War* (U.S. Naval Institute Press, 2001), about an F-14 squadron enforcing the no-fly zone over Iraq, into the sweepstakes. There are many more authors in the group. All these men bring something of value to their work—experience, craftsmanship, and in the case of one, downright abhorrence mixed with grudging understanding of military authority and its required organization. But in the end, there's something missing that would raise their work to the level of the novella from which we take our next excerpt.

The basic premise of this anthology is that it takes an extra measure of courage and dedication to make these massive weapons systems work. This book runs some 300 pages, yet James Michener accomplished the same end in barely half that amount in *The Bridges at Toko-Ri*. Nowhere is the story of carrier aviators so well and simply told. And surprisingly, the book generated a first-rate movie that won an Oscar for special effects when all there was to work with were models, lighting, and a camera—B.C., before computers.

The movie is also singular because it scrupulously adheres to the original book's story. The fear and planning for the strike on a Communist bridge avails the hero little. Lieutenant Harry Brubaker dies just at the moment when he might have gained an understanding of why he was in Korea. Read the book and rent the movie. Neither will disappoint.

In this excerpt, we meet Brubaker, a reluctant, recalled Naval Air Reservist from Denver, and many of the supporting characters, as he tries to bring his Banshee aboard the carrier *Savo*. That's one of the only differences between the book and the movie: the aircraft Brubaker's squadron flies is the McDonnell F2H Banshee. In the movie, perhaps for availability reasons, it is the Grumman F9F Panther. It's an acceptable exchange.

Michener's story recalls an important aspect of the air war in Korea: the role of thousands of recalled American reservists. Many of this group of participants, particularly the aviators, were not actively drilling—participating—in established reserve programs. After serving in World War II, they had mustered out in 1945 and 1946, thinking their military service was complete. However, whether they realized it or not, many of these veterans' names were on inactive roles, subject to recall when required. And by late 1950, they were definitely required.

While many drilling reservists volunteered, sometimes *en masse* in squadrons, to return to active duty and go to Korea, other inactive reservists were simply recalled, often with very little time to prepare their families or settle their business affairs. Thus, it was with anger and at least ambivalence that some of these men, in many services, reported to their duty stations. They had to find their own way in coming to terms with their second military careers. And that is the story of Harry Brubaker in *The Bridges at Toko-Ri*, which Michener dedicated to Marshal Beebe, a 10-kill Hellcat ace.

Michener drew from several sources and events to tell the story of the men of the *Savo*. Visits to the carriers USS *Essex* (CV-9) and USS *Valley Forge* (CV-45) helped generate his interest in carriers. An actual mission against North Korean bridges gave him facts and colorful personalities on which to base his plot.

James A. Michener was one of America's most successful writers of fiction. His best known works are Tales of the South Pacific, *on which the highly successful Broadway musical "South Pacific" was based;* The Source; Centennial; Hawaii; *and a host of other novelized "histories." He died in 1997.*

From *The Bridges at Toko-ri*, by James Michener

"We'll make it," Tarrant grumbled as his ships plowed resolutely on toward the crucial hundred fathom curve which he dare not penetrate for fear of shoals, mines and submarines. But he turned his back upon this problem, for he could do nothing about it now. Instead, he checked to be sure the *Savo's* deck was ready and in doing so he saw something which reassured him. Far aft, standing upon a tiny platform that jutted out over the side of the carrier, stood a hulking giant, muffled in fur and holding two landing-signal paddles in his huge hands. It was Beer Barrel, and if any man could bring jets surely and swiftly home, it was Beer Barrel.

He was an enormous man, six feet three, more than 250 pounds, and his heavy suit, stitched with strips of fluorescent cloth to make his arms and legs easier to read, added to his bulk. He was a farmer from Texas who before the perilous days of 1943 had never seen the ocean, but he possessed a fabulous ability to sense the motion of the sea and what position the carrier deck would take. He could judge the speed of jets as they whirled down upon him, but most of all he could imagine himself in the cockpit of every incoming plane and he seemed to know what tired and jittery pilots would do next and he saved their lives. He was a fearfully bad naval officer and in some ways a disgrace to his uniform, but everyone felt better when he came aboard a carrier, for he could do one thing. He could land planes.

He could reach out with his great hands and bring them safely home the way falconers used to bring back birds they loved. In the Pentagon they knew he broke rules and smuggled beer aboard each ship he served upon. Carrier captains knew it, and even Admiral Tarrant, who was a terror on navy rules, looked the other way when Beer Barrel staggered back after each drunken liberty, lugging his two ridiculous golf bags. The huge Texan had never once played golf and the two clubs sticking out were dummies. Once a deck hand, fearful that drunken Beer Barrel might slide back down the gangplank, had grabbed one of the outside golf bags to help, but the surprising weight of it had crumpled him to the deck. Beer Barrel, barely able to heft the bag himself, had got it onto his massive shoulder, whispering beerily to the boy, "Thanks, Junior, but this is man's work." And he had carried the bags full of beer into his quarters.

For he believed that if he had a can of cold beer in his belly it formed a kind of gyroscope which made him unusually sensitive to the sea and that when this beer sloshed about it harmonized with the elements and he became one with the sea and the sky and the heaving deck and the heart of the incoming pilot.

"Land jets!" moaned the bull horn.

"Let's hear the checks," Beer Barrel said to his spotters, staring aft to catch the first jet as it made its 180° turn for the cross leg and the sharp final turn into the landing run. Now the jet appeared and Beer Barrel thought, "They're always pretty comin' home at night."

"All down!" the first watcher cried as he checked the wheels, the flaps and the stout hook which now dangled lower than the wheels.

"All down," Beer Barrel echoed unemotionally.

"Clear deck!" the second watcher shouted as he checked the nylon barriers and the thirteen heavy steel wires riding a few inches off the deck, waiting to engage the hook.

"Clear deck," Beer Barrel grunted phlegmatically.

He extended his paddles out sideways from his shoulders, standing like an imperturbable rock, and willed the plane onto the deck. "Come on, Junior," he growled. "Keep your nose up so's your hook'll catch. Good boy!" Satisfied that all was well, he snapped his right paddle dramatically across his heart and dropped his left arm as if it had been severed clean away from his body. Instantly the jet pilot cut his flaming speed and slammed his Banshee onto the deck. With violent grasp the protruding hook engaged one of the slightly elevated wires and dragged the massive plane to a shuddering stop.

Setups like this won the movie an Academy Award for special effects. (Courtesy of Fred Freeman)

Beer Barrel, watching from his platform, called to the clerk who kept records on each plane, "1593. Junior done real good. Number three wire." Never did Beer Barrel feel so content, not even when guzzling lager, as when one of his boys caught number three wire. "Heaven," he explained once, "is where everybody gets number three wire. Hell is where they fly wrong and catch number thirteen and crash into the barrier and burn. And every one of you's goin' straight to hell if you don't follow me better."

From his own bridge, Admiral Tarrant watched the jets come home. In his life he had seen many fine and stirring things: his wife at the altar, Japanese battleships going down, ducks rising from Virginia marshes and his sons in uniform. But nothing he knew surpassed the sight of Beer Barrel bringing home the jets at dusk.

There always came that exquisite moment of human judgment when one man—a man standing alone on the remotest corner of the ship, lashed by foul wind and storm—had to decide that the jet roaring down upon him could make it. This solitary man had to judge the speed and height and the pitching of the deck and the wallowing of the sea and the oddities of this particular pilot and those additional imponderables that no man can explain. Then, at the last screaming second he had to make his decision and flash it to the pilot. He had only two choices. He could land the plane and risk the life of the pilot and the plane and the ship if he had judged wrong. Or he could wave-off and delay his decision until next time around. But he could defer his job to no one. It was his, and if he did judge wrong, carnage on the carrier deck could be fearful. That was why Admiral Tarrant never bothered about the bags of beer.

On they came, the slim and beautiful jets. As they roared upwind the admiral could see their stacks flaming. When they made their far turn and roared downwind he could see the pilots as human beings, tensed up and ready for the landing that was never twice the same. Finally, when these mighty jets hit the deck they weighed well over seven tons and their speed exceeded 135 miles an hour, yet within 120 feet they were completely stopped and this miracle was accomplished in several ways. First, Tarrant kept his carriers headed into the wind, which on this day stormed in at nearly 40 miles an hour, which cut the plane's relative speed to about 95 miles. Then, too, the carrier was running away from the plane at 11 miles an hour, which further cut the plane's speed to 84, and it was this actual speed that the wires had to arrest. They did so with brutal strength, but should they miss, two slim nylon barriers waited to drag the plane onto the deck and chop its impetus, halting it so that it could not proceed forward to damage other planes. And finally, should a runaway jet miss both the wires and the barrier, it would plunge into a stout nylon barricade which would entwine itself about the wings and wheels and tear the jet apart as if it were a helpless insect.

But it was Beer Barrel's job to see that the barriers and the barricade were not needed and he would shout curses at his pilots and cry, "Don't fly the deck, Junior. Don't fly the sea. Fly me." An air force colonel watching Beer Barrel land jets exclaimed, "Why, it isn't a landing at all! It's a controlled crash." And the big Texan replied in his beery voice, "Difference is that when I crash 'em they're safe in the arms of God."

Now he brought in three more, swiftly and surely, and Admiral Tarrant, watching the looming mountains of Korea as they moved in upon his ships, muttered, "Well, we'll make it again."

But as he said these words his squawk box sounded, and from deep within the *Savo* the combat intelligence director reported coolly, "1591 has been hit. Serious damage. May have to ditch."

"What's his position?"

"Thirty-five miles away."

"Who's with him?"

"His wingman, 1592."

"Direct him to come on in and attempt landing."

The squawk box clicked off and Admiral Tarrant looked straight ahead at the looming coast. Long ago he had learned never to panic, but he had trained himself to look at situations in their gloomiest aspects so as to be prepared for ill turns of luck. "If this jet limps in we may have to hold this course for ten or fifteen more minutes. Well, we probably can do it."

He studied the radar screen to estimate his probable position in fifteen minutes. "Too close," he muttered. Then into the squawk box which led to the air officer of the *Savo* he said, "Recovery operations must end in ten minutes. Get all planes aboard."

"The admiral knows there's one in trouble?"

"Yes. I've ordered him to try to land."

"Yes, sir."

The bull horn sounded. "All hands. We must stop operations within ten minutes. Get those barriers cleared faster. Bring the planes in faster."

The telephone talker at the landing platform told Beer Barrel, "We got to get 'em all aboard in ten minutes."

"What's a matter?" Beer Barrel growled. "Admiral running hisself out of ocean?"

"Looks like it," the talker said.

"You tell him to get the planes up here and I'll get 'em aboard."

So the nineteen dark ships of the task force sped on toward the coastline and suddenly the squawk box rasped, "Admiral, 1591 says he will have to ditch."

"Can he ditch near the destroyers?"

"Negative."

"Is his wingman still with him?"

"Affirmative."

"How much fuel?"

"Six hundred pounds."

"Have you a fix on their positions?"

"Affirmative."

"Dispatch helicopter and tell wingman to land immediately."

There was a long silence and the voice said, "Wingman 1592 requests permission stay with downed plane till copter arrives."

The admiral was now faced with a decision no man should have to make. If the wingman stayed on, he would surely run out of fuel and lose his own plane and probably his life as well. But to command him to leave a downed companion was inhuman and any pilot aboard the *Savo* would prefer to risk his own life and his plane rather than to leave a man adrift in the freezing sea before the helicopter had spotted him.

For in the seas off Korea a downed airman had twenty minutes to live. That was all. The water was so bitterly cold that within five minutes the hands were frozen and the face. In twelve minutes of immersion in these fearful waters the arms became unable to function and by the twentieth minute the pilot was frozen to death.

The decision could not be deferred, for the squawk box repeated, "Wingman 1592 requests permission to stay."

The admiral asked, "What is the absolute minimum of gas with which the wingman can make a straight-in landing?"

There was a moment's computation. "Assuming he finds the carrier promptly, about four hundred pounds."

"Tell him to stay with the downed man . . ."

The voice interrupted, "Admiral, 1591 has just ditched. Wingman says the plane sank immediately."

There was a moment's silence and the admiral asked, "Where's the helicopter?'

"About three more minutes away from the ditching."

"Advise the helicopter . . ."

"Admiral, the wingman reports downed pilot afloat."

"Tell the wingman to orbit until helicopter arrives. Then back for a straight-in landing."

The bull horn echoed in the gathering dusk and mournful sounds spread over the flight deck, speaking a disaster. "Get those last two jets down immediately. Then prepare for emergency straight-in landing. A plane has been lost at sea. Wingman coming in short of fuel."

For a moment the many-colored figures stopped their furious motions. The frozen hands stopped pushing jets and the yellow jeeps stayed where they were. No matter how often you heard the news it always stopped you. No matter how frozen your face was, the bull horn made you a little bit colder. And far out to sea, in a buffeted helicopter, two enlisted men were coldest of all.

At the controls was Mike Forney, a tough twenty-seven-year-old Irishman from Chicago. In a navy where enlisted men hadn't much chance of flying, Mike had made it. He had bullied his way through to flight school and his arrival aboard his first ship, the *Savo,* would be remembered as long as the ship stayed afloat. It was March 17 when he flew his copter onto the flight deck, wearing an opera hat painted green, a Baron von Richthofen scarf of kelly green, and a clay pipe jammed into his big teeth. He had his earphones wrapped around the back of his neck and when the captain of the *Savo* started to chew him out Forney said, "When I appear anywhere I want the regular pilots to know it, because if they listen to me, I'll save 'em." Now, as he sped toward the ditched pilot, he was wearing his green stovepipe and his World War I kelly green scarf, for he had found that when those astonishing symbols appeared at a scene of catastrophe everyone relaxed, and he had already saved three pilots.

But the man flying directly behind Mike Forney's hat wasn't relaxed. Nestor Gamidge, in charge of the actual rescue gear, was a sad-faced inconsequential young man from Kentucky, where his unmarried schoolteacher mother had named him Nestor after the wisest man in history, hoping that he would justify everything. But Nestor had not lived up to his name and was in fact rather stupid, yet, as the copter flew low over the bitter waves to find the ditched plane, he was bright enough to know that if anyone were to save the airman pitching about in the freezing water below it would be he. In this spot the admiral didn't count nor the wingman who was orbiting upstairs nor even Mike Forney. In a few minutes he would lean out of the helicopter and lower a steel hoisting sling for the pilot to climb into. But from cold experience he knew that the man below would probably be too frozen even to lift his arms, so he, Nestor Gamidge, who hated the sea and who was dragged into the navy by his draft board, would have to jump into the icy waters and try to shove the inert body of the pilot into the sling. And if he failed—if his own hands froze before he could accomplish this—the pilot must die. That's why they gave Nestor the job. He was dumb and he was undersized but he was strong.

"I see him," Nestor said.

Mike immediately called to the wingman: "1592. Go on home. This is Mike Forney and everything's under control."

"Mike!" the wingman called. "Save that guy."

"We always save 'em. Scram."

"That guy down there is Harry Brubaker. The one whose wife and kids are waiting for him in Yokosuka. But he don't know it. Save him!"

Mike said to Nestor, "You hear that? He's the one whose wife and kids came out to surprise him."

"He looks froze," Nestor said, lowering the sling.

Suddenly Mike's voice lost its brashness. "Nestor," he said quietly, "if you have to jump in . . . I'll stay here till the other copter gets you."

In dismay, Nestor watched the sling drift past the downed pilot and saw that the man was too frozen to catch hold. So he hauled the sling back up and said, "I'll have to go down."

Voluntarily, he fastened the sling about him and dropped into the icy waves.

"Am I glad to see you!" the pilot cried.

"He's OK," Nestor signaled.

"Lash him in," Mike signaled back.

"Is that Mike? With the green hat?"

"Yep."

"My hands won't . . ."

They tried four times to do so simple a thing as force the sling down over the pilot's head and arms but the enormous weight of watersoaked clothing made him an inert lump. There was a sickening moment when Nestor thought he might fail. Then, with desperate effort, he jammed his right foot into the pilot's back and shoved. The sling caught.

Nestor lashed it fast and signaled to Mike to haul away. Slowly the pilot was pulled clear of the clutching sea and was borne aloft. Nestor, wallowing below, thought, "There goes another."

Then he was alone. On the bosom of the great sea he was alone and unless the second helicopter arrived immediately, he would die. Already, overpowering cold tore at the seams of his clothing and crept in to get him. He could feel it numb his powerful hands and attack his strong legs. It was the engulfing sea, the icy and deadly sea that he despised and he was deep into it and his arms were growing heavy.

Then, out of the gathering darkness, came the *Hornet's* copter.

So Mike called the *Savo* and reported, "Two copters comin' home with two frozen mackerel."

"What was that?" the *Savo* asked gruffly.

"What I said," Mike replied, and the two whirly birds headed for home, each dangling below it the freezing body of a man too stiff to crawl inside.

Meanwhile Admiral Tarrant was faced with a new problem. The downed pilot had been rescued but the incoming wingman had fuel sufficient for only one pass, and if that pass were waved off the pilot would have to crash into the sea and hope for a destroyer pickup, unless one of the copters could find him in the gathering dusk.

But far more important than the fate of one Banshee were the nineteen ships of the task force which were now closing the hundred fathom mark. For them to proceed farther would be to invite the most serious trouble. Therefore the admiral judged that he had at most two minutes more on course, after which he would be forced to run with the wind, and then no jet could land, for the combined speed of jet and wind would be more than 175 miles, which would tear out any landing hook and probably the barriers as well. But the same motive that had impelled the wingman to stay at the scene of the crash, the motive that forced Nestor Gamidge to plunge into the icy sea, was at work upon the admiral and he said, "We'll hold the wind a little longer. Move a little closer to shore."

Nevertheless, he directed the four destroyers on the forward edge of the screen to turn back toward the open sea, and he checked them on the radar as they moved off. For the life of one pilot he was willing to gamble his command that there were no mines and that Russia had no submarines lurking between him and the shore.

"1592 approaching," the squawk box rasped.

"Warn him to come straight in."

Outside the bull horn growled, "Prepare to land last jet, straight in."

Now it was the lead cruiser's turn to leave the formation but the *Savo* rode solemnly on, lingering to catch this last plane. On the landing platform Beer Barrel's watcher cried, "Hook down, wheels down. Can't see flaps."

The telephone talked shouted, "Pilot reports his flaps down."

"All down," Beer Barrel droned.

"Clear deck!"

"Clear deck."

Now even the carrier *Hornet* turned away from the hundred fathom line and steamed parallel to it while the jet bore in low across her path. Beer Barrel, on his wooden platform, watched it come straight and low and slowing down.

"Don't watch the sea, Junior," he chanted. "Watch me. Hit me in the kisser with your left wing tank and you'll be all right, Junior." His massive arms were outstretched with the paddles parallel to the deck and the jet screamed in, trying to adjust its altitude to the shifting carrier's.

"Don't fly the deck, Junior!" roared Beer Barrel and for one fearful instant it looked as if the onrushing jet had put itself too high. In that millionth of a second Beer Barrel thought he would have to wave the plane off but then his judgment cried that there was a chance the plane could make it. So Beer Barrel shouted. "Keep comin', Junior!" and at the last moment he whipped the right paddle across his heart and dropped the left.

The plane was indeed high and for one devastating moment seemed to be floating down the deck and into the parked jets. Then, when a crash seemed inevitable, it settled fast and caught number nine. The jet screamed ahead and finally stopped with its slim nose peering into the webs of the barrier.

"You fly real good, Junior," Beer Barrel said, tucking the paddles under his arm, but when the pilot climbed down his face was ashen and he shouted, "They rescue Brubaker?"

"They got him."

Royal Navy Carriers in Korea

When the Korean War began on Sunday, June 25, 1950, only two aircraft carriers were on station to launch strikes against the invading North Koreans. Besides the USS *Valley Forge* (CV-45), Britain's HMS *Triumph* contributed its Spitfires and Fireflies of 13th Carrier Air Group to the Allied effort that helped fend off the initial assault. It's interesting that no

British jets flew from carriers in Korea. The squadrons involved operated World War II vintage aircraft, which had to be marked with distinctive D-Day-like black-and-white stripes on their fuselages and wings to distinguish them from Communist YAK fighters.

Although never reaching the numbers of the American fleet, the Royal Navy's carriers offered both discipline and spirit, and were always a welcome addition to whatever task force might be on the line. (Many of the postwar developments in carriers, particularly the flight deck, that we take for granted today, come from British dedication and imaginative skill. The steam catapult and angled deck that reshaped flight-deck operations in the mid-1950s were first developed in England.)

On October 5, 1950, HMS *Theseus* replaced *Triumph*. The new carrier arrived with the 17th Carrier Air Group, with Sea Fury FB.11s and Firefly AS.5s, perhaps the ultimate in carrier propeller-driven tactical aircraft. The *Theseus*'s aircraft went right to work, participating in many missions against Communist facilities before leaving for home the following April. HMS *Glory* came on duty, followed by the Australian carrier HMAS *Sydney*. HMS *Ocean* arrived in May 1951, and so it went. With never more than one British carrier—only five Commonwealth ships served in Korea—on station at any one time, the Fleet Air Arm gave a good account of itself throughout the war.

Little has been written about the contributions of British carriers in Korea, and it remained for such a book to come from the source. Whether it was sending strikes against Communist targets or fending off MiG attacks in propeller-driven fighters, the Royal Navy squadrons took a back seat to no one.

John Lansdown served in 821 Squadron aboard HMS Glory *in Korea during 1952–53. He left the Royal Navy as a lieutenant commander and went into private industry. He retired in 1986 and died in 2000.*

From *With the Carriers in Korea,* by John R. P. Lansdown

Captain Evans shifted *Ocean* 70 miles north of her usual operating area where he found entirely suitable conditions, the northern part of North Korea being clear. This well illustrated one of the inherent advantages of a carrier over an airfield as an operating base.

Weather delayed the start of flying on the 27th until 1000, and when it did a new twist was introduced into the war, best described in the words of 825 Squadron's diarist:

> 'Today for the first time since the Korean War started Russian built Mig-15 jet fighters attacked British naval aircraft. Lieutenant Hawkesworth's division was the one that was bounced. This attack was due primarily to the weather conditions prevailing in the operational area. A cold front lay east/west across Korea just south of the Chinnampo estuary. To the north the weather was good, 3/8ths cumulus with a base at 4,000 feet and good visibility; while to the south of the front conditions were such as to ground all 5th USAF.
>
> *Ocean's* aircraft were the only ones operating throughout the day on the west coast.
>
> 'In Event Baker four Furies reported seeing three MiGs in the Chinnampo area, but these did not attack the Furies. Also in Event Baker the CO [Lieutenant Commander Roberts] led his division to attack a warehouse, one of a group of three in the Hanchon area. The attack was successful and the warehouse completely gutted. On return to the ship the CO reported that the area had been thoroughly alerted and that his division had encountered more flak in the area than had been seen previously and that most of it was accurate. The majority of the flak was thought to be 37 mm type—bursts being seen up to 7,000 feet. This division saw no enemy aircraft.
>
> 'There were no Fireflies on Event Charlie. But on Event Dog Lieutenant Hawkesworth's division was briefed to attack a target in Kyomipo. Unfortunately during the briefing no mention was made of the MiG report from Event Baker. On arrival over the target area it was found that 8/8ths cloud, base about 2,500 feet, covered the whole area of Kyomipo. An alternative target in the shape of a large warehouse was found on the western outskirts of the town of Kangso to the north of Kyomipo. While pulling out of the attack, which severely damaged the building, the No. 3 Lieutenants Watkinson and Fursey were hit by the flak and began to lose coolant. The division then set course for the coast escorting the No. 3. In a position some 10 miles north of Chinnampo at a height of 4,500 feet three MiGs, there was possibly a fourth but it was not seen in the ensuing action, made a stern attack on the division coming from above cloud and out of the sun. It was rather a half-hearted attack except on the part of the MiG which had singled out the No. 4 Sub-Lieutenant Arbuthnot and Aircrewman Potter. This MiG fired one long burst closing the range while firing to approximately 100 yards, before breaking away over the No. 4 and in front of Lieutenant Hawkesworth. Another MiG made a firing pass at No. 2 who received a bullet through his starboard wing tip. This aircraft was flown by Mr. Brand with Aircrewman Dunmore as observer. The third MiG just flew through

the division without firing. In spite of the surprise, the No. 3 was able to fire a burst at the MiG which had attacked the No. 2; whilst Lieutenant Hawkesworth was able to fire a burst at the MiG which had attacked the No. 4. Unfortunately it is thought that the enemy was not damaged by these bursts. As a result of the MiG's attack on No. 4, No. 4's starboard tailplane, which received a direct hit by what was thought to be a 23 mm cannon shell, was severely damaged and the starboard after part of the fuselage was holed in seven places. The starboard wing was also hit in three places, one uncovering part of the wheel. After the attack was over, the MiGs did not attempt to renew the engagement although they remained overhead until all our aircraft had crossed the coast. On nearing the coast the Fireflies were joined by a division of Furies led by Lieutenant Hallam who unfortunately had no ammunition left. They had all seen the attack developing but had been unable to warn Lieutenant Hawkesworth in time. It appears that they saw 12 MiGs in all above the Fireflies in three flights of four, of which one flight only took part in the attack.

'On reaching the coast No. 3 of the Fireflies [Watkinson and Fursey] was forced to ditch off the west coast of Chodo and both the crew were rescued safely almost immediately. No. 2, after the ditching, was escorted back to the ship by a Fury as his starboard wing tip showed signs of falling off; whilst Lieutenant Hawkesworth and his observer Lieutenant Clancy, together with Lieutenant Hallam in a Fury, escorted No. 4 to the emergency beach at Paengyong-do where he made a safe landing, wheels and flaps up.

'In the following event, Event Easy, Lieutenant Williams' division attacked a coastal gun position in the Pungchon area. Three lots of RP fell in the target area, one batch fell in the sea. A small fire was started. Rail reconnaissance. No flak.

'Two further strikes each of three Fireflies were launched. Lieutenant Hawkesworth at 1625 attacked a Leopard target with Mr. Brand and Lieutenant Reynolds. The target, a village containing troops in the Ongjim area, was set on fire in one place. The aircraft continued to K.16 [Seoul]. In the last event Lieutenants Williams and Jacob with Sub-Lieutenant Hanson attacked a large municiple building in Haeju, severely damaging the east end of it. Cloud made reconnaissance of the area difficult.

'*Ocean*'s aircraft had one further brush with MiG-15s after Event Dog. In Event Easy four Furies led by Lieutenant Peniston-Bird were attacked by two MiGs at 4,500 feet one mile south of Chinnampo. The MiGs opened fire at extreme range. Two Furies retaliated, also at extreme range, and the MiGs made off.'

Sub-Lieutenant Arbuthnot's log book entry reads: 'Division bounced by 2 plus MiG 15s. My aircraft hit by 37 mm cannon fire in starboard tailplane . . .'

> The accident signal added more detail: 'Tail plane was severely damaged and fuselage and one mainplane received minor damage. . . . Subsequently when flaps were lowered to cruising position aircraft went into dive from which recovery could only be made after flaps had been retracted.'

Also in Event Dog, Lieutenant Hallam's flight had demolished a bridge and had used up all their ammunition; it was on their return from this attack that they were involved with the MiGs—a flight of four attacking his division and damaging his aircraft. Lieutenant Peniston-Bird's division out manoeuvred the MiGs and is credited with damaging the enemy flight leader before the action was broken off when the Furies entered cloud. And on the first event of the day, before all the MiG excitement, Lieutenants McKeown, Crosse and Graham had destroyed three rail bridges on the Chinnampo to Pyongyang line.

A Fairey Firefly F.R. 5 of 812 Squadron taxies up the flight deck of HMS *Glory* before a mission over Korea. The graceful Firefly also saw action in World War II, but never grabbed the headlines. (Courtesy of the FAA Museum)

Next day adverse weather again affected flying. On practically every mission there was 6/8ths to 8/8ths layered cloud from 500 feet to 25,000 feet. On Event Able at 0525, three Fireflies dumped their rockets on an old Leopard target near the coast south of Haeju because the weather prevented them from approaching either their primary or secondary targets. On the next event Lieutenant Williams' division could not reach their briefed target, but they found what appeared to be a stores dump and obtained direct hits on three houses. In two sorties, Lieutenant Commander Roberts first found and photographed a gun position near Haeju that had been firing at *Nootka*, and then on a later event attacked it, all rockets from the division landing in the target area. The Furies had an equally frustrating day. Their only good result was the complete destruction of a transformer station 15 miles north of Chaeryong by the division led by Lieutenant Carmichael—acting CO of 802 Squadron. Lieutenant Peniston-Bird, in a 'creeping beneath the weather' sweep, surprised a 5-ton truck driver and destroyed the truck. It was not a good day, but then Korea is notorious for the murkiness of its weather when influenced by the south-west monsoon.

On the 29th the weather was so bad that the Fireflies only flew one detail, led by Lieutenant Williams, to a village in the Ongjin/Haeju area. After 0800 the Furies flew 'weather test' CAP details only, on return from one of which Lieutenant Jenne entered the barrier. During the day an unfortunate accident occurred when PO/REM Jordan of 802 Squadron walked into the rotating propeller of a Fury whilst changing its radio set and received injuries from which he later died.

On the last day there was no break in the weather until 1630 when two strikes got off. The first one, a combined strike by Furies and Fireflies on a transformer station near Chaeryong led by Lieutenant Commander Roberts, caused explosions and sheets of orange flame. The second strike, and last event of the day at 1830, of four Fireflies was led by Lieutenant Hawkesworth on their secondary target of a village south of Chinnampo, causing primary and secondary explosions. *Ocean* left the area that evening, her place being taken by USS *Bataan.*

The 31st was spent on passage to Kure where *Ocean* arrived at 1000 on 1st August. That evening a concert was given in the after lift well by Bill Johnson from 'Annie Get Your Gun', and by a comedian and a pianist. On the Wednesday make and mend a ship's sports meeting was held at the main recreation field. The event was well supported and was won by the Royal Marines with the combined Chief Petty Officer's messes coming second.

RPCs were held for the military establishments in and around Kure and for *Crusader*, *Iroquois* and HMAS *Bataan*.

Three RNVR pilots, Lieutenants Adkin, Buxton and Clark from 1832 Squadron at RNAS Culham, joined 802 Squadron and spent the period at Iwakuni on familiarisation and ADDLs. They arrived in a blaze of publicity as the first RNVR pilots in Korea, but they had been preceded by Sub-Lieutenants Randall and Cook early in June who had already established the image of the RNVRs as perfectly normal squadron pilots.

Two replacement Fireflies were received from Iwakuni. Due to a shortage of rockets, 825 Squadron had to be rationed to 1,000 RPs for the next patrol. The inboard rocket launchers were removed from the Fireflies and their place taken by 1,000 lb. bomb carriers. The armament load was then either 2 × 1,000 lb. bombs or eight RP and 2 × 500 lb. bombs, but both bombs and rockets were never carried on the same sortie.

On the night of 12/13th August, the enemy massed in strength on the mainland opposite the island of Cho-do in the Haeju estuary, about eight miles north of Yonpyon-do. *Concord* broke up the attacks before they could be launched at the island, directed by a shore fire-control party on the island. *Concord* and USS *Strong* neutralised two guns, one each side of Haeju Gulf, on 21st August. Heavy air attacks in the area caused casualties estimated at 400–500.

On 14th/15th August, in the Paengyong-do area, a 'model' guerrilla raid was launched on Ongjin, near Kirin-do. Supported by *Rotoiti* and *Crusader*, 120 men under Lieutenant MacBride USN landed in junks and penetrated 4 miles inland, inflicted 80 casualties on the enemy, destroyed one gun and returned unscathed with four PoWs after spending five hours ashore.

Sixth Patrol

Ocean, with *Charity*, *Kimberley*, *Yarnall* and *Strong* sailed from Kure at 1700 on 8th August. The weather was uniformly good, 600 sorties were flown—the highest number then reached for a single patrol—at a daily rate of 75. The opening days were enlivened by the attention of Mig-15s operating mainly over the Hanchon/Chinnampo/Pyongyang triangle, strategically the most important area to the enemy in the whole of North Korea. As targets were attacked by *Ocean*'s aircraft at least twice per day, it was not long before air encounters occurred.

Flying operations started on the 9th, the early events well described by 802 Squadron's diary:

> 'Lieutenant Carmichael, Lieutenant Davis and Sub-Lieutenants Haines and Ellis started the ball rolling this morning by flying the first AR of the patrol.

Great Britain retained a viable traditional carrier force until the 1970s. British flattops saw action in Korea and several brushfire wars. Here, Korean War veteran, HMS *Ocean* plows the seas in 1952. Sea Furies and Fireflies are spotted on her flight deck. Note the large "O" on the forward flight deck. (Courtesy of the FAA Museum)

By 0600 they had entered the area and had commenced their Hanchon and Pyongyang to Chinnampo rail search. By 0630 they had reconnoitred as far south as Chinji-ri, a small village about 15 miles north of Chinnampo. As they meandered down the line, checking the bridge state as they went, they suddenly saw eight jet bogies to the north. Almost immediately the bogies were identified as MiGs—and they were closing. By this time drop tanks were fluttering earthwards and the flight had assumed proper battle formation and No. 4—Sub-Lieutenant Ellis—had noticed a shower of red tracer streaming past both sides of his fuselage. He cried "Break" over the R/T and the flight commenced a "Scissors". It was soon apparent that four MiGs were after each section of two Furies but by continuing their break turns our aircraft presented practically impossible targets to the enemy who made no attempt to bracket.

'On one occasion a MiG came head-on to Lieutenant Carmichael and Sub-Lieutenant Haines—they both fired—it broke away and proceeded to go head-on to Lieutenant Davies and Sub-Lieutenant Ellis—they both fired and registered hits. On another occasion a MiG pulled up in front of Ellis with its air brakes out and he was amused to find the range closing. He gave a long burst and noticed hits on the enemy's wings. The aircraft then proceeded northwards at a reduced speed with two other MiGs in company. Meanwhile the flight, still in its battle formation, managed a dozen or so more firing passes at MiGs head-on. The dog fight lasted 4–5 minutes and then the MiGs disappeared as quickly as they had arrived—as they departed an aircraft was seen to crash into a hillside and blow up. At first Lieutenant Carmichael thought it was one of his flight and ordered a tell-off. However when No. 4 came up "loud and clear" it was realised that the Royal Navy had shot down its first Communist aircraft. Lieutenant Carmichael as flight leader is being credited with its destruction officially but the rest of the flight are claiming their quarter as well.

'As a result of that five minutes fight one MiG-15 was destroyed and two others badly damaged—a remarkable feat achieved without a scratch to any of our machines.

'The ship was still humming with excitement about this when at 0800 a report came through that Lieutenant Clark, who was on a "Cook's tour" of the coast had been hit by MiGs in his starboard wing. His wing caught fire—but by side-slipping and releasing his drop tanks he put it out and brought his aircraft back to the ship. With him was Lieutenant McEnery who scored hits on the tail of one of the three attacking MiGs.

'By the time Lieutenant Clark had landed-on the ether was buzzing again with MiG reports. This time Lieutenant Jones reported that his leader, Bob Hallam, was bounced after attacking a rail bridge north of Chinnampo. By some trick Lieutenant Hallam was leading a weaving procession of one Fury, two MiGs and one Fury down the Taedong Gang toward Chodo. Eventually his aircraft was hit when he broke towards the enemy and the MiGs veered off. Bob Hallam was obliged to make a wheels-up landing on Chodo. On inspecting his aircraft afterwards, he found a large hole just behind the cockpit where a 37 mm had found its mark.

'This indeed was a fabulous start to our sixth patrol. On restricted sorties (nobody went north of Chinnampo) throughout the rest of the day, Lieutenant Carmichael and Sub-Lieutenant Haines destroyed a road bridge east of Haeju. Lieutenant Peniston-Bird obtained a 50% coverage on CAS with his flight, Lieutenant Jenne got a road bridge near Changyon and McEnery and Treloar got two more near Haeju. Lieutenant McKeown destroyed a transformer station north-east of Haeju with his flight and then got hit himself when flying over Chinnampo Waterways.

'In the forenoon Lieutenants Adkin and Buxton carried out DLP.

'What a day—Whew!"

Although the diary suggests that all four members of the flight claimed a share of the MiG, from a confused situation Carmichael as flight leader got the credit for its destruction; and Captain Evans had no hesitation in accepting Lieutenant Commander London's, the CO's, recommendation for the award of his DSC.

Indeed, what a day; a historic day for the Fleet Air Arm. To Lieutenant (now Commander) Peter 'Hoagy' Carmichael has fallen the unique distinction of being the pilot of the only piston-engined aircraft to shoot down a jet-engined aircraft, a formidable testimony to the Sea Fury's ruggedness and its excellent dog-fighting characteristics. In his own words:

> 'We, as usual, were flying at about 4,000 feet, and we always flew with gyro and fixed ring on the gunsight. Suddenly a MiG came down behind me: I turned towards him and as he flew past me I noticed he had his air brakes out. He made the fatal mistake of trying to dog-fight with us. I put my gyro sight on him and started to fire. At this point he realised he was in trouble and put his dive brakes in and started to accelerate like mad. I then switched to fixed ring and held him quite easily and my bullets started to hammer him. He started to roll over on his back and crashed into the ground with no attempt at baling out.'

CHAPTER 4

Times of Development and Transition, 1954–1963

The Rubber Deck

The decade following the Korean War was a period of great steps forward for carrier aviation. New landing systems, even the shape of the flight deck, changed the everyday—if there is anything "everyday" about flying from a 1,000-foot deck 80 feet above the sea in all types of weather and in the darkest night—operations. These new developments also greatly increased a carrier's mission capability and truly established it as *the* capital ship of any navy rich enough to support it.

There were a few experiments, however, that didn't go anywhere. The "flex deck" was one of these dead-end sidebars. Essentially a soft deck that ostensibly would permit aircraft to recover without the fuel-draining weight of heavy landing gear and related systems, it enjoyed a few years of testing until finally dropped.

Writer Martin Caidin quipped, "John Moore has more wheels-up landings than any other American pilot, landing the . . . Cougar, wheels-up into a rubber deck."

Retired Commander John Moore's memoir describes his full career in aviation in peace and war. During a combat tour in Korea, flying Panthers with VF-51 from USS Essex, *he was involved in a horrific fire generated by the crash of a Banshee on the carrier. Moore had been in another plane waiting to launch when the F2H vaulted over the barrier netting and struck Moore's aircraft, which quickly burst into flame. Somehow, he jumped off the deck into the water. Rescued by the ship's helicopter, he was horribly burned and spent the next several months in the hospital.*

After recuperating, Moore enjoyed a career as a Navy test pilot, flying many of the third-generation jet fighters, such as the North American FJ Fury and F7U Cutlass before they entered fleet service. Subsequent service included work in the U.S. space program and even a stint as mayor of America's space town, Cocoa Beach, Florida. Many of his friends and flying associates are on record as wondering how he survived all he has done.

From *The Wrong Stuff: Flying on the Edge of Disaster,* by Commander John Moore, USN (Ret)

The Air Force was making final preparations for arrested landings into their inflated Flexdeck reluctantly provided by the Goodyear Tire and Rubber

Company. They had selected the straightwing F-84G as their test vehicle, most likely because in the event of any loss, they had little to lose.

There was concern that damage would occur to the test vehicle's flaps following hook engagement with the arresting cable and subsequent deck/vehicle contact. Consequently, a microswitch was installed in the arresting hook face that would initiate flap retraction upon hook contact with the arresting cable, thus avoiding flap damage during landing. The result was that in the first attempted arrested wheels-up landing into the Goodyear deck, the hook struck the approach ramp, bounded over the arresting cable and returned to deck contact as the airplane flew by. The hook then encountered a puddle of water that had been sprayed on the deck for lubrication, the impact of which closed the microswitch in the hook and the flaps were retracted. The F-84 settled on the deck, dropped into the desert and ground to a halt in a large cloud of dust. The result was excessive damage to both pilot and test vehicle, putting both out of the program.

A second test vehicle was prepared and a second pilot willing to fly this mission was located, heroically armed with little knowledge of what had transpired to this point. Following a brief training program for the new pilot, the Air Force scheduled its second landing attempt.

In this landing, the hook successfully engaged the arresting cable, causing the F-84 to pitch violently into the deck on its nose for a series of three or four bounces become coming to a halt. Motion pictures showed that at first impact, the pilot, a tall lad who was strapped in snugly, disappeared completely in the cockpit. This left impressions of the pilot's lateral incisors in the top of the control stick and caused structural damage to his vertebrae. The result caused the hierarchy of a tailhookless Air Force to reconsider what the hell they were doing in this program anyway; wherewith they abandoned the whole thing, leaving further evaluation of this extraordinary concept in the capable hands of the United States Navy.

It should be stated that after observing these historic events of zero launch and wheels-up landings, this pilot did some soul searching for a reason as to why he should risk his cute little ass furthering this wacky program—and the answer was clear—stupidity.

An excellent introduction to the Flexdeck landing concept was offered by the British to the pilots who would fly the Navy program at Pax River. By the time our program had reached the hardware state, the British had made several hundred wheels-up landings into their deck at Farnborough, many of which were successful. Their test vehicle was the jet powered Vampire, a delightful little aeronautical innovation the English had hurriedly con-

John Moore sits in his F9F-7 during flex-deck tests in 1955. His VF-51 shoulder patch recalls his wartime service. (Courtesy of John Moore)

structed in World War II to chase down buzz bombs being launched by the Germans toward the United Kingdom. It was like a toy. You felt like you were flying a jet made by Radio Shack.

I was assigned to a three-week stint in England to fly the British deck and was chaperoned by the late Marsh Beebe who was Director of Flight Test at the Naval Air Test Center at that time. It was a bit like being chaperoned by Errol Flynn. Marsh had been Commander of Air Group Five on the aircraft carrier *Essex* during one of my VF-51 tours in the Korean War, and I had flown wing on him several times as we strafed high-priority haystacks and burned-out railroad cars south of Wonsan. Marsh was among the best of the best.

Our initial flights in the Vampire were made at Boscomb Down under the tutelage of a "batsman," the equivalent of our Landing Signal Officer (LSO), who stood at the edge of the runway with two "bats" (and a fire extinguisher) as we made practice runs, wheels-up, in the little toy airplanes. The best indication of being too low in a pass over the runway was the batsman's motion to arm the fire extinguisher, easily discernible with one's peripheral vision in the fly-by.

Following several days of practice at Boscomb Down, I ferried the Vampire to Farnborough as Marsh climbed into a sizable black automobile with darkened windows, piloted by a lady military type, and off they went in the direction of Farnborough. As it turned out, Farnborough was only about twenty minutes by Vampire and forty-eight hours by car.

Beebe and I were impressed by the courtesies extended to us by the British. At night, for example, before we retired in their Boscomb Down Officers' Quarters, we put our shoes outside the door and a gentleman called a batman came quietly by in the dark of night, picked them up, shined them gloriously, and had them ready for us in the morning as he came in to rouse us with a pot of tea. The similarity of names batsman and batman led Beebe to conclude we might broaden the responsibilities of our own batsmen (LSOs) at Pax River to include shoe shining, but the plan ran into some resistance from the likes of Bill Tobin, Sam Thompson and "Weakeyes" Boutwell, our own single-purpose Landing Signal Officers.

Related to the above, it is worth noting that shortly after returning from England, I was housed in the BOQ at Pax River for a few nights and I wondered what would happen there if I put my shoes outside the door as at Boscomb Down. And so I did. The next morning they were gone.

The English deck was configured quite differently from our own. It was higher in the middle than at each end, giving the pilot the impression he was landing on the hump of a bridge. The reason, of course, was that following arrestment, the deck in effect came up to meet the airplane, reducing the landing loads considerably.

The flight procedures at Farnborough were straightforward. One day run was made over the Flexdeck with the wheels and arresting hook up at an approach speed of about 90 knots. On the next pass, the hook was lowered and in a straightaway of about 3,000 feet a ten-degree glide slope was established with the airplane aimed at the center of the deck. Engagement was evident by the pitching down and rapid deceleration of the airplane followed by mild impacts and about a dozen bounces until the airplane halted, all in some 100 feet. It was a little like being dribbled by Magic Johnson.

The engine was then shut down and a cable was hooked to the nose of the Vampire allowing it to be winched off the deck onto a flatbed truck, the pilot remaining in the cockpit. The Vampire was then transported to some facility on the outskirts of the airport next to The Doxy Pub where a crane was off-loading sewer pipe from railway cars. The crane operator was gracious enough to pause the sewer operation and lift the Vampire from the truck. While suspended from the crane, I lowered the wheels and the airplane was lowered to the ground. The toy's engine was started, and the airplane taxied along the left side of the road (of course) back to the airport for another take-off and landing. Oncoming autos paid little heed to this ritual, simply pulling off onto the grass to clear the Vampire's wing. Marsh made six landings in this manner and I made eight, followed by an approved touch-and-go (bounce-and-go) landing, which I made to become familiar with that phenomenon, hopefully, never having to experience it on our own deck at Pax River.

The exercise proved quite beneficial as a training method in preparation for the American Navy program, although as it was to be subsequently determined, the speeds, energies, and kinematics involved with the F9F-7 and our deck were markedly different from those encountered at Farnborough.

Marsh Beebe did not fly the Navy deck at Pax River for reasons privy only to commanding officers but in all likelihood, because he found more important things to do. Smart man. Right stuff!

The first ten wheels-up landings into the Navy deck at Pax River were made by John Norris, Grumman test pilot and boy wonder, whose background was Navy aircraft carrier type with sound experience in arresting wires, barrier cables, and bent props on several aircraft carriers. John also availed himself of training on the British deck with the Vampire before beginning the tests in the F9F-7 Cougar at Pax River.

Prior to any landings in the Cougar, however, considerable thought was given to the results of the Air Force landings and the severe neck injuries their pilot had sustained as the airplane impacted the deck following arrestment. Analysis of this circumstance led us to believe our landings might well produce the same unsavory results. As a consequence, a sturdy aluminum device was fabricated to be worn by the pilot, designed as a back and head brace to protect his spine and neck from the kinds of injuries sustained in the Air Force landings. The protective metal harness extended from the cleavage in the pilot's buttocks to the top of his head.

A special helmet was also fabricated with a male probe in the back of the helmet, which was inserted by feel into a female receptacle in the protective brace, which allowed the pilot to lock his head to the back brace prior to an arrested landing. Both the back brace and helmet were molded to fit John Norris, but since he and I were about the same size, the special equipment also fit me fairly well except for the helmet, which seemed to be slightly square.

The operational procedure for using this equipment differed somewhat between Norris and myself. Norris would lock his head up on the downwind leg, then call in "turning base, wheels-up, head up and locked." I found that flying the base leg with my head locked in a fixed position was a little disorienting. It was as if the airplane were fixed in space and the horizon rotating around it. So I just flew a longer straightaway and locked my head up in the final approach. John Norris did not seem to find the approach as he flew it to be particularly disorienting, though in those memorable instances when I flew with John in other flying machines, he seemed to be disoriented a goodly part of the time anyway, so this seemed rather normal to him.

There was an anomaly associated with the protective harness related to its weight. It was calculated that with the harness on and the life jacket inflated (in the event of a water landing), the buoyancy was slightly negative. A simple procedure was developed to resolve this: Namely, in the event of a ditching and following the pilot's safe egress from the airplane, he had but to remove the life jacket, remove the parachute, remove the protective harness, reinstall the life jacket and inflate it. It was expected that this could be accomplished while the pilot was standing on the bottom of Chesapeake Bay.

Familiarization flights prior to arrested landings onto the rubber deck included making low passes over the regular runways at the Test Center with the wheels down at first, followed by wheels-up passes under the watchful eyes of our own "batsman," our Landing Signal Officer. These creatures could not contribute much to my wheels-up fly-bys which were made but three feet above the runway, because a high or low signal that related to a six-inch change in height above the deck was essentially meaningless to me. Considering the Cougar's close proximity to the ground in these passes, I was particularly disinterested in any flag waving by one of our LSOs, "Weakeyes" Boutwell. Bill Tobin was bad enough.

When the confidence level was as high as it was going to get, I commenced making wheels-up hook-up passes over the Flexdeck, concentrating on being at the correct height, on centerline, at 135 knots, wings level, again under the watchful eyes of our LSOs. They had resorted to shouting instructions on the radio because of my inability to respond to their frantic flag waving. The concentration required on line up, height control, et al., was too demanding to permit me to shift my vision to the LSO for signals, and regardless of all the semaphore he generated, it could not be recorded by my peripheral vision or saturated mind. The radio communications were helpful to tell me what I had done but of little help in telling me what to do.

Airplane height control across the deck was not too difficult in smooth air but, as expected, became increasingly difficult as turbulence increased. Crosswinds helped very little as well. In moderate turbulence, I was unable to control the height over the deck any closer than about plus or minus a foot, and as a consequence, we made no arrested landings onto the deck except under very stringent atmospheric conditions related to wind direction and turbulence. Remembering normal flying conditions in the Sea of Japan or in the Atlantic off Bermuda where the carrier deck is never still, this program seemed more ludicrous with each passing day.

It was interesting that in the final approach during which exacting speed control of 135 knots was required along with other constants previously mentioned, the airspeed indicator seemed to be inaccessible. As best I could tell, it took about one and a half seconds to shift my focus of vision from outside flight parameters into the cockpit to locate, read and mentally record my airspeed, then shift my vision back to deck line-up and other constantly changing variables. At 135 knots that amounted to more than 300 feet of travel, and as every "hooker" (tailhook pilot) knows, a lot can happen in 300 feet when you are close to the carrier deck.

In an attempt to improve this condition, we mounted a very large airspeed indicator on the instrument panel hood, almost directly in the pilot's line of sight during final approach. It was hoped I could read this monster with peripheral vision but I could not. It did, however, reduce the time spent with eyes in the cockpit by a fraction of a second and that was helpful.

Although a level final approach was required to minimize the possibility of an inadvertent touch down, it was determined that a flat approach—two degrees or less—resulted in an inadequate view of the deck. Consequently, my final approach was extended to about two miles from the deck at an altitude of 1,000 feet, which gave me time to get my head up and locked and provided an adequate view of the deck for alignment and speed control. The Cougar leveled off about 2,000 feet from the approach ramp at 135 knots in a flight path parallel to the deck to prevent inadvertent contact with the deck before reaching the arresting cable. Analysis had indicated that a touchdown into the deck without an arrestment would result in the Cougar being unable to become airborne again because of the pilot's inability to attain sufficient angle of attack for flight. It was projected that this would put the Cougar into the sand trap adjoining the fifteenth green at the Pax River golf course where I had been many times before but never in a Cougar.

During my initial wheels-up arrested landings, I was not aware of wire engagement and not certain of arrestment until the Cougar had traversed approximately 120 feet of the landing mat. Then as the airplane was decelerated by the constant runout arresting gear, a pronounced nose down pitch into the deck occurred. Following deck contact, I was extraordinarily aware of the longitudinal pitching into and out of the deck. In the first few landings, statistical analysis indicated that the severity of impact could be directly correlated with the size of the wet spot which appeared on the front of the pilot's flight suit. The lack of pilot sensitivity to the decelerating forces was attributable to the protective harness and head lock.

A unique aspect from this pilot's standpoint was the impression, as the Cougar pitched into and out of the deck during arrestment, that the airplane was motionless and that it was the horizon which was in motion moving up and down past the airplane's nose.

The most severe landing in the Navy program occurred on the twentieth landing. For this test, I was to engage the wire at the maximum height possible, about five feet above the deck with the Cougar at its maximum design gross weight for these tests as adjusted by fuel loading. On this landing, I let the right wing drop slightly after hook engagement so that deck contact was made with a bank angle of about five degrees right wing down. Vehicle penetration on contact was twenty-five inches into the deck, five inches from bottoming out.

Pitching out of the deck the first time was accompanied by a pronounced left roll followed by a roll to the right on the second pitch out as a result of the left wing/deck contact. The rolling and pitching continued until the Cougar came to a halt, which seemed like about four days. It was a wild ride. Black rubber deck marks were found on the upper surface of the right wing. If the Air Force Flexdeck landings did not disprove the concept, this landing certainly did.

Finally, I made the twenty-third and last wheels-up landing onto the deck at Pax River in this noble enterprise, thus releasing all the firetrucks and meatwagons to their normal duties at the Air Station.

From the flying standpoint, there was impracticality to the concept almost from the beginning. Pilot skills demanded for successful landings on a stationary deck were stretched close to practical limits and, although the British had landed the Vampire on a Flexdeck at sea under calm conditions, the possibilities of successfully deploying a squadron of higher performance sweptwing jets on a Flexdeck aircraft carrier by our Navy had to be considered unfeasible.

Since I was the only Navy pilot to fly the Navy Flexdeck (John Norris—civilian), I made that statement without fear of contradiction. Yet as any carrier pilot would surmise, it was fun to fly.

Young Pups and Old ADs

Earning Navy wings of gold is a tremendous achievement, a proud accomplishment to carry with you for the rest of your life. However, like any milestone, pinning the wings on merely gives you the chance to start at the bottom, in this case your first fleet squadron. Experience is only gained through time, trial, and endeavor. And in the case of carrier flying, one can never have too much experience.

By the early 1960s, the Douglas Skyraider was a veteran, powerful, strong, and fairly forgiving of occasional heavy-handed operation by young, newly minted aviators. But there was always something to learn, particularly for low-time ensigns. In his highly personal memoir, Captain Rosario Rausa delights in describing his first AD squadron, VA-85. Known far and wide, and with good reason, as "Zip," he struggled to get those wings of gold. With characteristic bulldoggedness and determination, he succeeded. He subsequently flew with VA-25 in the single-seat attack Skyraider's last combat cruise.

In this excerpt, Zip shows us how things were done in the old, pre-Vietnam AD Navy.

A native of upstate New York, Captain Rausa flew a combat cruise as a TAR, a type of active-duty reservist. The letters stand for "training, administration, reserve." It was unusual to find TARs in combat, but they were there. He later served two tours as the editor of Naval Aviation News, *the Navy's oldest publication, and then as Head, Aviation Periodicals before retiring. In between, he accumulated time in A-4s and A-7s, as well as his beloved AD Skyraider. He is a well-published author, with several books to his credit, including a full-length biography of the Skyraider. In 2000, he received the Admiral Arthur W. Radford Award, sponsored by the Museum of Naval Aviation Foundation. He is currently the editor of* Wings of

Gold, *the quarterly magazine published by the Association of Naval Aviation. His son, Zeno, continues the tradition and flew FA-18Cs with VFA-94, and the FA-18E and FA-18F with VFA-122.*

From *Gold Wings, Blue Sea: A Naval Aviator's Story,* by Captain Rosario Rausa, USNR (Ret)

Interval in the landing pattern around the carrier was important. It was a sign of professionalism to be able to "recover" large numbers of aircraft aboard ship efficiently and with a minimum number of wave-offs. Wave-offs usually meant delays and forced the captain to keep the ship into the wind, which was undesirable, because even in the spacious Mediterranean the fleet had tactical boundaries inside of which it was supposed to remain.

So, each of us tried to keep a proper distance from the plane ahead and the one behind while in the approach and landing sequence. For purposes of uniformity it was essential that we fly precisely on the proscribed altitude and airspeed. The jets were intermingled with us in the pattern and flew at different speeds, but we compensated for them. Squadrons competed against each other to demonstrate their proficiency, and the air wing commander railed when timing was off and for one reason or another the pattern was botched up. The CAG (as he was called in reference to Carrier Air Group, former title of Carrier Air Wing) would get on the backs of the squadron COs, and the COs would get on the backs of their pilots.

Skipper Lee never seemed to get a wave-off, and one evening in the ready room—surrounded by a group of us junior pilots—he was asked why that was. "Even when you look as if you're too tight on the next guy and a wave-off for close interval seems inevitable," one of the officers said, "you get aboard."

"Well, boys," he allowed, sipping from his mug of coffee, "you can do it, too." *Like pups at the master's feet, we listened.*

"Airspeed is the answer," he began. "Let's say you're coming around the corner at eight-seven knots and Dilbert is ahead of you." (Dilbert is a cartoon character created by eminent illustrator and satirist Robert Osborn during World War II when Osborn was an officer assigned to training duties. A chronically error-prone pilot, Dilbert's goof-ups were depicted on posters as safety aids. The main message was: "Don't be a Dilbert." Osborn also helped create Grampaw Pettibone, a character who has been promoting safety in aviation since 1943 when he first appeared in *Naval Aviation News* magazine. He still appears in that publication monthly, and Mr. Osborn still draws him.)

"He's long in the groove as usual, messin' everybody else up. But he's not so long that Paddles will wave him off and send him around. What you do is ease the nose up just a tad, tickle the throttle a bit, and slow to eighty-five, maybe eighty-four knots. That's below what the book says so be on your toes."

"Hold that speed for a second or so," he went on, "and keep an eyeball on your pal Dilbert up ahead. You'll gain precious feet of interval. Get your nose back down, massage the throttle again, get back on speed and drive her on in. Fly the bird with everything you've got. Concentrate. Stay with it all the way. You can hack it."

I had a couple dozen traps under my belt by then and tried his system one day. It worked! It was neither a recommended nor an authorized procedure, but when the master talked we listened—and learned.

The early sixties were played out under the somber shadow of the Cold War. Therefore, tactical planners inserted frequent long-range, simulated nuclear strike exercises into the training scenarios. For us in the ADs it meant flying lengthy, low-level navigation flights called "Sandblowers." Masked by the terrain from the probing sensors of radar, or skimming low over the ocean to avoid detection, we sped to simulated targets like airfields and oil storage facilities, made practice bomb delivery maneuvers, and returned to the carrier. These journeys were normally flown at less than 200 knots and often required six to ten hours to complete. A few warriors exceeded that extreme, logging marathon distances involving a dozen hours and more in the saddle.

Although these excursions were ruinous to the backside, they set us apart from all other air wing types. If our flight suits were more greasy and unkempt than our counterparts who flew the air-conditioned jets, and if our flying boots were scuffed and lusterless because we spent our spare time in the air rather than with a buffing cloth, (or elected not to promote our image through glossy shoe tops, which was more likely the case), then it was more power to us. This behavior gave us and the fraternity of Skyraider pilots throughout the Navy a special identity. Other aircraft drivers flew low-levels, to be sure, but none remained aloft and alone and in such delightful agony as we did.

We were off the Sardinian coast steaming in the Mediterranean night well along on my first cruise when a typical strike exercise was under way. The sequence of events went like this:

0200: The stateroom telephone rang and I fumbled for it in the darkness, knowing already who would be at the other end.

"It's me," a tired voice said. "Your friendly air intelligence officer." Lieutenant (junior grade) Chuck Watson sounded as if he hadn't slept in days. He was a Virginian with impeccable manners, a keen intellect, and a tremendous capacity for hard work. He knew his stuff and was highly respected in the wing. On such exercises he played a key role in that he gave us overall briefings and coordinated much of our activities with the strike planners a level above in the chain of command.

"When's the launch?" I asked bluntly.

"Oh four-thirty. You're off to Italy," he replied.

I plodded my way to the shower, which partially awakened me, shaved quickly, and pulled on a flight suit. The passageways were bathed in the dull red light of "darken ship" conditions, a precautionary environment in warships. I stumbled my way to the air intelligence spaces. My body was still in first gear but my mind was shifting to second. I had an hour and thirty minutes to plan the route, have breakfast, and perform the one-hundred-and-one functions necessary for a carrier launch.

Watson was nearly punch-drunk with fatigue, having been on the go since the exercise began nearly a day and a half ago. But he dutifully had charts and kneeboard cards waiting, and I set to work building a map on a table cluttered with pencils, scissors, plotters, trimmings, and other remnants of target planning. A few pilots and aircrewmen assigned similar missions crossed the area, shuffling between tables, asking questions, and examining the overlay chart of NATO's southern flank and the Mediterranean, which covered one bulkhead.

0245: After a final glance at the map I was satisfied I knew where to go and how to get there. The route would take me over miles of ocean and a good portion of Italy. I would make a landfall at Sorrento and then swing southeast to the heel of the boot before turning north and slicing up through central Italy. My target was a small, abandoned airstrip east of Florence and after hitting it, I would go "feet wet"—that is, depart land—at Livorno, cut to the island of Elba and then proceed south to rendezvous with the ship. The route promised mountains, valleys, and flatlands. It would be a respectable challenge to navigational ability. Presumably there would be some easy-on-the-eyes scenery as well.

There was a hitch, however, The flight would be a butt buster; time en route would be nine hours and fifteen minutes. I thought of the jet jockeys, especially the fighter pilots who seldom flew more than single-cycle, one and

Then-Lt. Zip Rausa in his VA-85 Skyraider. (Courtesy of Rosario Rausa)

one-half hour hops, and for a moment despised them. Most of those fellows will have launched and recovered twice in the time it would take me to complete a single mission. I reconciled myself with the thought that there is a distinction in being able to solo an airplane for a period as long as an average businessman's eight-to-five day at the office.

0310: In the wardroom I loaded up with scrambled eggs, bacon, home fries, and fresh coffee. But there was no time to savor an extra cup. Mentally, I was in third gear and cruising. Check-lists were cycling through my mind.

0315: The ready room teletype pecked out the ship's position and area weather as I flipped through flight logs once again. Three other VA-85 pilots plus the duty officer were milling about, preoccupied. Chatter was limited. A messenger appeared with a stack of box lunches and at oh four hundred on the button, the squawk box came alive. A voice from air operations ordered, "Pilots for the oh-four-thirty launch, man your aircraft." Then, in deference to our friends who flew the E-1 Tracers, also known as the Willie Fudds, "and fill the Fudds!"

I zipped up the G suit, checked my pencil flares and other emergency equipment, and pulled on the yellow Mae West. Systematically, I scooped up the hard hat, the nav bag, and the checkerboard-sized plotting board, tucking it under my arm. I was through the doorway when the duty officer tugged at my shoulder and shoved a box lunch under my other arm. On the escalator that led to the flight deck, the same old jibes flowed forth from the jet pilots riding with me.

"Where's the picnic?" boomed one.

"AD pilots lead the world in hemorrhoid disorders!" cried another.

The brisk night air whiffed away some of the perspiration as I emerged onto the catwalk. There was no moon, and against the grey-black sky the Skyraiders on the fantail were awkward silhouettes, their folded wings like inverted Vs over the canopies. A twinge of envy went through me as I looked at the slim-lined Skyhawks, Skyrays, Skywarriors and Crusaders. But it was quickly gone. The Able Dog would win no beauty contests, but it was still the most versatile airplane on the ship.

0415: The preflight took longer than usual because of the darkness and having to poke into joints and crevices with a red-lensed flashlight. By the time Airman Prothro (pro' throw), my plane captain, strapped me in, the air boss spoke through the bull horn and cracked the hush of early morning.

"Start the ADs," he ordered, "stand by to start the jets!" We fired up the engines and the crescendo of sound became a powerful roar. The flight deck came alive, like a waking animal.

0430: The jets were poised on the catapults, and a green beacon outside the island pri-fly (the primary flight control) tower flicked on. The launch officers at their stations on the bow and waist (angled deck) catapults swirled their slim wands and the launch began. The entire carrier felt the rocking, thrusting force of the steam-powered sling shots as jet after jet was kicked off into the darkness. The heat from their exhausts, as engines labored at full power, cascaded down the deck and flooded over my machine. The temperature soared momentarily in the cockpit.

0440: I was connected to the cat and the launch officer waved his wand. I slowly moved the throttle full forward and reviewed the red-lit instruments. Everything was vibrating, the aircraft, the gauges, me. All the needles were where they should have been. I snuggled back into the seat and pressed my helmet firmly against the head rest. I felt like a bronco straining at the gate. I flipped on the master light switch, located outboard of the throttle quadrant, illuminating the Skyraider's exterior lights, which signaled that I was ready to go. Then I waited the one or two seconds that preceded my "shot."

Then it came, like a good solid sock to the behind, and I was hurtled into the night. I raised the gear handle and the wheels folded obediently into the plane. At 500 feet I leveled off and throttled back to a cruise power setting, which brought the tumult of sound down to a reasonable level. I transmitted to the ship using the squadron call sign, "Buckeye five oh four, departing on course."

0530: The horizon was a blue-grey swath between the dark of both the sky above and the sea below. I could make out wave patterns, however, and changed heading five degrees, correcting for the wind. Landfall, the "feet dry" point, was an hour away, and except for an occasional freighter there was little to see. Fuel pressure from the external tanks was steady, the autopilot was holding nicely on a northeast heading and I settled back to enjoy the solitude, my hand loosely on the control column. The engine hummed deeply.

0620: The sun was up a good angle so I descended to the daytime altitude limit over water, 100 feet. Discomfort was already setting in, especially in the buttocks area, but once land was sighted the distraction of having to navigate carefully over the hills and plains would lessen the pain, if only in mind. I twisted in my seat, the Skyraider pilot's massage.

0634: Sorrento passed under the left wing in all its crystal blue and white beauty. I had lost two minutes along the way, but that wasn't bad considering that the ship's position may have been off a few miles at launch time. The real estate was post-card pretty, especially along the coastline, but it was soon enough behind me and I was over land pointed southeast toward higher ground. Up ahead mountains seemed to have sprouted unexpectedly, so I added power. The engine rose anxiously, drawing the aircraft with it. I shook away complacency, replacing it with aggressive concentration. I remembered the stories from low-level training about flyers who found themselves boxed in by canyons with no way out. I did not want that to happen to me.

0700: I flew into a wide valley between jagged mountain ranges that paralleled the course. It was as if the earth had opened up. On the right appeared a small city at the edge of a finger-shaped lake. I checked map. So far, right on course.

0750: I hit the southern coastline of Italy a minute late. No sweat, though, I could make it up later. There was no chase pilot behind me to check my expertise, and this was only an exercise, but pride is a fine motivator and I wanted this strike to be as real as possible. Timing accuracy was vital to that realism, so I maintained my navigation log with close attention to the elapsed time of the mission in minutes and seconds.

0930: Before me was a plateau, flat, sienna-colored, desolate. Not even a farm house was in sight. The sun grew hotter and beat down through the plexiglas of the canopy. The cockpit seemed smaller. My anatomy ached. Same old symptoms. A can of pineapple juice from the box lunch was warm but refreshed me a little bit anyway. I tried not to think of the endless hours ahead but it was unavoidable. My hind quarters would remind me anyway. If this were a road trip, I'd pull over and walk around for a few minutes. One of the pilots told me he unfolded a map and draped it over his head on really hot days as a shield against the sun. "It cooled the temperature a few degrees," he said. I did not consider myself dexterous enough to try that, especially at such low altitude.

0945: I was two minutes overdue at a checkpoint. A railroad bridge should have been easy to see, even from my height, the proscribed 200-foot limit over land. There was a city dead ahead, but according to the chart there should have been two of them united by a highway. OK, I figured, maybe I was lost. A thumb rule hammered down our throats ever since we started Sandblower training came to mind: "Hold your course and if the checkpoint doesn't come up, turn on time anyway."

So I swung left twenty degrees to my planned heading and anxiously checked back and forth between land and map, map and land. I was over rolling hills and to the left saw two sharp peaks that formed a crude V rising from the earth. I couldn't pinpoint them on the chart. Then suddenly, directly below, appeared a strip of brown in sharp contrast to the green of the countryside. Running diagonal to my course and projecting vertically from the strip like enormous robots, were steel towers supporting a network of power lines. These I could locate on the map. I found myself ten miles east of course and banked the Skyraider to intercept the route.

1050: Smooth going now, I thought to myself. I had picked up the lost minutes by eliminating a short leg and was a few seconds from the crucial run-in to the target. There could be no mistakes from here on.

I passed a U-shaped bend in a river and accelerated. The ground swept by faster and faster as the AD bucketed along at 240 knots of airspeed. I could feel the growing pressure on the machine as it sped through the air mass. The road I had selected to follow in the approach led to the Initial Point, the IP, a seventy-five-foot-high water tower, conveniently visible because of its bright, white coat of paint. I made final checks. Harness tight, mixture rich, fuel boost pump on, all other switches set. Then I glanced at the timer setting on the automatic bombing gear to ensure it was correct. My speed was now 260 knots.

The tower, a huge ball on struts, whipped by the port wing. I depressed the pickle on the control stick and held it down. The carefully calculated and preset sequence of seconds began clicking away, and a warning tone sounded in my earphones as the red timer light illuminated on the instrument panel. This was the moment of truth, the time of attack.

I pulled back on the stick and hauled the bomber upward, my eyes shifting to the gauges. Airspeed fell off rapidly and the stress of four and a half Gs inflated air pockets in my anti-gravity suit. These held my blood in check, preventing excess flow of the precious fluid from the brain. Otherwise, I might black out. The light and tone stopped as I passed through the near-vertical position simulating release of the bomb. On top of the maneuver the altimeter pegged at 1,900 feet. The world was upside down but I pulled on through the horizon, completing three-quarters of a loop before righting the aircraft and descending to tree-top level for my escape out the run-in line.

I had performed many of these maneuvers, called medium-angle lofts, but each always paid off with its own special kick. On this day I was pleased with my on-top altitude and the G schedule. I was nice and tight and believed that my "hit" would have been reasonably near the target.

1230: The Italian coast was behind me and I took a long look at Elba. It was the last land I would see for the day. Discomfort began to dominate once again. (Try sitting in a chair without standing up for eight hours some day.) To offset the gnawing fatigue, I finished the box lunch. There was another hour or so to go.

1320: I was still at 100 feet when I saw the grand profile of the *Forrestal* in the distance. Coming home to the carrier is one of the great and exhilarating sensations in naval aviation. Minutes later I joined three other Able Dogs in the holding pattern above the ship. I sucked some 100 percent oxygen directly from the supply tube to help revive my mental faculties in the struggle against exhaustion. I began to feel a bit punch-drunk like Watson. But it was uplifting to be in formation with the other squadron flyers who had gone through the same evolution.

1355: The jets finally got aboard and I double-checked gear, flaps, and hook down as I turned off the abeam position. On final, the ball on the signal mirror stayed frustratingly high and I corrected downward too abruptly, close in. This was not going to be a good pass. Safe enough, but no cigar. Our LSO, Kit Saunders, knew that we had been up in the air for hours and, as a pilot himself, identified with us. He was slightly more tolerant than usual.

I drove the Skyraider down the slope and bumped onto the mobile runway catching the number four wire. The firm tug of the harness straps against my shoulders told me I was home. With an economy of movement that never ceased to amaze me, the yellow-shirted flight deck directors guided me forward to the bow where Prothro waited, tie-down chains draped around his neck. When the last aircraft slammed aboard a few minutes later, the bull horn wailed "Recovery complete!" By the time I climbed out of the plane, the carrier was quiet but for a few commands barked by the directors and the sputtering of a tow truck here and there.

I took a moment and faced into the wind. There was satisfaction in this endeavor, the sort that is appreciated silently, personally. As much as I believed that my partners and I could do the job if the real bell rang, however, I hoped that time would never come.

Carriers in the Middle East: The Suez Campaign

Aircraft launching from carriers was a fairly rare thing in the Middle East until the 1991 Gulf War, where eventually six American ships were engaged in the intense six-week campaign. Most fighting, both on the ground and in the air, occurred over vast stretches of desert far removed from any body of water large enough to accommodate such a large ship and its escorting task force. During World War II, even when naval aircraft were engaged in operations in the area, they usually staged from front-line airfields on what today we call detachments, or "dets." American and French carriers had engaged in limited operations during the civil war in Lebanon, and had launched patrols and strikes against rebel positions in 1982 and 1983. The October–November 1956 campaign against the Egyptians by British, French, and Israeli forces was also an example of carriers at war in the Middle East.

Fed up with Egyptian posturing, and finally by Egyptian President Gamal Nasser's nationalization of the Suez Canal, Britain and France mounted an expedition to the eastern Mediterranean and, in concert with Israeli units, started a series of punitive strikes against Egyptian airfields and facilities on October 31, 1956. While several of the strike aircraft launched from land bases, many of them were based aboard British and French carriers.

The carrier planes met intense antiaircraft defenses, and occasional airborne threat of MiGs. It's worth noting that only the British carriers operated jets. The French used veteran Corsair prop fighters they had obtained from the U.S. after Korea. Their F4U-7s were similar to the U.S. Marine Corps', AU-1 ground-attack variant of the Corsair. The Royal Navy also used, for the first and only time, Westland Wyverns, dedicated ground-attack fighters powered by turboprops. The first carrier strikes launched early on November 1 and their allied aviators soon had their hands full.

This trio of researchers and historians has a long history of interest in Middle East military aviation. Brian Cull and David Nicolle have written extensively on the subject in several books and magazine articles. Shlomo Aloni, an Israeli writer, has made his mark recently as a chronicler of his country's air force's short but proud heritage.

From *Wings Over Suez,* by Brian Cull, with David Nicole and Shlomo Aloni

Two fully armed Avengers of 9F Flotille were launched from the *Arromanches* at 0349 to carry out an anti-submarine patrol, followed by a further two at 0405. The latter pair sighted an Egyptian destroyer, the *al-Nasr*, near Nelson Island but were greeted by heavy anti-aircraft fire as they approached. At 0520 the Aeronavale patrol encountered and attacked with rockets the Egyptian frigate *Tarek*, the French crews claiming two hits. When news of the attack reached *Arromanches*, *La Fayette* was instructed to launch a strike of eight Corsairs, each armed with two 1000lb bombs. The fighter-bombers were launched at 0700, led by Lt deV de Saint-Quentin, the deputy commander of 15F Flotille. After 50 minutes, the wakes of two ships were detected and the leader ordered drop tanks to be released, but first Ens deV Caron (son of the Task Force's Admiral) was sent down to confirm that the ships were not part of the U.S. Sixth Fleet. On observing Caron's aircraft being greeted by AA fire, de Saint-Quentin ordered an attack, having been briefed to bomb from a height of 8,000 feet—far too high to ensure accuracy against moving targets. However, both Egyptian vessels did suffer minor damage during these attacks although they were able to withdraw to Alexandria under cover of a smokescreen. Rocket-armed Corsairs of 14F Flotille were also launched during the morning, Lt deV Cremer (flying 14F-1) briefed to attack any military convoys encountered north of the Nile Delta. Although one such target was located, the pilots were unable to attack without the possibility of inflicting casualties among the many refugees also on the road, so they returned to the carrier having first discharged their rockets into the sea.

At 0520 the Fleet Air Arm commenced its assault, 36 Sea Hawks and Sea Venoms being launched from *Eagle*, *Bulwark* and *Albion* to attack the airfields at Cairo West, Almaza and Inchas, which were the priority targets; a further eight fighters (four Sea Venoms and four Sea Hawks) were sent off to fly protection patrols over their respective carriers and escorts, while Whirlwinds and Dragonflys from the carriers took off to undertake planeguard duties in case of emergencies.

The Sea Hawks of 897 and 899 Squadrons reached Inchas without incident, led by the four Sea Venoms of 892/893 Squadrons, as recalled by Lt Cdr Henley (WW193/096):

> I was Strike Leader [with Lt Ian Gilman as observer], which involved four Sea Venoms leading 12 Sea Hawks on a strafing attack against Inchas. The launch and form-up were pre-dawn, the plan being to use the Sea Venom's superior navigation capability (radar, observer) to arrive at the target at first light, closely following the high-level bombing attacks by the RAF [Canberras from Cyprus]—almost too close as a stick of bombs exploded across the airfield as we were on the run-in. Thereafter, the Sea Hawks and Sea Venoms operated separately.

The Sea Hawks rocketed and strafed a number of MiGs and other aircraft seen on the airfield and claimed four MiGs destroyed, five probables and three other aircraft damaged. Lt Cdr Rawbone of 897 Squadron (flying Sea Hawk 201) claimed one MiG destroyed and two possibly damaged:

> There were several craters from RAF bombs but no damage to runways, hangars or airfield primary targets. We strafed aircraft, hangars and the control tower, claiming damage only on those aircraft which actually burnt. Three flamers for Flight.

Another pilot of 897 Squadron, Lt Mills (XE448/191), graphically described his attack:

> We had no trouble in finding the target and started letting down to begin our attack. We could see the hangars and runways with rows of parked aircraft which we attacked with three-inch rockets. As I climbed away from the attack, following the Boss, I suddenly realised the sky was full of little black puffs of smoke, and I thought "My God, they're shooting at me!" On my second attack, this time using my 20mm cannon, I saw a lot of winking lights coming from the airfield perimeter, but it was much later when it penetrated that these came from anti-aircraft and small-arms fire. Then we re-formed in good order and started the trip back to the ship expecting to be attacked by fighters at any minute. As we crossed the coast, "tail-end Charlie" turned to make sure we were not being followed, giving away the position of the Fleet, but all was clear and we landed on without incident. At the debriefing, it was clear that we had destroyed or damaged a large number of parked aircraft, but how many was a different matter. I had seen a MiG blow up when hit by one of my rockets, and claimed it, but I think the score was much higher.

As the 802 Squadron section approached Almaza they encountered a flight of MiGs, as recalled by Lt Cdr Eveleigh (WM911/133):

> Almaza was the main operational airfield of the Egyptian Air Force. We were surprised by a flight of MiGs which climbed away from Almaza towards us, and as we were only straight-winged compared with their swept-wing high performance, I can still recall the feeling of my Squadron closing up tight on me on sighting them. We were apprehensive to say the least, but they swept straight past and did not attempt to mix it with us. We can only assume they were being flown away to some nearby safe haven.

Lt Cdr Ron Shilcock of 809 Squadron led the strike in Sea Venom XG670/220:

> Strafed line of small aircraft—thought to be MiG-15s. Encountered some light flak. Results of strike not observed.

His observer, Lt John Hackett, however noted:

> Attacked Almaza at 0601—four Sea Venoms. Five MiG-15s damaged on ground.

Another of the Sea Venom pilots, Lt Bob Wigg (in XG620/226, with Lt Eddie Bowman) who had flown operations during the Korean War, recalled the strike in more graphic detail:

> On our way in to the primary target, Almaza airfield, the sun came up, bathing the Egyptian landscape a stunning rose red. We had heard at the final briefing that the Canberras, flying direct from Malta, would bomb the runways at the Egyptian airfields and generally soften them up for us, so that we could get on with destroying the aircraft on the ground with rockets and cannon. As we made our long descent towards the target, it became obvious that the Canberras had not been there ahead of us. Almaza was untouched, apparently still asleep. Apart from a couple of unidentified transport aircraft [which he claimed damaged], there was no sign of the Egyptian Air Force's MiGs and Ilyushins. With the lack of targets, there was no problem in deciding where our rockets would do most good. The hangars on both sides of the runway erupted as 64 assorted HE and AP rockets exploded inside them, and then it was time to switch to cannon and concentrate on hitting something worthwhile with them and at the same time avoid hitting the ground with the aircraft.
>
> The Egyptian anti-aircraft guns must have been at maximum depression. Sited between the hangars on both sides of the runway we were currently streaking along, all they had to do was squeeze and wait for us to fly through, and that's what they did. The white hot balls of tracer criss-crossed the runway in front of us and I remember shouting a warning, not that any of us could have taken avoiding action. A few more seconds of the sort of terror I hadn't known since Korea and we were all clear.

Sub Lt Tony Yates and his West Indian observer, Sub Lt Charlie Dwarika (XG677/225), observed an Il-28 on the airfield which they duly strafed and claimed damaged. The leader of the Sea Hawks, Lt Cdr Eveleigh, continued:

> However, we did find some MiGs on the airfield, and some Ilyushins, to strafe. We were overflown by RAF bombers who actually bombed through our formation when we were over Almaza at medium level (8,000 to 10,000 feet). Apart from a little friendly barracking there were no problems.

Flt Lt George Black (WN118/137), 802 Squadron's 'tame crabfat' (the Navy's endearing colloquialism for RAF aircrew serving with the FAA), added:

> The flak we encountered had lived up to the intelligence briefs—even in excess in certain areas of the target complex—and we had sighted the first enemy aircraft in the air. I had gone to war and survived my first operational sortie.

The Sea Hawks strafed and one MiG was seen to go up in flames and two more of the jet fighters were damaged, while a Vampire was also hit; strikes were reported on an Il-28 and an Il-14 transport by Lt Jack Worth, who was leading the 800 Squadron flight in XE400/107, while a C-46 of 7 Squadron EAF and a 3 Squadron EAF Dakota were also damaged. On their return to *Albion*, the pilots generally agreed that they had failed to notice any damage to the airfield from the nocturnal bombing attacks.

A Sea Venom launches from HMS *Eagle* during the Suez campaign as the aircraft's tow bridle drops away. The Sea Venom was the Royal Navy's first all-weather jet fighter. (Courtesy of the Royal Navy)

The last bombs from the Valiant and Canberra attacks were also dropping on Cairo West as *Bulwark*'s Sea Hawks—eight from 804 Squadron led by Lt Cdr Randy Kettle, another Korean War veteran, and four from 810 Squadron—went in. Many Il-28s were seen in their blast pens and a small aircraft believed to have been a Vampire (probably a 2 Squadron EAF machine) was seen on the airfield. Four of the jet bombers were claimed destroyed by 804 Squadron and two more by 810 Squadron; the latter unit also destroyed a Lancaster. Four of the jets and two more of the elderly Lancasters were damaged during the strafing attacks. Lt Cdr Peter Lamb (XE403/238), another veteran of the Korean War, leading the 810 Squadron quartet (one of whom was Lt Graham Hoddinott in WV796), claimed one Il-28 destroyed and a second damaged. One pilot reported:

> Every pilot tensed, concentrated on his own target. The Beagles [Il-28s] and Lancasters took shape and grew larger beneath the fixed centre cross. Fire and fairy lights mottled the silver fuselages in the cold grey dawn. Suddenly a sheet of flame arched skywards, etched with black billowing smoke.

Two swept-wing aircraft, presumably MiGs, were reported to be over the airfield at 6,000 feet, although the Sea Hawk pilots failed to sight these and they were not attacked. Sqn Ldr Mohammed Nabil al-Messiry,* commander of a MiG squadron at Cairo West, recalled:

> I was in a state of readiness and could take off and get into action if the enemy were to come. However, I was attacked while I was on the ground without any warning. We were at Cairo West. I had only a very narrow chance to jump from my jet and run from the attacking aircraft. The jets were turning to the left toward me so I ran to the right to get away from their guns. I got only 20 or 30 metres before my aircraft was destroyed. We had four aircraft near the start of the runway in an open area. My No2 asked me, "What should I do?" I yelled, "Jump!" I didn't see him after I ran and saw that all four aircraft were in flames. After a while I saw this young pilot come out of the flames not touched or burned and he was still struggling with his parachute which was going left and right . . . it was very funny at the time. No one saw us get out and they thought we were killed. There were repeated attacks and most of the aircraft at our base were destroyed on the ground. Some of the Il-28 bombers were able to take off between attacks and escaped but many of the aircraft were destroyed.

* Major-General al-Messiry served later as Deputy Commander of the EAF during the Ramadan War in 1973.

Eight Sea Hawks (four from each of 800 and 802 Squadrons) and two Sea Venoms from 809 Squadron repeated the attack on Almaza, departing *Albion* at 0720, while two other Sea Venoms flew CAP. Leading the 800 Squadron quartet was Lt Cdr Tibby (XE435/104):

> We carried out a strike on Almaza and strafed aircraft. Light flak was coming up but it was not very accurate. An interesting experience because it revealed that despite the training you give your young pilots, you can never be quite sure how they are going to react in an emergency. Leading my Flight in the dive, my No4 suddenly shot right past me and as I was about to press the button to fire my rockets, he was in my line of sight. So that was an aborted attack. He realised that he had panicked and overshot the target. That does show that despite all the training there is no substitute for operational flying. He learnt the lesson pretty quickly after that.

As a result of the strafe of the airfield one Il-28 and one Vampire were claimed destroyed, and two MiGs and two Il-28s damaged. The pilots reported that the north-south runway was in use by MiGs during the strike (although there was no mention of these fighters actually taking to the air), and that Cairo International airport showed signs of being bombed. One pilot wrote:

> Throughout the day the strikes continued on a one-hour-five minute cycle. Eight or twelve aircraft launched; cut out at 20,000 feet; let down on track; attack with cannon; retire on the deck at 400 knots to the coast; then throttle back, up to about 5,000 feet, and back to the ship via the picket. The 'one pass and away' principle was adhered to strictly at first.

Wyverns of 830 Squadron were sent into action at 0800. Seven aircraft from *Eagle*, each carrying a 1,000lb bomb, attacked Dekheila airfield near Alexandria. Three bombs were reported to have exploded on the intersection of the north-east/south-west and east-west runways, and another two on the intersection of the north-south and east-west runways, leaving the north-east/south-west runway probably still usable. Lt Cdr Bill Cowling*, the Squadron's colourful American-born Senior Pilot, recalled:

* Lt Cdr Cowling left the U.S. in 1938 to join the RAF as a pilot but lacked the necessary education, so enlisted as ground crew until sent to Canada for pilot training in 1941. He finally qualified in 1943 and was commissioned but, to his chagrin, was retained as an instructor. With the war over, he transferred to the Royal Navy in 1945.

> The first Wyvern strike was against Dekheila airfield; six [sic] Wyverns in two vics attacked with 500lb bombs. We concentrated on runway intersections with a view to the MiGs getting airborne.

As the Wyverns withdrew, 11 Sea Hawks of 895 Squadron arrived from *Bulwark*, the sections led by Lt Cdr John Morris-Jones (XE396/167), Lt Ted Anson, the Senior Pilot in XE375/239, and Lt Eric Palmer; the pilots subsequently reported damage to two Beechcraft Expediters of 4 Squadron EAF seen on the airfield. Two large crates, assumed to hold aircraft, were damaged, as were hangars to the north and east of the airfield. Lt Cdr Morris-Jones noted:

> No other aircraft seen on airfield, hangars appear empty except for aircraft crates.

A priority non-airfield target was the blockship *Akka*, which was lying in Lake Timsah. She was a derelict craft loaded with 3,000 tons of cement, scrap iron and heavy rubbish. It was hoped to sink it before it could be placed in the Canal. The first strike at 0845 by four Sea Hawks from 897 Squadron armed with 500lb bombs, and eight from 899 Squadron armed with rockets, failed to sink her, as Lt Cdr Clark recalled:

> I was briefed to take eight aircraft armed with solid shot anti-submarine rockets to knock holes in the bottom of the blockship. Ray Rawbone was to follow dive-bombing. I objected to attacking with R/P on the grounds that there was only two feet of water under the hull and my rockets would stick in the mud. Or, if they managed to miss the mud they would not penetrate the concrete. I also suggested that dive-bombing was unlikely to damage a hull full of concrete. I suggested an alternative which was to simply take two aircraft and write off the tug using 20mm front guns. I should point out that a small tug was alongside in Ismailia waiting to tow the blockship into position. This suggestion was refused on the grounds that the tug was manned by civilians.

Lt Cdr Rawbone of 897 Squadron (XE448/191) added:

> We hoped to sink it in the lake before it could be moved. Intelligence reports positioned it accurately but we were led to believe it would be heavily defended by AA and patrolling MiG fighters. We therefore planned to

> attack immediately it was sighted and then re-form as swiftly as possible. On approaching the target there was some broken cloud but we could see the blockship quite clearly. We dive-bombed in succession, had several near misses but no one had a direct hit. We re-formed swiftly but were amazed that there was little AA fire and no enemy fighters!
>
> The attack was a waste of time. I could see the tug raising steam and again asked (by radio) if I could attack her but was refused.

Although Lt Cdr Rawbone had not sighted any Egyptian fighters, one of his pilots, Lt Mills (XE388), did, but not the feared MiGs:

> As I pulled out of the dive, I saw two Vampires flash in front of me. I turned after them calling on the radio, but there was so much chat about the attack on the blockship that nobody herd me. Eventually, bearing in mind the strictures about wandering off alone, I gave up the chase and turned to rejoin the Squadron. When we landed on and debriefed, Ray [Lt Cdr Rawbone] said I had almost certainly missed a chance to down two Egyptian aircraft but he was very understanding about my reasons for not doing so.

The Egyptian Vampires were probably returning from a strike against Israeli forces in Sinai, or were involved in the brief attempt to disperse aircraft to forward airfields in western Sinai.

On their return to *Eagle* the 897 Squadron pilots reported the near misses during their attack on the *Akka*, and recommended an immediate restrike. However, they were held at readiness pending a reconnaissance report. Four camera-equipped Sea Hawks of 810 Squadron were sent at 1050 to photograph the *Akka*.

While *Eagle*'s Sea Hawks were involved with the blockship, seven more from *Albion*'s 800 and 802 Squadrons carried out a further strike against Almaza, in company with two Sea Venoms from 809 Squadron led by Lt Wigg (with Lt Bowman) in XG669/224:

> I believe the idea was to check the damage and to surprise any last-minute aircraft movements. As we started our dive I saw a MiG being towed along the perimeter track. I saw the tractor driver sprinting for cover, and then our 20mm shells began striking the MiG. By the time we had passed overhead, I was sure the MiG would not be bothering anyone for a while. The flak gunners must have been on their tea break because we had a clear run across the airfield and out over the desert. And then I saw something that I believe none of us noticed on the first sortie. The run out was directly over a mili-

> tary camp; khaki-clad figures were running in all directions, except for one small group who were lying on their backs with their weapons pointing upwards—shades of Korea again! A split second of bum-crinkling panic and we were clear again, heading east towards the Canal and our second part of the brief, a recce north along the Canal to see if we could spot any interesting movements along its banks.

During the attack one MiG was seen to burst into flames and another was claimed damaged. One Sea Hawk returned slightly damaged. This attack was followed by another against Inchas by ten Sea Hawks from *Bulwark*'s 804 and 810 Squadrons (two aircraft having failed to join up due to catapult failure). A MiG was damaged during the attack and other aircraft strafed a hangar and the control tower. *Bulwark*'s three Sea Hawk squadrons were now sharing aircraft.

Cairo West airfield was targeted again at 1035 when seven Sea Venoms drawn from 892 and 893 Squadrons from *Eagle* were launched, led by Lt Cdr Malcolm Petrie, 892 Squadron's New Zealand-born commander. A successful strafe of the airfield was carried out which resulted in claims for the destruction of four Vampires (2 Squadron EAF), one Il-28 and two Lancasters, plus damage to another Vampire and a second Lancaster. The bombers were claimed by 892 Squadron pilots, the Vampires by 893 Squadron. Anti-aircraft fire was slight and none of the Venoms was damaged.

Albion's Sea Hawks (800 and 802 Squadrons) carried out a third strike against Almaza, eight aircraft, plus two Sea Venoms from 809 Squadron, departing at 1045. Flt Lt Black, 802 Squadron's Raf pilot, recalled:

> I was sent to Cairo West on the second sortie of the day, and I saw Pan-Am Dc-4s where I had expected to see Egyptian Air Force aircraft. I therefore flew to the secondary, which was Almaza.

A MiG and a Fury were seen to catch fire during the attack, and of each type plus two Dakotas and an Il-28 were claimed damaged. Flak was light but accurate. Returning pilots reported small-arms fire from a military camp to the south-west of the airfield. Almaza was the base for Egypt's new MiG-17s commanded by Sqn Ldr Shalabi al-Hinnawi (who had earlier that morning led an attack on Israeli forces in the Sinai—see Chapter Seven), who recalled:

> I was giving a briefing to my pilots and, at this moment, we saw four aircraft—Sea Hawks—come in and strafe. Unfortunately our MiGs were sitting in the open. The attack met with very weak anti-aircraft fire. Many of

> our aircraft were wrecked and I remember seeing one Il-28 jet blow up. A shell splinter from the strafing hit me in the leg and I was ordered to go to the hospital for surgery but I refused. Some eight of our MiGs were covered by nets near the hangars and weren't hit in the first attacks. We tried to move the surviving MiGs but we couldn't get approval from headquarters and they were destroyed in later raids. When the Sea Hawks started their attacks the two Soviet advisers to our squadron drove away in their car. Later they came back after the war. When we challenged them they said, "those were our orders."*

Midday saw four Sea Hawks of 802 Squadron launched from *Albion*, two of which carried cameras. Led by Lt Cdr Eveleigh (WM996/135), photographic runs were carried out over Almaza and Cairo International, initially at 10,000 feet, then at 1,000 feet:

> The photo-recce was in response to fear that some other flight might have attacked a civil airliner by mistake. These were not in the target brief!

Two Il-14 transports of 11 Squadron EAF were seen at Almaza and were immediately strafed and set afire by Lt John Bridel and Lt John Carey, while at Cairo International the two pilots observed and photographed cratered runways and two burning aircraft—one identified as a civil airliner, which was probably Viscount SU-AIC of Misrair, and a transport aircraft, although this may have been SU-AFF, a Fiat G212 of Air Orient, which was also destroyed during an air raid. The damage at the civil airport was presumably inflicted during the night attack by the Canberras from Cyprus.

At midday *Bulwark* sent 11 Sea Hawks from 804 and 810 Squadrons to Inchas, where four MiGs were claimed destroyed, one by Lt Cdr Lamb (flying XE 396/167 of 804 Squadron) of the latter unit. Reconnaissance photographs taken by *Bulwark*'s Sea Hawks of the *Akka* reached *Eagle* at midday and these revealed that the blockship, towed by a tug, was moving towards the Canal. 897 Squadron's bomb-laden Sea Hawks were launched again at 1240—eight aircraft led by Lt Cdr Rawbone (on this occasion flying WV907/190)—and found the blockship still under tow:

* Sqn Ldr Shalabi al-Hinnawi had flown Macchi MC205Vs during the closing stages of the 1948–49 War, and had been shot down by an Israeli Spitfire. Severely wounded, al-Hinnawi would carry a steel strengthening plate in his skull for the rest of his days. During late 1967 he was made Commander of the EAF, tasked with rebuilding this force following the catastrophic June War.

> Such were the political implications, however, that our briefing strongly emphasised that under no circumstances was the ship to be attacked if it was actually in the Canal when we arrived. Much to my dismay, the tug and blockship were just entering the Canal as we approached and when they saw us arriving the tug manoeuvred to turn across the Canal and scuttle our target in the centre of the stream. We were now out of range of *Eagle* on the R/T and I decided to position the Squadron ready to bomb whilst climbing to altitude in the hope of regaining R/T contact with the carrier, thus obtaining further instructions. Clearly it was better to sink the blockship at the edge of the Canal rather than have it scuttled in the main stream. The Squadron aircraft were now nicely poised to attack and there were no enemy fighters. However, by brief had been so specific that I could imagine the outcry if I decided to bomb and any damage to the Canal resulted. In the event I decided to bomb because I could not contact *Eagle.* Sub Lt [David] Prothero was first in position, and this bombs narrowly missed the blockship and blew a large hole in the west side of the Canal! Not very encouraging! But the gods must have been with me as the next bomb (delivered by Lt Mills) was a direct hit and blew the blockship into two halves, each of which sank near the west back of the Canal.*

The successful pilot, Lt. Mills (XE379), added:

> For once, I got it right and sent my bomb more or less down the funnel.

With the task apparently completed—although reconnaissance photographs later revealed that the Egyptians had nevertheless been able to move the severely damaged and half submerged vessel across the Canal—Lt Cdr Rawbone led his Sea Hawks to Abu Sueir, their secondary target, where those with bombs remaining attacked the hangars:

> At least two hangars and their housed aircraft were destroyed. We all strafed lines of MiGs parked on the hardstandings and were surprised at the almost complete lack of defensive fire.

One pilot reported seeing an Egyptian soldier, armed with a machine-gun, standing firmly between two parked MiG-15s, hosepiping at the swooping fighters. Eight MiGs (probably aircraft of 30 Squadron EAF) were claimed

* Some four years later when newly promoted Cdr Rawbone took his first sea command, the frigate HMS *Loch Killisport,* through the Suez Canal, he was very aware of a large hole in the west bank. It was repaired but obviously there.

destroyed during the attack and a further three damaged, of which Lt Cdr Rawbone claimed two flamers and two possibles. Sqn Ldr Farouke al-Gazawy was at Abu Sueir as the strike came in, and commented later on his experiences:

> I was at Abu Sueir which was a former Royal Air Force base. It was a lovely place full of beautiful gardens. We were just finishing lunch and I was about to go to the Squadron when we were caught. I remember seeing four Hawker Sea Hawks come in and strafe our aircraft. I was face down in one of the gardens and looked up and watched the jets hitting us.

Syrian Air Force personnel training alongside their Egyptian comrades at Abu Sueir inevitably became involved in the fighting, and a number of SAF training mission MiG-15 fighters and trainers were destroyed on the ground by British and French aircraft. On his return to *Eagle*, Lt Cdr Rawbone was much relieved to have Admiral Power's backing for his decision to attack the *Akka*, which was no doubt vindicated further by the sinking in the Canal of some 50 other blockships later in the operation.

The Wyverns from *Eagle* were off again at 1140, five aircraft completing a second bombing and strafing attack on Dekheila, where the runways were again targeted. Four 1,000lb bombs were seen to straddle both the north-south and east runway, and the east/south-west runway. Additionally, an aircraft observed parked in the centre of the airfield was strafed and damaged. Six Wyverns returned at 1455, when four bombs were seen to explode on or near the runways as a result of their attack.

Avenger crews from *Arromanches* had earlier reported seeing an Egyptian destroyer entering Alexandria harbour and seven rocket-armed Corsairs were despatched, in two flights, from *La Fayette* to hunt for further movement of Egyptian naval units, albeit without success. Another Avenger was sent to identify a vessel detected on the radar of a patrolling aircraft and discovered a U.S. Navy minesweeper which was not displaying its national colours. However, as the Avenger approached, its crew spotted an American sailor hurriedly hoisting the stars and stripes.

Shortly before 1100, *Albion* launched a dozen Sea Hawks (four flown by 800 Squadron pilots and eight of 802 Squadron) to carry out a strike against the EAF Flying School at Bilbeis, where an estimated 100 trainers were observed within the airfield perimeter. Lt Cdr Eveleigh of 802 Squadron, who led the attack in WM938/131, wrote:

Bilbeis was a training station but we were asked to destroy planes on the ground in case any could be used offensively.

Flt Lt Black (WN118/137) led one section of four Sea Hawks:

To my surprise I can remember going into the dive and seeing a variety of static aircraft types which had all been pushed out of the hangars and left scattered in the centre of the airfield. There were Harvards, Yaks, Chipmunks, a Sea Fury trainer etc. It seemed too good to be true. I sighted on the first target, a Harvard, and opened fire with a short burst down to minimum range. With hindsight, it was exhilarating to experience 20mm cannon shells rip into static aircraft, followed almost immediately by a small explosion as the aircraft fuel tank caught fire. The 20mm ammunition we carried had every third round tipped with an incendiary head, the other being semi-armoured piercing rounds—the effect was quite dramatic as the tracer ammunition showed instantly whether the aiming point was being achieved.

Time did now allow a total count of static aircraft, but I estimate it was in excess of 90—clearly there was no shortage of targets for all of us to have a field day! This gave us tremendous confidence and I can remember hitting one trainer before inching the sight onto the next target and claiming two more 'flamers'. This type of sortie felt unreal and reminded me of the more mundane range-work which was routine for ground attack pilots. I note that my logbook records I claimed five aircraft destroyed and two damaged on this mission. Ammunition exhausted, we headed north leaving behind large palls of smoke from burning aircraft.

800 Squadron diarist noted:

Here [at Bilbeis] a strafing roundabout was set up as pilots made attack after attack on aircraft parked literally nose to tail.

Lt Cdr Tibby (XE435/104) leading the 800 Squadron quartet added:

We were strafing training aircraft—a bit of a "turkey shoot" since there was no flak at all. So at one stroke we probably destroyed the training capability of the Egyptian Air Force.

Following the strafing passes, five Harvards, three Chipmunks and two Yaks were claimed destroyed in flames, and a further 15 aircraft heavily damaged, of which Lt Worth (XE411/108) claimed one of each type destroyed. Lt Cdr

Eveleigh reported seeing two aircraft on fire and a third obviously damaged as a result of the strike. All aircraft returned safely. Flt Lt Black continued:

> Following the landing back on board *Albion*, the next surprise was to be told to go back to Bilbeis for another strike and finish off the remaining undamaged aircraft! This we duly did, although I can recall the task of finding undamaged machines quite difficult; indeed, I remember opening fire and, as I got closer, noting that the target had already been partly damaged. All in all, the 'turkey shoot' at Bilbeis, as it was subsequently referred to, proved an excellent means of sharpening our aiming procedures whilst adding greatly to the confidence factor for what was to come in the following days. Moreover, I was to add a further three aircraft to my score!

As a result of the second strike, pilots of 802 Squadron claimed ten trainers and left a dozen others damaged. During the two attacks—in addition to Flt Lt Black's eight destroyed—Lt Paddy McKeown (the Senior Pilot and another Korean war veteran) and Sub Lt Carl Clarke (the Squadron's most junior pilot) each claimed the destruction of three trainers and a further six damaged jointly.

The crew of one of *Albion*'s helicopters, patrolling in case of an emergency ditching by returning aircraft, reported seeing two unidentified aircraft at 1350, described as silver with black star and red stripes, armed with rockets; these were believed to have been USN Skyraiders from one of the American carriers. Six Sea Venoms from *Eagle*—two from 892 Squadron and four from 893 Squadron (including WW154/448, an aircraft of 892 Squadron crewed by Lt Ben Neave and Flg Off Duncan Watson, and another flown by Lt Cdr John Willcox with Flg Off Bob Olding)—followed up the strike on Bilbeis two hours later. The crews reported the destruction of a further nine Harvards and one Chipmunk, and damage to four of each type, of which 892 Squadron's share comprised four Harvards destroyed and three damaged, plus one Chipmunk damaged.

Eight Sea Hawks from *Bulwark*'s 895 Squadron, the sections led by Lt Cdr Morris-Jones (XE394/165) and Lt Palmer, were launched at 1410, this time joined by three more from 810 Squadron (including XE375/239 flown by Lt Hoddinott), the pilots briefed to attack Almaza. On their return to *Bulwark* they reported the destruction of one MiG, two Dakotas (of 3 Squadron EAF) and three C-46s of 7 Squadron EAF, the probably destruction of an Il-28 and a C-46, and damage inflicted on four Dakotas, two MiGs and a Fury. Two of the transports were claimed by the 810 Squadron section, and a further one damaged. A general report of the strike noted that aircraft were seen on the runways and dispersals all over the airfield: on the western side were transport aircraft, and on the eastern side MiGs, Furies and Il-28s, while up to 50 soft-skinned vehicles (referred to as SSVs) were observed to the south of Almaza.

Three 800 Squadron Sea Hawks from *Albion* visited Cairo West at 1500, flying over the airfield at 1,000 feet. The pilots reported seeing six Lancasters on the ground south-east of the airfield, of which four had been destroyed. Near a hangar six Vampires (apparently aircraft of 2 Squadron EAF) were observed, of which four appeared to have been burnt out. A lone Il-28 was seen in the same area, and another of the jet bombers was seen on the ground south of the airfield and was adjudged to have crash-landed. This was strafed and destroyed. The pilots also reported a ditch was being constructed about four miles south of the airfield alongside which appeared to run a light railway. As *Albion*'s aircraft returned, they were passed by four Sea Hawks of 895 Squadron from *Bulwark*—a section led by Lt Palmer—which carried out a further attack on Cairo West and claimed two Il-28s and five Lancasters damaged. On receipt of this report, *Eagle*'s last strike of the day, launched at 1600, was also directed against Cairo West, Lt Cdr Clark (XE457/487) leading eight Sea Hawks of 899 Squadron to destroy one Il-28, one Lancaster and two Vampires. A second Lancaster was claimed as probably destroyed and a third Vampire was damaged. In addition, the control tower was strafed, as were three trucks, anti-aircraft positions and other buildings.

Albion's last strike of the day consisted of a dozen Sea Hawks, six each from 800 and 802 Squadrons being sent to Inchas at 1515, where the pilots reported no sign of life on the airfield. The wrecks of seven burnt out MiGs were observed on the runway and, although wheel tracks were clearly seen leading from the hangar area to a plantation to the north-west, no aircraft could be seen concealed there. Flt Lt Black of 802 Squadron, on his fourth sortie of the day, noted:

> We were off back to one of the more heavily defended airfield targets near Cairo. Here, life in the Sea Hawk cockpit became much more lively and sporting.

With dusk rapidly approaching, *Bulwark* launched six Sea Hawks flown by 895 Squadron pilots led by Lt Cdr Morris-Jones (XE396/167) and Lt Anson (flying XE378/168) for a final strike against Almaza, where two MiGs were claimed destroyed, one in flames. Aircraft damaged during this attack included a Vampire, a Meteor, a Fury and a Dakota. The Sea Hawks returned at 1720. During one of the raids on Almaza, a Mallard amphibian of the former King's Flight was strafed and destroyed.* Bombs also destroyed the EAF's Museum which housed, amongst other types of aircraft, a Spitfire, a Hurricane, a Tiger Moth, an Avro 626, a Hawker Audax, a Hawker Hart, a Lysander, and a Gladiator.

With the end of the day's carrier operations, the captain of one of *Bulwark*'s escorts, HMS *Decoy*, sent a signal to the carrier:

> I hope you will allow a fish-head [a non-flying naval officer] to say that I have never seen such consistently first class landing-on and flight deck drill as your team have put up today.

However, other signals were not in such a complimentary vein, as witnessed by the one sent by Admiral Durnford-Slater to the Admiralty regarding the activities of the U.S. Sixth Fleet:

> Sixth Fleet are an embarrassment in my neighbourhood. We have already twice intercepted US aircraft and there is constant danger of an incident. Have been continually menaced during the past eight hours by US aircraft approaching low down as close as 4,000 yards and on two occasions flying over ships.

Sqn Ldr Maitland of 249 Squadron, who had served with the Americans in Korea, had been less perturbed:

> The Americans from the Sixth Fleet frequently intercepted us outward bound but never anywhere near the war zone. They would come up alongside, wave and depart.

A view not entirely shared by Flt Lt Black of 802 Squadron, since the carrier pilots were called upon to carry out interceptions of all intruders approaching the Fleet:

> It was not uncommon for aircraft to be scrambled to intercept US aircraft coming to look at what was going on. This hampered operations because the carriers had to keep turning into wind to launch intercepting aircraft, often at the most inconvenient time.

* The second Mallard of what had been the King's Flight survived, and could still be seen, no longer serviceable, at Almaza in the early 1970s.

Lt Cdr Eveleigh of 802 Squadron similarly recalled:

> Knowing the Egyptians had MiGs, we had to intercept all incoming fighters and, until identified, US swept wing fighters appeared as a threat. The wing destroyer of their Fleet even got inside the screen of our own Fleet, occasioning an interchange of signals which probably cannot be repeated. Our Wing CO asked the American whose side he was on. On receiving a rather negative reply, our Wing CO made back in plain language, "Then why don't you f . . k off!"

However, at least one American unit had offered its services, albeit without official sanction, as recalled by one particular Fleet Air Arm observer:

> We had been working-up the Squadron at Hal Far in Malta in mid-October, and had become very friendly with the US Navy squadron of Neptune aircraft (Squadron VP-24) which was also based there at the time. If I remember rightly, their Commanding Officer was a Lt Cdr Cook. When we started our strikes on Egyptian airfields, the Neptunes quite unofficially did some high-level spotting for us which was most useful. Their involvement was certainly at the very beginning of the operation either at the time of, or shortly after, the RAF bombers had bombed Egyptian airfields from high level. We were in air-to-air communication with the Neptunes and they gave us information of what targets had been hit. I was personally in communication with one of the Neptunes and reported accordingly. I do not believe they were in the vicinity during the strikes. I understood later that Commander Cook incurred the displeasure of Mr Dulles [the U.S. Secretary of State] and the US Navy, and was relieved of his command. I am unable to corroborate or verify the sequel to this story, but the Neptunes certainly helped us, and Commander Cook was no longer in Malta when we returned there on 30 November.

French Hellcats in Vietnam

Before America became entangled in the mire of Southeast Asia, another country had painfully extricated itself from the bloody civil war that had spread over the region by the mid-1950s. While America and its United Nations allies fought the Communists a thousand miles to the north in Korea, France found itself heavily engaged with another Communist enemy in the steamy jungles of Vietnam, or Indo-China as it was then called.

Using one American and two former British carriers (HMS *Biter*, renamed *Dixmunde*, and HMS *Colossus*, now *Arromanches*), the Aeronavale sent squadrons of Grumman Hellcats and Curtiss Helldivers into action against the insurgents. Then-Lieutenant de Vaisseau (the equivalent of the American rank of lieutenant) Roger Vercken was the commanding officer of 12e flottille (Deuxieme Flottille or 12F), sailing in the carrier *Arromanches*, then the *La Fayette* (formerly USS *Langley* [CVL-27]). In 1952–53, he flew strikes against enemy positions in Vietnam, Northern Laos, and Northern Cambodia. He had assumed command at such a junior rank because two preceding COs had been killed in action.

Often the weather was as foul as it could be only in that region of the world. In this excerpt, now retired Vice Admiral Vercken describes the problems surrounding recovery aboard the small escort-class carriers. It is a good reminder that another country tried to save Southeast Asia long before America's longest, most costly, and ultimately unsuccessful war.

Born in 1921, Roger Vercken entered l'Ecole Navale, the French naval academy, in 1941, serving during the occupation period of World War II. He gained his wings in 1947 and flew Seafires, Hellcats, and Corsairs in several fleet squadrons. As noted above, he commanded 12F aboard Arromanches, *then as part of the air group in* La Fayette. *After retiring in 1979, he wrote his memoir from which this excerpt is taken,* Au-Dela du Point d'Envol (Beyond the Flight Deck). *He is vice president of the Association pour Recherche de Documentation sur l'Histoire de l'Aeronautique Navale (ARDHAN), dedicated to the history of French naval aviation.*

From *Au-Dela du Pont d'Envol* [Beyond the Flight Deck], by Vice Admiral Roger Vercken, FN (Ret)

Excerpt translated by Captain Robert Feuilloy, FN (Ret)

French Hellcats Over the Gulf of Tonkin, 1953

Gray, everything around was gray. I was wrapped up in the grayness, in turn diaphanous, opaque, transparent or blurred, as would glint from some sort of cameo or lavish with an endless variety of shades. And the dark and polished surface of the South China Sea was developing at 2,000 or 3,000 feet underneath.

Once in a while, I cleared the mist to discover a clearing, stretched between two pale white tones, which were rejoining in the distance like two sheets on an infinite bed of desert, or, elsewhere, an iced lake suddenly scintillating under a lost sunbeam falling on it. Then, my vision vanished, while I kept gliding on an immaterial mattress, turning darker and darker. I was alone, totally alone in my aircraft and in this environment which was drifting as in a dream, I could not take the time to wonder if I wanted to stay alive.

On the earphones, a distant spoken voice, from Robin or Vetillard, the best radar officers aboard the carrier, was tearing me out of those visions to bring me back to reality. Using a metric radar, which was partially broken and not designed for this kind of Hertzian aerobatic use, they were doing their best to get us down to the mother ship, my wingman and me, that meant to bring us in sight of the flattop. The bloody *La Fayette*, that we had left at the end of the morning, managed to entangle herself in a mass of wet air, which had coagulated at once and from which the ship could not slip out.

Visibility at low level must not have been better than inside the ill-fated stratus. The ship must have been optimistic when they were broadcasting a 200-foot ceiling! Every minute, that ceiling was descending closer to the ocean. But our good old Hellcats could not be considered seaplanes!

"Numa leader, this is Figaro. Take left heading zero six zero. Stay level. Over."

"Figaro, this is Numa leader, to turn left to heading zero six zero. Stand by."

Caron and I had already split from each other, 10 minutes ago. Our controllers had opted for individual letdowns rather than a close formation penetration, which was always a strain for the wingman and could become dangerous if he lost sight of his leader. I could not recall which one of us had started first to leave the milky brightness of the dry mist to dive into that soaked-through pillow. But the ship's commanding officer would probably have asked to have the senior pilot come down first just to see how it goes. It was me, of course, not only in my section but also among the 20 pilots of the squadron I commanded.

Therefore, each of us was having his own trajectory, guided by the ship on separate radio frequencies. If we couldn't land, we would have to try to divert to Haïphong at a bare 250 km in the north. For sure, the weather up there was probably the same. A letdown above the Tonkin delta and the bay of Along, guided by an antique radar, above invisible peaks of limestone, was like playing Russian roulette. Did we have enough fuel to even get there? At least, in our situation here, above the sea, we were sure not to crash into some swelling of the Earth, before reaching altitude zero.

One thing was certain: we were the last to come back that day. The other flights had recovered a long time ago. We were the flight which had been sent the farthest and that was four hours ago. The ship must have come into the wind, some time before, to prepare the recovery, and the air boss must have broadcast the ritual wording "Free deck to recover two Hellcats." This sentence had the effect of bracing all personnel involved in a recovery process; it was a strange ballet of colorful jerseys. That sentence also implied that the aft part of the deck was clear for the landings. But waiting, the skipper of the *La Fayette*, who had recently taken command in Indochina, was certainly getting anxious, angry at not having been in a position to recover all aircraft before the ship was engulfed in the fog.

Caron and I also wanted to finish up quickly. We had left our Noah's Ark just after the morning haze had dissipated. The ship was then off the northern coast of Annam, abeam of the city of Vinh. Initially, we were to participate more in the south, in one of those amphibious operations that higher command was fond of. Those operations were supposed to ease our ground forces from the pressure of the Viet Minh, by creating a diversion in the center of Annam. But the Viets had, themselves, diverted their interest toward Laos. This is why we had been sent out to see what was happening inland, 350 km to the west. The concern was that a Franco-Laotian post was keeping silent; they had left Sam Neua, attempting to escape southward on uneasy tracks through a breathtaking maze of mountains, in order to rejoin the Plain de Jarres. They were hounded by a Viet Minh division, sweeping down from Tonkin. Our task was to locate the column Maleplatte, bearing the name of the commanding lieutenant colonel; eventually we had to protect their retreat.

From long ago, the army guys had made it their custom to name their elements by their CO's name—group X or battalion Y. This "personifying" was not at all in line with navy traditions. Our ships bore an official name and the Aeronavale units were happy with just a number, which was replaced on the radio by a call sign, easy to pronounce and to understand. For example,

Roger Vercken after landing at Glyeres, June 10, 1953, from carrier *La Fayette*. Note his suitcase resting against his Hellcat's chock. (Courtesy of Roger Vercken)

in the routine life, my squadron was called 12F and the tactical call sign was Numa, a name probably picked up in a Latin dictionary by a distinguished member of the navy staff in Paris. One of them, fond of Beaumarchais, the 18th-century French author of the opera "Figaro," or a fan of Pierre Brisson, the chief editor of *Le Figaro* newspaper, had elected to allot the *La Fayette* the call sign "Figaro."

"Figaro, from Numa leader, steady heading zero-six-zero, altitude two thousand feet, over."

"Numa leader, this is Figaro. Roger, out."

Strapped for a few hours in my aircraft, sitting on a pack containing my folded dinghy and parachute, as comfortable as a public bench in the street, I was starting to feel the ache. To get some relief, I was balancing from one hip to the other, or folding one leg, just keeping one foot on the rudder pedals, like a one-legged man. I was barely looking outside, electing to keep my attention on the instrument panel, mainly on the altimeter. The engine was purring gently, and the dials showed the different parameters, the still needles giving normal indications. We had to say that, for the last seven months, we had been very careful with our engines. The times of the great circus of the carrier fighters, where we were shifting carelessly from full power to idle, were gone. For our present missions as truck drivers going to the end of the world, we were gently pushing the throttle and adopting a long-endurance regime. Our propellers were turning at the speed of those of a merchant ship. We were accordingly obtaining the maximum range.

Generally speaking, our engines were turning like well-tuned clocks. Those were Pratt & Whitney R-2800-10W with 18 cylinders, cooled by air and capable of 2,000 horsepower. They were jealously maintained by our mechanics, whose devotion was unlimited. Since 12F had started operating in Indochina, we had not had any serious malfunction, even when the Viets were shooting at us with their AAA, fortunately limited to small caliber at this period. The pilots had complete confidence in their aircraft as well as in their mechanics, and therefore were able to concentrate on their mission.

In addition, there was no precise assignment of an aircraft to a pilot. Everybody, including me, was flying any given aircraft. So, all birds received the same attention, which meant a lot of care. The safety of the pilots was at stake, and I have always been stringent on this point, despite the pressure from a few pilots who would have preferred to have their personal "taxi."

"Numa leader from Figaro. By the left, come heading two-four-zero. Resume your descent to one thousand feet and hold."

"Roger, wait."

The gray clouds were repeatedly licking the fuselage and were increasingly darkening. At least, here, above the sea, I was sure not to be swallowed by a cumulonimbus, like those we had seen in the distance on our climb to Laos, emerging above the dry haze at about 20,000 feet.

Just after the cat shot and once the rendezvous was made, Caron and I found ourselves wrapped in a reverberating, white atmosphere, which rapidly put the sea and the land out of sight. Heading west, I had hoped to be able to climb up to rejoin a clear sky that seemed close. In vain. After a 45-minute climb, at 15,000 feet, I had given up going higher. We then came down, always to the west, trying to catch sight of the ground, which refused to show up. We also knew it was becoming more hilly.

A large enough hole finally permitted us to see a patch of land and we spiraled down, ignoring the elementary rules of safety. We did not have a choice, anyway. Resuming our course at low level, we spotted by sheer luck a French flag floating on the mast of a small post, whose name was written on the ground in capital letters using white painted stones. It was Nong Het. That's a brilliant idea from those infantry guys to help lost aviators update their navigation! In the past, in France, many young pilots had found it convenient to read the names of locations on the roofs of train stations.

Nong Het was at the border of Annam and Laos on the road colonial No.7—the R.C.7—nicknamed "road of Queen Astrid" in memory of the queen of the Belgians who took it in 1932 to travel from Luang Prabang to Vinh. Those were happy times!

"Figaro from Numa leader, steady on course two-four-zero, at one thousand feet.

"From Figaro, roger, wait."

Looking at the altimeter, I was 1,000 feet above water and still couldn't see anything underneath or forward. It was about 4:00 P.M., it was getting darker and darker, and there was no hope that the visibility would improve. The more difficult was to come: catch the boat in sight and trap on the first pass.

The worst was that we had strained ourselves with little to show for it! After Nong Het, we had clung to the R.C.7 road that swung like a dragonfly, winding for a thousand curves, climbing on a pass then tumbling down into smiling valleys. But at Ban Ban, we had to release the thread of Ariadne to head north and find, at 60 km out, the post of Hua Mong, lost in the mountains, through which the Maleplatte column must have passed. The landscape had become chaotic, bristling with interwoven ridges and peaks with abrupt slopes, overcast by a dense coat of greenery, once in a while torn by the scintillation of a stream in the deep of a ravine. It was an illusion to pretend to follow a track or read a map in such a country.

We were not out of maps, however. We carried with us in flight a full bag of them, going along with other baggage like an emergency medical kit, a machine gun, and a Colt pistol. The full collection of maps covered the north of Indochina at various scales—1/1,000,000, 1/400,000, and finally the 1/100,000, with one centimeter equal to one kilometer. This more precise scale was on more than one hundred different maps, and was too detailed to be of any use. The main lines of the relief were lost in a jumble of level curves, each a different design, and there were just too many topographic names. Each sheet was covered with small local names, whose approximate spelling must have been hastily established by zealous cartographers. The same word could be found several times inside a 10-km-diameter circle that revealed a geographic fantasy and deep innacuracy. To avoid the errors of homonymy, the maps had been overlapped with a kilometric grid to permit designating the targets by rectangular coordinates, just as in a crossword puzzle. But in those crossword games, thousands of lives were in danger. We had the impression we were to strike the black squares . . . and to strike the empty ones.

After about a quarter of an hour, we had at last found a naked hilltop, with remnants of houses still smoking and close by, the relics of a Morane Criquet, a spotting aircraft, on the edge of a road. The post of Hua Mong had been evacuated, then set on fire, or vice versa. But who had set the fire, our troops or the Viets? We couldn't see anybody. How do you to know if the Maleplatte is up to the north, south, east, or west? It's like searching for a needle in a haystack. They certainly had no desire to be seen, for fear of being spotted by some Viet posted on a nearby hill. Another Criquet, taking off from the Plain de Jarres, had finally joined us, and asked us to strafe a strip of the track. But even flying at deck level, we had not spotted anything. By itself, this information was valuable. Then we turned back to the ship.

"Numa leader, by the left, come to heading zero-six-zero. Perform vital checks. Current altimeter setting twenty-nine, ninety-two inches. At completion of your turn, take the approach speed and start your descent at three hundred feet per minute. Over."

"This is Numa leader. Roger, three hundred feet per minute. Wait."

In our airman jargon, those were the vital checks—lowering the wheels, the flaps and the arresting hook, setting a certain RPM on the engine (propeller on low pitch, carburetor on rich) and diverse technical verifications. Usually, in clear weather, we did the checks at 1,000 meters abeam the ship, on the downwind leg, before initiating the left U-turn into the wake of the carrier. But in this fog the controllers were having us let down in a straight line, approximately on the axis of the deck, without any other possible maneuver than to adjust the heading on final approach. The foremost thought was to have the ship in sight. The radars were not precise enough to guide the aircraft within a 10-meter bracket, and it was out of the question to land without seeing anything. Five more meters to the right, and the aircraft was going to hit the island. Five more meters to the left, it was going into the catwalks where the flight-deck personnel were sheltered during landings. Thirty meters too short and we would have a ramp strike. Seventy meters too long and it was a guaranteed crash into the parked aircraft on the bow. Off those limits, it was water all around.

"Figaro from Numa leader, steady heading zero-six-zero. Wheels, flaps, hook down. Descending at three hundred feet per minute. Speed ninety knots. Final checks performed."

"From Figaro, roger."

As expected, the fog was becoming thicker. I could no longer see the trails of clouds passing by and, outside, all perceptable sensations of motion had vanished. The needles on the dials were motionless, except the one on the altimeter, which had resumed its rotation counterclockwise. The song of the engine had simply become harsher when I had reduced the pitch and increased the RPM as I started my descent. To fight the numbness, I was instinctively moving stick slightly. The aircraft responded instantly, and the instruments reacted. It was reassuring. However, I was arriving at 600 feet, still descending, and when glancing outside, it was yet the same opacity created by millions of water drops.

Robin was now sending me distances and altering my heading.

"Come to heading zero-six-two. I have you at three miles."

"From Numa leader, roger. Heading zero-six-two. Still in descent, crossing five hundred."

He was now hooking me and he was right. It was not the time to wander around.

"Come back to zero-six-one. Distance two nautical miles."

He was struggling hard to tune the precision of the approach. It was certainly not easy with a radar that was painting the smallest aircraft on the CRT like a big banana of 15 degrees of angle. But it was his job and his talent, and I was trusting him. He had been controlling aircraft for years, and along with Vetillard, he was one of the more competent radar interception officers in the navy. But this time it was not an exercise, and the game was being played at very low altitude.

The radars fitted on the *La Fayette* couldn't give the altitude, and it remained for me to watch the height. I also knew that, within 1,000 meters around the ship, the radars were blind. The plot of my aircraft would then mix with the clutter of the sea return so often described by the radar operators.

The long needle of my altimeter had just passed through 100 feet, and still nothing was in sight. I decreased my rate of descent to only 100 feet per minute. In the moments to come, four things could happen. I would see the boat and one way or another, manage to land aboard; I could overtake her without seeing her at all, and the duo Robin-Vetillard would get me another circuit or divert me to Haïphong. I could hit the water, or I could hit the back of the ship.

This last possibility did not give me a good feeling. On the contrary, in case of a ditching, there was not a second to lose, because my mount would rapidly sink. I would have to unstrap from the seat, disconnect the radio cord from the helmet, stand up, exit the cabin, open the CO_2 bottle of my Mae West jacket, slide into the water, then inflate the dinghy located beneath me, and simultaneously separate from the now-useless parachute pack. That was a lot of things to do in a few seconds. Meanwhile, I opened the canopy. The wet air, braced by the propeller, whipped my face and the noise of the engine went up.

Damn it! The three needles of the altimeter had joined at the fatal zero mark. Instinctively, I pulled the stick to level off.

Figaro was calling me. "Your distance is one mile. Continue your approach."

I answered, "Indicated altitude is zero. Confirm the barometric setting on the ship."

Robin sent me back a new setting which now credited me with 80 feet. It was not a thick mattress! Carefully, I was now descending by small steps, a little push on the stick, then a little pull to level off. I did that four or five times, bracing myself for a possible touchdown on the water. The long needle of the altimeter was coming back to zero, and this time, the ship would not send me another cushion to please me.

"You are now at one thousand meters from the ramp," Robin radioed. "Keep steady on course. You should see me by now."

All of a sudden, something jumped up out of the grayness, a mass darker than the mist, whose distance was difficult to grasp, but something on which

I could fix my attention. Breathlessly, I made the call, "OK, Figaro, I see you."

I was barely above the deck level, and coming in on a little bit of a slant from aft port, which was, by the way, giving me a good view angle on the LSO, all dressed up in his yellow suit, his forearms elongated by the multicolored rackets. That was probably Castelbajac, and he must have had no desire to have me going around, up in the clouds again.

I throttled back a little bit and cocked the aircraft gently up to settle it into slow flight, practiced only by the naval aviators, at a few knots above the stall speed. The batman had extended his arms horizontally, meaning that my altitude and speed were correct, then he bent slightly sideways to signal me to come left and line up on the centerline. Everything was now happening very fast, and this carrier, so much longed for, was now jumping at me. I was entering the short final. I received, very late, the signal "too high," immediately followed by the "cut." I idled full back, dropped the nose, aimed at the forward section of the centerline, then pulled the stick to land on three points, main wheels and tail wheel together.

I felt the cheerful thump of the landing gear on the deck and the drag of the hook catching one of the arresting wires, thrusting my chest forward in the straps. On my right, the island was crowded with spectators. From the deck, spread in echelon to the fore, emerged one, two, three, four, five tall yellow devils, the flight-deck directors, who would guide me safely into the chocks.

On a signal from the first one, I raised the hook and the flaps and unlocked the wings. Then, two men in blue shirts manually folded the wings along the fuselage. Dripping with sweat, I felt completely washed out. I had spent more than four hours in such dreadful conditions. I would not do it again soon. I gave a good throttle up to roll over the barriers and free the aft portion of the flight deck.

Once the aircraft stopped in the chocks, engine shut down, propeller stopped, I waited for my plane captain to come to get me out of my various impedimenta and then to help me get out of the cockpit. Suddenly, I felt I weighed a ton. Standing on the inboard section of the wing, I stayed there a moment, facing aft to gaze at the flight deck, already green to recover Caron, my wingman. He too, was coming out of the mist, at less than 1,000 meters from the ship. He landed like a flower. He must also have been struggling. We were going to be able to go to the officer's mess and buy a drink for Robin and Vetillard. They deserved it. That was on the 16th of April of 1953.

CHAPTER 5

The Vietnam Crucible, 1964–1975

The Skyhawk's War

The Douglas A-4 Skyhawk was one of the workhorses of the Vietnam War. In the thick of the fighting from beginning to end, the little attack bomber designed by American genius Ed Heinemann ranged from Hanoi and Haiphong in the north to Cambodia and Laos, and the Delta in South Vietnam. Flown by Navy and Marine Corps aviators, the A-4 flew thousands of day and night attack missions, even though it was basically a clear-weather aircraft with limited radar equipment. An A-4 even managed to shoot down a North Vietnamese MiG-17 in May 1967 as its pilot, Lieutenant Commander Ted Swartz of VA-76, was preparing to dive on an enemy airfield.

During the first two years of the war, following the 1964 Gulf of Tonkin Incidents, Skyhawks flew from small and large carriers. The "big decks" accommodated big air wings that included A-4s and A-6s as the attack portion of the wing. By 1968, however, the A-4 had begun to give way to the newer, more sophisticated Vought A-7 Corsair II. By war's end, the A-4 could be found only aboard the USS *Hancock* (CVA-19), the last *Essex*-class, 27-Charlie flattop engaged in combat operations. *Hancock*'s sister ship, the *Oriskany* (CVA-34) had changed its A-4s for A-7s by war's end. (The Marines still flew the Skyhawk from land bases, making some of the war's last missions in August 1973.)

But in the early war years, the Skyhawk was king of light attack. Maneuverable, tough, capable of a wide variety of missions, the so-called "bantam bomber" was without a doubt one of the finest American combat aircraft ever designed.

Charles Brown commanded VA-112 during its 1967 combat cruise in USS *Kitty Hawk* (CVA-63), a big-deck carrier. Although a less-publicized A-4 squadron, VA-112 had an eventful cruise in company with the other squadrons of Air Wing 11.The war had quickly become an exercise in frustration with restrictions and micro-management affecting how the men of the air wings fought.

Although his book has a much larger story, namely the development of all-weather, carrier-attack squadrons, Captain Brown gives us an excellent view of the A-4 pilot's war during the intense middle period of the Vietnam conflict. As commanding officer, he also had special considerations regarding responsibility to his unit and to carry out his superiors' orders. It wasn't easy.

A member of the class of 1952 of Annapolis, Captain Brown flew his early squadron tours as a night-attack pilot in AD-5N Skyraiders with VC-35. After a tour with VX-5, a test and development squadron, he transitioned to the A-4 and served as executive, the commanding officer with VA-112 during two combat deployments.

From *Dark Sky, Black Sea: Aircraft Carrier Night and All-Weather Operations,* by Charles H. Brown

The target was the Haiphong thermal power plant. The strike group included twelve light and medium attack bombers loaded with 1,000-pound bombs, eight other A-4s as flak suppressors and SAM suppressors carrying Shrike antiradiation missiles (ARMs), and four F-4 fighter escorts. An E-2 Hawkeye AEW aircraft was airborne to relay information between the strike group and the *Kitty Hawk.* KA-3 tankers were airborne for refueling any aircraft, if necessary. (By that time, the A-3s had transferred their bombing mission to the A-6s.)

The ship launched a weather reconnaissance airplane about one and a half hours before the strike's scheduled launch time. At launch time, the weather airplane reported several cloud decks extending from the middle of the Gulf of Tonkin inland as far as he could see. The clouds definitely covered Haiphong. The bottoms of the clouds were at about 2,000 feet, with the tops at about 20,000 feet. Believing the target weather would improve, the *Kitty Hawk*'s captain ordered the strike to launch. The E-2 was off the deck first, followed by the fighters, the A-4s, and the A-6s. The KA-3s launched last.

My A-4 was about the tenth airplane launched. I pulled the gear and flaps up, called departure, and disappeared into the clouds. Scanning across the altimeter, airspeed, turn-and-bank, attitude, and rate of climb indicators provided a comfortable feeling that all was going well on the climb-out. Departing airplanes had enough interval between them to prevent midair collisions while climbing through the clouds. I had set the strike group rendezvous to the north of the ship at 20,000 feet, above the reported cloud tops. When I skimmed out of the clouds, I could see airplanes in front of me already joining and setting up the strike formation. The fighters were above the center attack group. The flak and SAM suppressors were on the flanks. Joining on my A-4s, I moved into the lead.

Then-Cdr. Charles Brown, CO of VA-112 in March 1968. (Courtesy of Charles Brown)

When the A-6 division leader radioed that all his airplanes were formed up, I reported that we were departing for Haiphong and switched to the ship's weather reporting frequency. The weather scout continued to report marginal conditions near the coastline, but it seemed that the cloud decks were breaking. As I continued toward the point at which we would come into SAM range, I switched back to my group's tactical radio frequency to see if there were any problems. All was quiet. Soon I had a good picture of the weather between my strike group and Haiphong. I switched again to the ship's frequency, checking for further instructions, but the ship had no guidance. To proceed or not was my decision. From where I sat, completing the strike was worth the risk. I called the group, saying something like, "We're going in!"

I began a slight descent, dodging thicker clouds and following the landmarks toward our planned dive roll-in point by the power plant. A few SAMs, which looked like flying telephone poles, flew up through the lower cloud deck into the formation, but none hit. Either the electronic countermeasures were working or the Shrikes were forcing the SAMs' guidance radars to shut down after the SAM sites fired. As the flak suppressors turned away to look for flashes of active antiaircraft gun sites, a clump of clouds drifted over the spot where I wanted to start our dive on the power plant. Sliding around the clouds, I rolled in, followed in turn by the other eleven bombers, each varying its dive heading a bit from the airplane ahead. The spread of airplanes in a fan pattern created a more difficult target for the gunners. The antiaircraft fire was heavy, but we were lucky on that strike. Perhaps the gunners had not expected us to get through the breaking overcast. The fighter escort was alert but inactive—no MiGs that day.

After the attack, the group collected itself in sections of two airplanes, with each section leader reporting when overwater to the E-2. The E-2 reported to me and the ship when all the aircraft cleared North Vietnam. One of the F-4 fighter escorts had taken a hit in his starboard wing but was apparently in good enough condition to attempt a carrier landing. One of the tankers joined with the damaged F-4's section in case he needed fuel.

The rest of the strike group returned directly to our marshal. At the appropriate time, the ship's controller started the strike airplanes' approaches. Because the overcast was above 1,000 feet, we let down below the overcast in divisions and entered a visual weather condition landing pattern. Without the all-weather training that the group's aviators had, the strike would not have been possible.

On 21 January 1967, about a month and a half after my first combat cruise began, I flew my first night interdiction mission into North Vietnam. There had been no Alpha strikes scheduled during the day, but the winter monsoon's overcast rose to about 15,000 feet late in the afternoon. My section's mission was to locate and destroy trucks or truck parks along North Vietnam's Route One, the major road connecting North and South Vietnam, located a few miles inland from the coast. I briefed my wingman in the red light of the ready room. Squadrons still briefed under red ready room lights at night, although aircrews lost most of their night vision under the flight deck's moon lights.

Coming out of the island hatch, the floodlights cast shadows around the crowded flight deck, but it was easy to pick out my aircraft. The VA-112 Broncos' airplanes, distinctive with a black stenciled horse's head on the fuselage, were as usual parked tail outboard along the deck edge toward the after part of the flight deck. A thorough preflight check, using my flashlight, of the airplane's exterior and the six 500-pound bombs preceded the seven-foot climb up the ladder to the cockpit. Illuminated by the floodlights, I settled into the cockpit, tightened the seat belt to my harness, and connected the g-suit. I put on my helmet and plugged in the radio cord and oxygen mask. I was ready to go. I had flown several combat missions before I realized that by the time I'd gotten strapped in my airplane, ready for engine start, most of my anxiety about the coming mission was gone. By that time in my career, the night was not so mysterious.

After the engine start and aircraft system checks, I checked in with Prifly and departure control. The E-2 "Hummer," so called because of the loud, humming sound its turboprops made at idle, launched first at every cycle to relieve a squadron mate on the AEW station. The A-4s followed the fighters. The taxi director guided me smartly with his wands onto one of the bow catapults. I ran the engine up to 100 percent, turned on my lights, and trusted the catapult officer. With my head braced against the headrest, elbow against my stomach, and hand resting behind the stick, the A-4 jumped off the cat easily despite a heavy load.

On the gauges going off the catapult, I checked the plane's altitude and attitude, raised the gear and flaps, and proceeded to rendezvous with my wingman, who had gotten off the deck before I did. His lights appeared where I expected them to be, west of the ship about ten miles away below the overcast. I had decided it was not worth the time and trouble punching up and down through the clouds before going "feet dry" over the North. The *Kitty Hawk* was not that far from the coast.

I took the lead with my wingman in a loose trail, keeping track of my airplane's dorsal and dim blue formation lights. When we reached the beach at a point near the middle of our route, we turned out all lights and my wingman began dropping back. I had chosen to let each of us search independently, but as leader I called all route and altitude changes. We had enough airspace below the overcast to see and evade missiles if the enemy fired SAMs at our airplanes. It was clear and dark underneath, but with no horizon. At 1,000 feet above the terrain, I could barely make out the winding path of Route One through the trees below. I turned north.

We bypassed Thanh Hoa, patrolling north toward Ninh Binh. If either my wingman or I located a promising target, the one finding the enemy would drop flares and call for the other to join the attack. I could see bursts of rapid-fire 23-millimeter tracers around the Ninh Binh bridges, but nothing close to us. Those 23-millimeters at night really attracted my attention. That gun's rate of fire was so fast that the stream of bullets, always with tracers at night, looked like burning water from a fire hose. At the end of the stream's trajectory, it sort of petered out, tracers dropping like a fancy fireworks burst.

I reached the northern end of our route and swung west. When my wingman called his turn at the same spot, I headed back to Route One and started south. Neither of us had seen anything. Near where we had started our interdiction mission, I saw flickers of light. There were moving trucks coming toward me, probably empty, but at least moving. Checking that I had armed my bombs and 209-millimeter cannon, I pulled back to idle power and started a strafing run on the lights. The guns jarred the little A-4. I went on the gauges during the pullout from the dive, then climbed, making a tight left turn, and released two flares.

I called my wingman, reporting my luck in discovering the trucks. He answered that he understood and soon reported that he saw my flares and was joining me. I had seen no gunfire. I ordered two passes apiece, telling him to start his bomb run from south to north after I called that I was off the target. My wingman and I would be aiming only at a spot on the road under the flares, for I had seen nothing after my strafing run. After my wingman called rolling in, I waited for him to call off target—and waited. Nothing! He never called.

By that time, the flares had extinguished themselves. I had seen no secondary explosions from our bombs or any other explosion that might have been an impact from an airplane. After dropping my other four flares, I circled just above the light in the hopes of seeing something under them. I climbed and reported my wingman's disappearance and possible loss to the airborne E-2 controller. He sent a tanker toward my location and vectored an F-4 section to help search. In the darkness, I had them take altitude separation on me and we flew over the area for about thirty minutes. We saw no fires on the ground and heard no emergency radio transmission.

I called the KA-3 tanker, letting him know that we needed fuel. As the F-4s and I approached the refueling position, I saw the tanker's bright and flashing lights about ten miles away as it proceeded on the refueling course. The F-4s refueled first, so I stayed away from the tanker and watched the lights of the F-4s take position behind the A-3. After the second F-4 finished refueling, I accelerated, slid into position, and took my turn refueling.

Refueling from an airborne tanker's hose at night requires as much precision as any maneuver in an airplane. The refueling hose, called a drogue, drops behind and below the tanker. It has a basket at the end that has the fueling fitting at the joint of the basket and the hose. The pilot of the refueling airplane must put his airplane's fuel probe into the bobbing and weaving basket, making a firm connection with the fitting. At night there are only four lights on the rim of the basket to guide the refueling pilot. The A-4s had been tanking for each other on buddy strike missions for years. Tanking from an A-3 was easier because the larger airplane was a much more stable platform. An A-3 tanker's drogue basket was almost still, except in very rough air. After leaving the tanker and reaching marshal, the approach, CCA, and landing were routine.

That mission was typical of the many light attack night missions flown into North Vietnam with one terrible exception. A pilot and an airplane had been lost. I had probably been looking in another direction when my wingman flew into the ground in his bomb run. Explosions from airplane impacts sometimes do not leave lingering fires. Resuming the search the next day, we saw no evidence of what happened to the airplane or pilot.

Paying the Price: Frank Elkins of VA-164

The 1966–68 period of the Vietnam War was perhaps the most intense portion of the nine-year conflict. Carriers rotated on and off the line, serving full four-to-six-week periods of operations, relieved only by occasional side trips to Hong Kong or the Philippines. Most of the missions Navy flight crews flew were "up North," over Hanoi and Haiphong, which had become the most heavily defended cities in the history of air warfare. Often flying more than 100 missions during a six-month deployment, including several of these line periods, naval aviators, particularly those who flew almost daily strike missions, developed their own ways of getting through their time away from home and families.

Keeping a diary was an outlet and a way to record one's thoughts in this most difficult, yet exhilarating and personal time in one's life. Lieutenant Frank Elkins was an A-4 pilot with VA-164, aboard USS *Oriskany* (CVA-34) in 1966. This particular cruise was one of the most eventful of the aging carrier's eight deployments to the war zone. Besides considerable air action and corresponding losses, it also included a devastating fire on October 26, which took the ship out of action for nearly a year.

Living aboard a ship at sea for an extended time is hard. Personal space is at a premium, and always there are the sounds and smells of a ship underway. Loudspeakers, man-overboard drills, launches and recoveries right above your head, and the smell of jet fuel permeating everything from water to clothing. The lucky ones were the aviators who could leave the ship for a few hours as they flew missions, dangerous as that might be.

In his diary, Frank Elkins describes living conditions and the missions he flew before he went missing in action on October 13, 1966, two weeks before the fire. His natural exuberance and love for flying and his wife are always evident in his writing. But like many young aviators of the time and place, he is angry at having to fight the war hamstrung by so many official restrictions levied by people half a world away with no concept of what is really needed.

Frank's loss began a struggle for his young wife, Marilyn, as she tried to find out what had happened to her husband. Was he alive in a North Vietnamese prison? No one could, or would, tell her. She spent several frustrating months trying to get information, even going to Paris to talk to the North

Vietnamese delegation. It wasn't until 1990, after relations between the U.S. and Vietnam had thawed, that more definitive data was released, and American teams were allowed to tour parts of the country to find crash sights. Remains at one site were eventually identified as those of Frank Elkins, allowing his family to know what had happened to him. He had died in the crash of his A-4.

Marilyn Elkins moved on with her life and earned a Ph.D. in English from the University of North Carolina. She is now a professor in the English department at California State University, Los Angeles.

A native of Bladenboro, North Carolina, Frank Elkins earned his wings of gold in 1962. He married Marilyn Roberson in 1966, shortly before he sailed for his combat cruise in Oriskany. *He was posthumously promoted to commander.*

From *The Heart of a Man,* by Frank Elkins

June 27, 1966, under way to Dixie Station off South Vietnam

Goddam boat! They're cycling the dadblamed catapults—your bones rattle and your ears ache with the loud, hollow echoing resonance.

I'm in the wardroom now because it's the only comfortable place on the whole goddam ship. I just left my hot-ass room because even the damned fan doesn't do anything for the humid heat beneath the cats. Shit, what a fat life! It's so hot that I sleep naked with no cover and still wake up soaking wet. I sweat so much that my mattress is wet; even my ankles sweat.

The shoals here are beautiful. The rock formation and shallow waters make a color from the air. It's funny; you get the impression you're over land there. Actually, you know you're still over water, but you feel safer around the shoals than you do out over the open sea. It's strange because the breakers would probably make short work of any survivor who parachuted into those treacherous waters.

On my second hop today, Bost and I were in the lead section with Don and Greg in second. We bombed the shoals and the wrecked ship, but this time we did so at night using MK-24 pyrotechnic flares to illuminate the target. From the very first, I was uneasy and uncomfortable, seeing the sun go down, knowing that I hadn't made a night landing on the carrier since the fourth of April—almost three months—and the weather was, in a word, shitty: low, blowing puffy stuff and black hairy lightning and thunderstorms in all quadrants. The bombing went okay though, and I forgot my uneasiness till we were off target and en route back to the ship. Then I started dreading making the landing on the ship—plus the added attraction of flying Don Bost's wing, which is always a thrill at night. He pumps the stick like he is drawing water. Even the wings of his plane seem to have a nervous twitch.

He finally left me at marshall near another storm at 140°, 44 nm from the ship at 29M, with a good, strong, serious case of 30°-left-wing-down-equals-vertigo. I fought and fought that sensation for four or five minutes. Finally my balance started believing the instruments, and I got the vertigo under control. I put my hook down and commenced my letdown at 20:19, set the power, and checked and rechecked for things I might have forgotten. Radar picked me up at twenty miles and brought me in. I've never had such a sensation of being able to think so much so fast as when I was on the ball, in close, just prior to landing. Things were buzzing through my head like lightning: "Kid, if you bolter after fighting that vertigo, you need your tail kicked—easy don't be rough with the power pole." I got aboard, settling to the ramp to a number-one wire—piss-poor driving—but I smiled when I felt the tug of the gear and really wanted to holler, "MADE IT"—cheated death again.

I decided that maybe I'm cool enough to go to my tiny-ass room, so I climb the ladders, cross the goddam hurdles, rearrange things in the room, get out my typewriter, and reach for this sheet, and the sonofabitchin electricity goes out again. Such a miserable existence—and I have a far better lot than most, being a Lt.

And today we're really off to war. I feel as unprepared as I have ever been. Four days from now we'll be dropping bombs on people; that's difficult to believe. I don't know how I'll feel about it; I hate to even wonder because taking someone's life and then wondering how *you* feel about it is harsh, to say the least.

We'll be flying close air support for ground operations for about a week. This is quite safe, actually, since South Vietnam has no great antiaircraft system of guns. It's sort of a transition phase, leading up to the more dangerous North Vietnam work. I'm glad, since I'm still not used to the idea.

What a damn hot-ass boat! Sweat in bed, sweat in your chair, I even sweat in the shower!

Well, the damned cats are shaking the whole room again. *Goddam boat!*

July 3, 1966, Dixie Station

It's after midnight now and the end of my fourth day on Dixie Station.

On my second flight, the FAC led us to a wooded area, where we orbited a few minutes till some low clouds blew by. Then he marked an area of woods south of the river using a smoke rocket. Again we bombed without actual visual contact with the enemy. Then the fun part of the day began. We agreed to meet the Spads, single-engine, propeller-driven planes from the AD squadron of the ship, at 040°, 30 nm, 10M for a big twilight dogfight. (One of the Spad's main functions is to fly protective cover for helicopter rescues of downed pilots.) It was dark by the time we rendezvoused. Radios sounded something like this:

"Hey, I'm rolling in at 160° on a low speed."

"Okay, is that you above me? Flash your lights."

"Holy hell, who just went by me?"

"Was that you? I thought you were the Spad. Chip, do you hold us?"

"I think so; flash your beacon—hey, look out, they've got a sleeper up high—Who's that above me?"

"Goddam, I just met him head-on. I thought he was going the other way."

And so it went. If you think 150 knots in the tip of a high yo-yo with 120° bank and 1.5 G's is a hairy trick in the daytime, try it in the black of night with three ADs and three A-4s, and a sky filled with stars that look like aircraft lights until you try to rendezvous with them and stall and fall. Whee, but it was fun—Chip came back white!

Everyone seems to consider me the pilot that I feel I am. I've raised pure hell in almost every flight I've flown and impressed most of the people with whom I've flown. If there were a who-is-the-best-pilot-in-the-squadron contest, I think that I would win. I have to continue to prove myself to myself, but as long as I succeed, I guess I'm okay.

I'm mighty undecided about leaving in September. I want to see Marilyn and do all the things we could do, but I would also like to get all the air medals, etc., and more particularly I feel that I need the combat experience. Of course, I am speaking without having flown against enemy fighters, enemy AAA, and enemy SAMs. As George Johnson used to say, "Those that want to go are those that ain't been." Now I fall into that group. Still . . . I am undecided. The real advantage I'd like to gain is that extra year of seniority that I would have on my contemporaries by "dragging my feet" and getting an extra year of sea duty. If I stay on sea duty for another year, I'll be rotating along with people a year my junior, and I'll always have that one-year jump on them as far as seniority in billet choices. I want to give the whole thing some more thought after I've seen the flying steel and lead. Then I think I'll know.

Lt. Frank Elkins in a relaxed moment. (Courtesy of the U.S. Navy)

I must have made friends with Tim. He just brought me a cold San Miguel Filipino beer which, aboard ship, is worth at least a small fortune. I have a fifth of bourbon and gin, but beer is bulky and, therefore, logistically impractical—and all the more wonderful a surprise.

Independence Day, Dixie Station

Turns out that Richard is the one really funny guy in our outfit, and he is game for anything—dogfight to poker game. He's always making some false, raving charge against someone. Today he tore into me for getting so much mail from Marilyn. He's really good company, and we played gin a lot until we hit the combat zone. It's been nothing but max thrust since we arrived.

Joe's been in on a couple of hops on which BDA (bomb damage assessment) gave him credit for several people's death. And, like me, it's a different case when you know that you've killed for sure 'cause now you've got to be right in what you're doing. Killing reaches way inside you and searches for the truth. I see Joe feeling this now. We both believe in what we're doing, but still there's something to face every time it happens.

This morning I led Bob and Ralph across the country over to the Cambodian border, where there had been a major battle yesterday between the Charlies and the U.S. Infantry. The VC were in retreat and were in a wooded area near a friendly rubber plantation. We each had eight 250-pound bombs, and we pounded the wooded area with our ordnance. It seemed just like flying down to Stevens Lake and "bombing" for a while.

What a big zero Bost is! Tonight I really went out with the tigers: Bost-Toastie, Ralphie, and Jerry. What an enthusiastic challenging hop! I lost my oxygen at the start of the flight, and when I mentioned it on our way back to the ship, I thought Bost would spin in. Ralph lost his radio, and Bost loused up bringing him aboard so badly that when they landed they had messages to see both the CAG and the captain right away. Bost is the third officer in command. What a sterling example for the rest of us to live up to!

For me flying is just pure fun—danger or not—and I can see how some people consider Navy pilots as members of a refined motorcycle club. Damn, though, I love to fly. I zoomed up a vertical cloud today and in my heart the words of Gillespie's "High Flight" came welling: "Oh I have slipped the surly bonds of earth/ and danced the sky on laughter silvered wings."

I spent some time today reading and rereading Marilyn's letters. She's it as far as the kid is concerned. There can't be but one, and I decided six months ago that for the rest of my life, however long that might be, she's the one. If

I've ever made a decision that paid me back in the deep hours of the lonely night, this was it. In my life I've loved three entities that deserved it: my mother, because she unselfishly wanted only what I wanted for me; my flying, because it gave me joy and self-respect; and Marilyn, because I've never felt so close to another human being. Her letter describing the kittens and rain displays just the sort of feeling and sensitive understanding which so endears her to me. She says that she bungles, but what she manages to do is hit the very heart of things I know that I've felt only after she tells me about them.

July 8, 1966

What a hot SOB this whole tub is! Outside my door in the passageway, the thermometer says 94 degrees, and it's 10:00 at night. At noon today it was 105 degrees here in this room. I'd ten times rather be on a combat mission in my nicely air-conditioned cockpit with people shooting at me than supposedly getting a day of rest on this furnace.

One of the things the Air Force does occasionally is to spray wide stretches of jungle with defoliates which kill the leaves and vines and rob the Viet Cong of their hiding places. When we were flying in the south, we flew over one of these areas and dropped fire bombs on it to set it on fire and maybe smoke out some of the VC. Well, the air is so wet and volatile that when the heat from the fire started rising above the fire, it caused a small thunderstorm right over itself and put the fire out within fifteen minutes. At first we didn't know what was happening, but when we tried it again the next day, it happened again; the heat caused the air to rise and form rain clouds. That's the best example I know of how wet and humid it is in this country. Add that humidity to the steam of the cats, and it's no wonder that all I do is bitch about the heat.

It was our first day to fly over the north, and I was handed a tanker hop so I still don't know what the big action is like. Don and Richard went out on the same launch as my tanker and destroyed nine river barges they happened to find.

The staff captain in the War Room explained how the North Vietnamese continue to supply their troops in the south. Three main routes the NVN use are all almost directly parallel and within twenty-five miles of each other and run practically the length of North Vietnam. These routes are the railroad, the main highway itself (the one we call "Highway 1-A"), and the coastline waterway. Supply troops carry something by bicycle for a distance, then transfer it to the railroad and run it perhaps thirty miles by rail, then transport it to the road if a railroad bridge is out, or put it on a barge. You can't blow up rivers, and we sink barge after barge and still they come. All this river traffic—junks, barges, flatboats, etc.—are cheaply made, so there's no real problem to replace them. To that extent it's a game of patience; who'll get tired of the war first, a test of endurance.

July 9, 1966

There was a special meeting in the ready room at 09:00 this morning. All aviators were forced to get up to attend the meeting. The purpose of the meeting? The skipper wanted to warn us not to get complacent. Can you get that? We fly until after midnight, and the skipper routs everyone out of bed to tell us not to get complacent. How about getting rested, skipper, how about that, how about a little sleep on this hot-ass boat where twelve hours of sleep equals eight because of the goddam heat!

At 11:15 I briefed for my first "combat" hop over North Vietnam. What an abortion! The skipper (to whom Heaven would be a thirty-plane echelon with himself in the lead going in parade formation into Hanoi) led the flight with Chip on his wing; Joe and I were in the second section. We approached the beach. We were supposed to make our run from the south, down Highway 1, dropping one instantaneous and three chemical delay fused ANM 64 (500-pound GP's) on a little piddly-ass bridge north of Vinh. Approaching from the southeast, do we make our run from the south, up the road? No, we fly over the target (waking everybody up so they can get a chance to fire at us), then do a 270° turn to the left. If anybody in town had been awake with a gun, some of us would not have come back.

Tonight I go out with Darell on a night armed reconnaissance hop. Bost-Toastie spares the flight, and if he gets to go, I'd almost not go—except that I like to consider myself somewhat daring—and to go anywhere with that zero is to just dare death. What a complete ineffective!

Combat jitters are not too noticeable about the ship except in isolated cases. The XO of one squadron turned in his wings on July 6, having had enough. He was to be sent off on the COD [Carrier-on-board-delivery, an S2-F] and flown to Cubi Point in the Philippines. When you are a passenger in a COD, you are strapped in your seat backward for the catapult shot. When he was catted off, his straps came loose, and he pitched forward and smashed his head into the bulkhead. He cut his head from the back of his scalp across his head and all the way down the right side of his face. They brought the COD around and landed again, and he has been in the operating room for hours now. He had quit because he had premonitions of a forthcoming crash. Louie had the watch, saw the whole thing, and said, "When I turn in my wings, I want to go ashore in a long boat."

October 11, 1966

Wahoo, hurrah, and yippee! We're off the mid-to-noon for a whole week. Now it's noon-to-midnight, and that's so sweet it's like a chance to fly other than completely exhausted.

Sign on the ready room chalkboard: "Only 30 more Bombing Days till Christmas, get yours done early!" Sign number 2: "Flight Surgeon Sez: Get your annual physical. Those not complying will be grounded (upon arrival in San Diego)!" Sign number 3: "Yossarian Sez: The Ironhand that was scheduled to be cancelled will brief 30 minutes early."

Night before last I took B-10 out on a night bombing hop and found and bombed a convoy of trucks. Stumbled around for twenty minutes not knowing where the devil we were, getting shot at from everywhere, looking for checkpoint 38. Finally gave up and decided to go out and get oriented and find something else. Heading out, I saw two lights, dropped two flares to mark the spot, and found the convoy.

On a later hop yesterday, I took Jerry out and scared the devil out of us both. I led him through a couple of double-Immelmans, stalling out over the top of each one. Then I did some zero-airspeed, hammerhead climbs that always scare me. But then I did the grand finale which I never admitted to him that I had never had the guts to try before; I let down to low altitude over the water, getting lower and lower, 100 feet, then 50, then 20, then 10. Then finally I let my hook down and eased it down till the hook touched the water twice, and that was at about 350 knots. Great fun! I took Barry out today and did the same thing; but Barry did it too. I not only never did it before; I never heard of anyone's doing it. And when I saw Barry's hook touch the water, I decided that maybe I'd never try it again. When your hook hits, your drop tank is only about eighteen inches above the water.

It seems that we never go into AI in the morning anymore but what there's another report of a Navy pilot downed. Just every day. Just count them off.

But now, glorious day, we're off the mid-to-noon for a week.

Night before last I wrote Marilyn, while I was lying in bed for that six hours, I lay there and hated that night hop and hated that night hop and hated the strain of flying where you have to look in every direction at once and look out for yourself and somebody else and hated the Navy and mostly hated that doggoned night hop and the fact that I couldn't go to sleep so long that the very idea of a whole career in the Navy was so remote to me that I decided that I wanted out of all this, decided that the thing for me to do was go through three years of shore duty and then go back to North Carolina and do something: law, teach, anything, start a business, run the farm, and get out of all this. As far as that period of tossing and turning, rolling, forcing myself to lie still while I took first one hundred deep breaths in one position and one hundred deep breaths in another position in an effort to make my body go to sleep, as far as that four hours went, it was certain that the only thing for me to do was to get out. I thought of all the times when I was growing up, even in high school, that I was afraid of something and took the easy way out, of all the times I was scared and took the coward's way out, of all the times I didn't think I could look the other guys in the face and feel that I was quite the same stuff they were, fights I backed down from, loafing at football practice, loafing on the farm sometimes, or not putting everything I had into any one of a thousand things. I could think of all that, knowing that if I turned in my wings, that I'd have to live all my life in that same feeling of shame, having the medals but secretly knowing that I had given up because I didn't think I had the stuff to keep going when it got rough. And yet, it wasn't quite enough.

I had all that on the one hand, and on the other hand was that damned night hop coming up in the ink black that seems to belong exclusively to the night over here, the uncertainty that the airplane was going to make it into the air off the cat, the terror of not knowing—or of having to find—the exact right spot to coast in, of having to watch out not only for myself but to coordinate and control, to maintain order, to keep track of that all-important wingman, and to get back and find the ship and make sure that he got back and got aboard, and then to fly that ink-black approach and not break my neck on the ramp or fly into the water or fly into the sea or into a mountain or into the ground and see to it that my wingman didn't.

All these things can be done easily enough taken one thing at a time, but lying in bed knowing that you'll be required to do it all, it all lies on your mind at once, and you can't sleep and every minute goes by and you know that that's one minute more of precious sleep you lost and need to be safe and get the job done safely and yet you can't sleep and you just lie there and lie there and you can't stop thinking about it all and you toss and it never quits.

It would almost be a relief to get up right then and brief and go and do it all, but you can't even do that; you just lie there knowing that when it's really time to go that then you'll be sleepy and tired out and unsafe, and there's nothing, absolutely nothing, you can do about it. It's horrible. And it happens every night, every night until you're off that awful schedule.

And now we're off that schedule for a week. The feeling of relief is inexpressible. Now I can know that although it'll be night flying and Vinh and Nam Dinh and Cam Pha and Haiphong and all the rest, it'll be a schedule on which I face things one at a time, doing each thing as it becomes necessary, living one day at a time, and getting enough sleep and rest to do it sanely and as safely and as much under control as possible.

Still in the back of my mind is the knowledge that sooner or later, a week, eight days, ten, and we'll be back on it again. That's what I came to find out: whether it was really just in my mind, or if I could really hack a combat flying billet, and what it was really like. So now I know. To tell the truth it's about what I expected—harder'n hell.

Scott and I had a discussion once. We're pretty honest with each other, and the subject of his having to prove himself to himself all the time and my having to do the same thing came up. I remember his laughing once and saying something like, "Well, it's true, I don't believe in myself enough to just say, 'Now Scott, you can do as much or more than anybody else and you don't have to prove a thing,' I've got to show myself." And then he said, "But dammit I do show myself, and do prove things to me and as long as I can keep doing things other folks can't do and things I don't really believe I can do, where's the harm except in the fact that someday I'll fail to prove something—and I've done that before too—where's the hurt if I manage to successfully prove myself?" And I guess that's me over here.

I hate night hops, but almost only when they're the first hop of the day. And every time I walk up on that deck knowing what's coming up, it's like facing death. Hell, more than that, it is facing death; but I think I face it sometimes more heavy-hearted than other folks. I think I'm sometimes more cowardly about it than others, more hesitant. But dammit, I do it. So where's the hurt as long as I manage to get it done?

Yankee Station, 12 October 1966

My Darling Marilyn,

I don't really have time to write a decent letter, but I just received your letters, and I wanted to write a thank you note. Or better, an I love you and am thankful for your note. I do and I am.

I'm sitting here in the room eating geedunk because I didn't feel like getting dressed and then back into flight gear. That's because Robert McNamara, General Wheeler, and a bunch of other admirals and Army generals and captains and colonels are aboard for the night.

The captain came over the intercom today and said, "Now we're just going to show them a normal day's operation," and now the aviator's chow line is closed.

I missed the big action this morning. Down to the south, Darell was hitting a truck park and was using his bomb computer. The computer fouled up and threw a bomb out into the boonies, and they got all sorts of gigantic secondary explosions. So everyone kept bombing that area and kept getting fires and explosions. It was probably a stockpile area for the DMZ forces. The next three launches were diverted into that area and got all kinds of results. Really a big day. I missed it all. My hop was traded with Joe so that he could escort the VIP's tonight, and I got a tanker hop which aborted. I didn't care; I'd rather give up the hop than do that escort duty.

Another bit of excitement occurred today when a number of motorized craft resembling PT boats were approaching the ship's steaming area from the north. We all figured that maybe we'd get some antisurface action, but they turned out to be Chinese motorized fishing trawlers. The *Constellation* found them first and was all set to sink them when pictures revealed that the single gun mount on the bow was really a crane.

Another AD pilot down back in the woods. One a day.

I read and read again and again the things you said about the autumn, and it sounded so good I couldn't stop thinking about it. You can't know how much I wish I were there. Won't be long. Even so it seems so long. Tick-tick. It seems like a million years since I saw you last.

I'm looking forward to our cross-country trip to California. Maybe we can borrow sleeping bags and camp out a couple of nights. I love you so much I don't think we'd need but one sleeping bag or even use the other one if we had two.

Well, brief time for my flight. Taking Chip out again tonight. All my love, darling, and thank you again for your lovely thoughts in those two letters today.

I love you. Frank

WESTERN UNION TELEGRAM

To: Mrs. Frank Callihan Elkins, Route One, Dunlap, Tennessee

I deeply regret to confirm on behalf of the United States Navy that your husband, Lt. Frank Callihan Elkins, 658100/1310, USN is missing in action.

This occurred on 13 October 1966 while on a combat mission over North Vietnam. It is believed your husband was maneuvering his aircraft to avoid hostile fire when radio contact was lost. An explosion was observed but it could not be determined whether this was hostile fire exploding or your husband's aircraft. No parachute or visual signals were observed and no emergency radio signals were received. You may be assured that every effort is being made with personnel and facilities available to locate your husband. Your great anxiety in this situation is understood and when further information is available concerning the results of the search now in progress you will be promptly notified. I join you in fervent hope for his eventual recovery alive. I wish to assure you of every possible assistance together with the heartfelt sympathy of myself and your husband's shipmates at this time of heartache and uncertainty.

The area in which your husband became missing presents the possibility that he could be held by hostile force against his will. Accordingly, for his safety in this event, it is suggested that in replying to inquiries from sources outside your immediate family you reveal only his name, rank, file number, and date of birth.

Vice Admiral B. J. Semmes, Jr.
Chief of Naval Personnel

Fire on the Oriskany

There is no more dangerous hazard than a fire aboard ship. Unchecked, fires spread quickly through confined internal areas, roasting everything and everyone. A fire aboard an aircraft carrier is especially dangerous because of the presence of highly flammable jet fuel and ordnance. During the Vietnam War, there were three major fires on carriers. The first, on October 26, 1966, occurred on the USS *Oriskany* (CVA-34), in the middle of her second cruise to the war zone. The second blaze occurred aboard USS *Forrestal* (CVA-59) on July 29, 1967. The third erupted on board the nuclear-powered carrier *Enterprise* (CVAN-65) on January 14, 1969, while she was undergoing a predeployment exercise. Each of these fires resulted in great destruction and loss of life. New fire-prevention and firefighting procedures were instituted, as well as the provision of new personal equipment for individuals to use to escape a fire.

Oriskany's fire started, in part, because of the mishandling of a flare by two young sailors. The fire quickly spread throughout the ship and ended with many of the ship's company and air wing members coughing and blinking in the sunlight of the flight deck, having just escaped from the conflagration below that claimed many of their friends' lives.

Lieutenant Andre Coltrin was an RF-8G pilot with *Oriskany*'s VFP-63 det. Throughout the war, three-plane, fifty-man detachments of this photo-reconnaissance version of Vought's famed Crusader flew dangerous, unarmed missions over heavily defended areas of North Vietnam. As courageous and experienced as these aviators were, however, facing a raging fire in their small staterooms was for some the ultimate challenge.

Andre Coltrin flew photo-reconnaissance missions in three tours during the Vietnam War, in 1966, 1967, and 1972. He retired from the Navy as a TAR commander and went into private industry in California.

From *RF-8 Crusader Units Over Cuba and Vietnam*, by Peter Mersky

Besides the dangers of flying combat over heavily defended areas of North Vietnam, aviators also faced danger even when supposedly safe aboard their carriers. Fire is a constant threat for any ship's crew, and aircraft carriers have had their share of conflagrations. One of the most devastating was the fire that broke out in the early morning of October 26, 1966, on board *Oriskany*. Forty-four officers and men died. The veteran ship was on the second of seven combat cruises during the Vietnam War.

Only nine months later, in July 1967, USS *Forrestal* (CVA-59) had a fire that killed 134. In January 1969, a fire on board USS *Enterprise* (CVAN-65) took 27 lives.

The *Oriskany* and *Forrestal* tragedies resulted in a major overhaul of firefighting and survival training, including the use and maintenance of oxygen breathing apparatuses (OBAs).

Lieutenant Andre Coltrin had faced enemy flak and SAMs, but he found he now had to deal with what was potentially the most fatal enemy of all.

> It was early morning. The predawn launch had been scrubbed, and the ordies were downloading flares from the A-4s and preparing the aircraft for

daylight operations. As two red shirts [ordnancemen] were taking the flares to the lockers, they decided to play catch. The fuse of one flare got pulled, and the flare ignited. Although the edge of the ship was not far away, in his panic, one of the airmen threw the burning flare into the flare locker instead of overboard. Before the sailor could shut the locker door and dog it down, the explosions started. [It was never determined that the ignition of the flare started with a haphazard game. Rather, a mishandling of the Mk-24 resulted in the flare's lanyard catching on something followed by ignition. (Ed.)]

The first blast ripped the door open with such force that the dogs were torn out of the edge of the door like large bites. The door threw the sailor who was setting the dogs well clear before the first fireball emerged. The door then crushed a metal ladder behind it. Every few seconds, fireballs—with temperatures in the thousands of degrees—shot out the locker doorway like huge Roman candles and raced through the passageways of the ship. Oxygen and burnable material fed the fireballs as they went.

I was asleep in my stateroom, one deck below the flight deck, when the fire alarm went off. When the alarm was called, my roommate and I thought it was another drill. We had been having a lot of those during the cruise, and I rolled over to go back to sleep. As a member of the air wing and not ship's company, I would let them play fire drill on their own. Shortly thereafter, my roommate got up saying, "I heard something that sounded like an explosion." He left the room.

Still not too excited, I continued to try to sleep. Later, I heard the door open as someone came in. I heard a strange voice ask, "Is anyone here?" I asked who he was and he replied, "Seaman so-n-so." (I can't remember the name.)

I said, "Seaman, this is Lieutenant Coltrin. What are you doing in my stateroom?"

He replied, "Sir, the ship is on fire and we are trapped!" I decided to get up! I went to the door and opened it. I could see nothing but thick smoke. I put my arms and hands out into the passageway, and they disappeared in this white haze. I could feel many people lined up trying to get out of the area. It was too crowded to try to squeeze in. I went back into the room and shut the door.

At the start of the cruise, my roommates and I had installed an air conditioner in our room. We vented it into the head next door. It was very inefficient so we had sealed every air leak we could find in the room to keep the cool air in. This now paid off by keeping the smoke out.

I put on my Nomex flight pants and shirt. [At this time, several squadrons had non-regulation gear obtained through swaps with Army units. *Oriskany*'s squadrons sported camouflaged flight suits, as well as unofficial Nomex outfits and flight gear. (Ed.)]

Lt. Andre Coltrin in a VFP-63 RF-8G Crusader. (Courtesy of Andre Coltrin)

The sailor went to the phone to call for help, and I went back to the door and checked the passageway. He said no help was available, and I found the passageway still jammed with people.

The floor was starting to get hot and the tile was burning my feet, so I put on my flight boots. The smoke was now thick and heavy, and the room was very hot. We were being driven to the floor to breathe. I went to the sink and soaked two towels. I gave one to the sailor and I wrapped the other, a black one, around my head and said, "Let's go."

As I opened the door, my companion darted to the left. I yelled to him that was a dead end, but he disappeared into the smoke. I went to the right. The passageway was now clear of people. The passageway led fore and aft. In about 15 feet it fed into a cross-ship passageway. I knew I was there when I ran into the knee-knocker. By this time, I knew the main fire was on the starboard side of the ship, so I turned right and headed to the port side. I did not turn fast enough and ran into a locker and smashed my hand. I later found out this locker contained OBAs, but it wouldn't have mattered because I didn't know how to use one. I wanted to continue down this passage until I reached the ladder that I knew went down to the hangar bay.

At about the point that I should have been coming to the ladder, I became confused because the port-catapult room was now on fire. The flames from the catapult room mixed with the smoke and disoriented me. This problem turned out to be one of my luckier moments. A friend of mine, the air-wing flight surgeon, found the ladder I was searching for. As he reached the first landing, the helicopter that was parked just below caught fire and roasted him on the spot.

Unable to find the ladder, I continued to the end of the passageway knowing that there was a hatch there that led to a catwalk outside. I reached the hatch and found it dogged down. Scared, disoriented, and finding it very hard to breathe I tried to undog the hatch. I thought I had undone all the dogs but the hatch would not open. I heard screams—I do not know whose, maybe mine. Frustrated and thinking I was going to die, I angrily kicked the hatch. To my great surprise, it opened. I stumbled out onto the catwalk and laid there a few seconds, breathing the air.

I stood up and looked across the flight deck. There were only a few inches of clear air between the flight deck and the smoke layer. The skipper had turned the ship, putting the wind from the starboard side, blowing the smoke to port to increase the visibility on the starboard side to fight the larger fire more effectively.

As I looked between the flight deck and the smoke, I saw the feet of two main groups of people. One group was near the island and the other near the fantail. I started crawling toward the group near the island. Suddenly to my right, I saw the CO of VF-111 crawling toward me.

Oriskany crewmen assemble on the stricken carrier's flight deck. In the group in the lower portion of this photo, facing the camera with a crew cut is Cdr. Dick Bellinger, CO of VF-162, and the first Navy pilot to shoot down a MiG-21, barely a month earlier. Losses during the fire included the air wing commander, which required Bellinger to step in as CAG for a limited time. (Courtesy of the U.S. Navy, PHAN W.E. Warren)

"Andre," he gasped, "do you have a cigarette? I'm dying for a smoke." I couldn't believe what I heard.

I said, "Skipper, if you just raise your head three inches, you can get all the smoke you want."

He said, "No, no, I need a cigarette," and crawled off. I continued crawling toward the island, where I checked in with the mustering officer and tried to see about the status of our detachment. Most everyone from the det had been accounted for. My roommate, who always wore a red-and-white striped nightgown to bed, looked very much out of place, but no one else seemed to notice.

He had reached the cross-ship passageway but did not turn right or left. He continued straight ahead, and ended up in the admiral's quarters. He, the admiral and a member of the admiral's staff pulled off the air-vent cover. They gathered around this 4-to-6-inch opening, breathing what air they could. Since the admiral was the admiral, a rescue team soon arrived and led them all to safety.

As I stood around, I noticed a lot of activity just forward of the island. Crews were pushing bombs and other explosives overboard. Men with OBAs were entering smoke-filled compartments to retrieve those explosives. They had wire-lead cables attached to their backs in case they became incapacitated. People outside could then drag them to safety.

I was very impressed by the bravery of these men. One of them was a young officer from VF-162, Lieutenant, j.g. Jim "Flaps" (he had big ears) Andrews. He was willing to go anywhere and do anything. More than once while I was standing there, one of these brave men had to be dragged out of the compartment by his safety wire because he had been overcome by smoke. It was here that I learned how to put on, test, and use an OBA. Unfortunately, a very large percentage of them were not useable and were being thrown overboard.

After my lesson, I joined a Marine and two white hats [enlisted sailors] on a search party. There were still a lot people missing. I grabbed an untested OBA and followed them down a couple of decks. When I put on the OBA, I found it almost useless. Although it did produce oxygen, the eyepiece was so scratched I could barely see through it.

We entered the dark, smoke-filled passageways carrying a metal stretcher. It was very hot and muggy, and we were in water above our knees. It was worse than being in an oven.

Every now and then, we would pass live electrical wires sparking off the water and bulkhead. As long as the wires were on our side and we did not go between two of them, we were supposedly safe. In water up to my knees, carrying a metal stretcher, with live electrical wires sparking around, and unable to see—I was not too confident.

We found the stateroom of one of the missing officers. At first we thought no one was there but looking more closely, we found a lump of something we realized had been a man. The heat had been so intense he had melted into an unidentifiable heap. His dentures could not even identify him. (The process of elimination later identified him.)

I helped carry a man out whom I did not think I knew. I thought he was black. He turned out to be a good friend that I had flown with many times. The fire had charred his skin and singed his hair. Only when we rolled him over and I saw his unburned side did I know who it was. This same pilot had survived two ejections earlier in the cruise.

In one of the fighter ready rooms, the SDO received a call from some of his shipmates who were trapped in their stateroom on the third deck. He told them that all the rescue teams were very busy and that they would have to find their own way out. Shortly thereafter, there was a loud, horrible noise that sounded like uncontrollable screaming over the phone, then it suddenly went dead. We didn't know if it was the sound of the men being burned to death or a strange sound made by the phone lines as the wires melted. It made the hair on the back of every man's neck in that ready room stand on end. One man out of eight did escape from that room.

As a couple of the men succumbed, an LDO [limited duty officer, a former enlisted man who had been commissioned in his area of specialty] from VF-111 decided he was not going die this way and busted out of the room. He raced to the ladder that took him to the second deck. As he was running blindly across the deck he passed out from the smoke. Falling to the floor, he regained consciousness, breathing the smoke-free air close to the deck.

Getting to his feet, he again started running across the deck to the port side of the ship. Passing out again, he fell down a ladder to the third deck. At the bottom of the ladder, he accidentally slammed into the stateroom door of one of his shipmates, who had been able to sleep through all the previous commotion. The loud bang against his door woke him up.

As he opened the door, he saw his friend and realized the dangerous situation. Throwing his unconscious shipmate over his shoulder he carried him to the safety of the flight deck. They had saved each other's lives.

Over on the port side of the ship, the helicopter crew was directing a fire crew on the flight deck on where to lower a fire hose to an open porthole a few decks down. In this stateroom was a pilot from one of the VA squadrons. As he had tried to exit his room he was met with huge balls of flames. He tried to escape a couple of times to no avail. He stuck his head out the porthole to breathe and to put the fire hose that had been lowered to him over his shoulder, spraying down his back, while his stateroom burned around him. The only injury he received was a bruised knee caused by him banging it against the bulkhead trying to escape the heat.

A man was trapped in one compartment when the water used to fight the fire flooded it. The lights were out, and it was pitch black. As the water rose, the man shimmied up a pipe in the corner of the room to escape the rising water and to breathe. He stayed in this dark, torturous situation for hours before he was rescued. He suffered psychological trauma, but later recovered.

Some time during my roaming around the ship, to my surprise I came across the sailor who had come to my stateroom. I had not expected to see him alive again. I asked him what happened after he left my stateroom and how he got out. He had tripped over the body of an officer who lived across the hall, which is why he disappeared so fast. He checked the downed officer to confirm he was dead and then proceeded to the end of the passageway to a compartment where some of his friends worked. As it turned out they had already evacuated the area and he did likewise.

During the fire, we received a lot of help and supplies from other carriers and ships that were on the line. And after what seemed like a very long time, the fire was under control and eventually put out completely. We headed for the Philippines. En route, we had a burial-at-sea ceremony. That was tough. It still raises goose bumps. The trip to the Philippines was very somber. We couldn't escape the smell of death and burned flesh.

Lieutenant Coltrin made another cruise in *Oriskany* in 1967. After serving in several other squadron assignments, he got the chance to join VF-194, flying F-8Js, in 1970. Coltrin made yet another deployment to the war zone. For personal reasons, he eventually resigned his regular Navy commission for one in the reserve, and soon became the OINC of a VFP-63 photo detachment aboard USS *Franklin D. Roosevelt* (CVA-42). This time, the cruise went to the relatively peaceful waters of the Mediterranean. But the *FDR*, normally assigned to the Atlantic Fleet, had made a single combat cruise early in the war, with a det from AIRLANT light-photo squadron, VFP-62's, as part of Air Wing 1.

America *and* Liberty: *Passing Ships*

By virtue of its larger fleet, the U.S. Navy quickly assumed the role of the western world's patrolman, sending roving deployments of task forces into the world's oceans as situations demanded. One of the regular "beats" was the Mediterranean, particularly the eastern Med that bounded the volatile Middle East coasts of Israel, Egypt, and Lebanon. Periodic flare-ups became a way of life in the region, but in June 1967, full-scale war broke out between Israel and Egypt, Iraq, and Syria. Lebanon and Jordan, aided by other Arab and Muslim states, also joined in.

Propelled by a massive first strike on the early morning of June 5, Israel soon became complete master of the conflict, quickly finishing off Egypt by June 8, and turning its attention to the northeast to deal with Syria. In this highly sensitive period of the short war, an American surveillance ship, USS *Liberty* (AGTR-5) appeared off the north coast of the Sinai Peninsula between Egypt and Israel. It was not a good place to be.

Fearing a Soviet or Arab presence, the Israelis sent Mirage fighters to scout the contact. However, their reconnaissance was inconclusive, and soon, a large Israeli force of aircraft and torpedo boats attacked the hapless American ship, killing 34 and wounding 171 of her crew. The captain, William L. McGonagle, though badly wounded on the bridge, remained at his station and ultimately received the only Medal of Honor to be presented for action not involving the Vietnam War during this timeframe.

To this day, 35 years later, there remains considerable, often heated, discussion as to whether the Israelis truly knew they were attacking an American ship. (Brassey's recently published an in-depth look at the incident, written by retired Captain Jay Cristol, that sheds considerable light on the entire matter.)

Ships in the general area tried to respond to the stricken *Liberty*'s distress calls. Among these was the carrier *America* (CVA-66), on only her second deployment since her commissioning in January 1965. Her captain, Donald D. Engen, was a highly experienced naval aviator with considerable combat time in World War II and Korea. He had been an SB2C Helldiver pilot with VB-19 aboard USS *Lexington* (CV-16) in October 1944. He had received the Navy Cross for his role in sinking the Japanese carrier *Zuikaku*, the last survivor of the six carriers that had attacked Pearl Harbor three years before. Six years later, he flew some of the first carrier strikes of the Korean War as a Panther pilot with VF-51 in *Valley Forge*. After that, he became one of the Navy's premier test pilots and flew many of the new jet fighters then approaching fleet introduction.

On June 8, 1967, however, he was the CO of an American carrier desperately trying to help a sister ship under attack. He had the air wing and was ready to send it against *Liberty*'s attackers.

Donald Engen was one of this country's most experienced naval aviators. He participated in many of the Navy's most important fleet actions and also helped develop many of the carrier systems and aircraft that made the fleet such a vital instrument of national policy and pride. Retiring as a vice admiral in 1978, he took up a second career with the National Transportation Safety Board, and then as administrator of the Federal Aviation Administration.

In 1995, he became the director of the Smithsonian's National Air and Space Museum. It was an arduous assignment, but as he led the museum from a period of extreme controversy regarding a sensitive exhibit of the Enola Gay, *the B-29 that dropped the first atomic bomb on Japan, he also prepared the institution for the new century and especially the upcoming centennial of the Wright Brothers' 1903 flight. These preparations included the complicated plans for the new Steven F. Udvar-Hazy Center at Dulles International Airport. A visitor's tower at the center that will give a commanding view of operations at the airport will be named the Donald D. Engen Memorial Tower.*

Engen still found time to fly gliders over the Sierra Nevadas that border California and Nevada, and it was over these mountains, at age 75, that he met an airman's death in July 1999.

From *Wings and Warriors: My Life As a Naval Aviator*, by Donald D. Engen

At about 1400 on June 7, Bob Goralski was visiting on the navigation bridge when CIC called me on the squawk box to say that a U.S. ship, the USS *Liberty*, had been attacked by unknown naval motor torpedo boats and aircraft. Up to that time I had no knowledge of any U.S. ship named *Liberty*, let alone that it was in the Mediterranean. I later learned that our covert intelligence gathering ships were in a different "navy," and subsequently, because of the *Pueblo* and *Liberty* incidents, the Navy would learn not to have two distinct lines of authority and responsibility for signals intelligence operations and line command. Up to that time, each was not aware of what the other was doing. Bob Goralski's eyes became as big as saucers, and I asked him to leave the bridge. At that time *America* was conducting a CTG 60.1–directed no-launch weapons readiness drill, which involved the movement of many different weapons about the flight and hangar decks.

Thus the notification about *Liberty* came at the exact wrong time for *America*, thus adding a certain amount of additional complexity. I found Rear Admiral Geis in his war room and informed him that unless otherwise directed I was stopping the training exercise to strike below those weapons and to arm the air group airplanes with more flexible and appropriate loads of ammunition. He was so busy that he hardly answered me, but his staff agreed. Along the way, Rear Admiral Geis, as CTF 60, directed *America* and *Saratoga* to launch a specific number of airplanes against whoever was attacking *Liberty*. In *America* there ensued about an hour's effort to move some bombs below and bring up others and to launch four A-4Cs fully loaded along with F-4B escorts. That group was judged to be the correct size for a response—not too large and warlike, but still large enough to protect *Liberty*. As our airplanes departed the task group, I heard the flight leader announce on the departure frequency, "We are on the way. Who is the enemy?" Our knowledge was such that no one knew yet who had attacked *Liberty*.

Within minutes the national command authority—President Johnson—was on the radio telephone from the White House direct to Rear Admiral Geis directing that the aircraft be recalled. The president told Rear Admiral Geis that the Israelis had said they had believed *Liberty* to be an Arab intelligence ship. That recall of our airplanes presented an additional but far less serious problem. The A-4Cs were armed with Bull Pup air-to-surface missiles with highly volatile hypergolic rocket fuel that could not safely be brought back on board ship. Rather than jettisoning perfectly good and perhaps soon-to-be-needed Bull Pup missiles, the A-4Cs were directed to land at the NATO airfield at Souda Bay, Crete. The airplanes were recovered later; the missiles, much later.

Captain Joe Tully, commanding officer of *Saratoga* in TG 60.2, which was positioned some 100 miles east of us, was able to launch his airplanes before we in *America* could. But the airplanes from both ships were recalled before ever reaching *Liberty*. Vice Admiral Bill Martin issued a public denial to those nations that claimed U.S. Navy airplanes were supporting the Israeli forces. In a way, the Soviet navy ships, through their presence, could confirm that to their own government and as a result may have helped reduce U.S-Soviet tensions.

During the evening of June 7 *Liberty* was directed by Vice Admiral Martin to steam as best as the ship could north out of reach of any further attack, and he sent USS *Davis* and USS *Massey* to escort the intelligence ship. Commander John Gordon, senior medical officer in *America*, remained on board to prepare our eighty-bed hospital to receive the wounded. Lieutenant Commander Peter A. Flynn, second most senior medical officer in *America* and an accomplished surgeon, and two medical corpsmen were sent in *Davis* to help *Liberty*'s heavily taxed ship's doctor and two corpsmen.

The two destroyers rendezvoused with *Liberty* at 0600 on June 9, transferred the medical personnel, and then assumed the difficult task of following in *Liberty*'s wake to pick up any bodies that floated from the ship and to destroy any highly classified papers that the commanding officer of *Liberty* was concerned might be coming from the underwater holes in the ship's sides. The Soviets were definitely interested.

At 1030 on June 9 two *America* helicopters rendezvoused over *Liberty* and

Formal photo of Capt. Donald D. Engen. He wears six rows of decorations and campaign ribbons typical of his generation of naval aviator, who saw action in World War II and Korea. Foremost is the Navy Cross, first on the left, top row, the highest Navy combat award, and second only to the Medal of Honor. (Courtesy of the U.S. Navy)

began transferring the more seriously wounded to *America*, where Commander John Gordon and his doctors and corpsmen were well prepared for their arrival. After the helicopters touched down on the flight deck, the patients were carried in Stokes litters directly to nearby bomb elevators already prepared to take them down five decks to the hospital. At 1130 *America* rendezvoused with *Liberty*, and we put our boats in the water to facilitate damage assessment and the transfer of the dead. I took *America* slowly down the port side of *Liberty* from bow to stern and about 200 yards away. Two thousand *America* crewmen lined *America*'s port side flight deck, and there was dead silence as reporters and crew alike visually assessed the major damage that *Liberty* had sustained. As we slowly passed abeam, I called Nick Castruccio, coordinating helicopter traffic in Primary Fly, and said, "Let's give them three cheers!" Nick then led the two thousand officers and sailors on the flight deck in a cheer by using the flight deck speaker system. "Let's hear it for *Liberty!* "Hip, hip," keying them to a thunderous "hooray." Three times that rousing cheer of two thousand voices rose from the flight deck of USS *America* and roared across the 200 yards of water between the two ships, saluting those in *Liberty*. I will never forget that sound. It made the hair stand up on the back of my neck and made me proud to be in the Navy. Thirty-four *Liberty* crew members lost their lives, and seventy-five were wounded, fifteen seriously, when *Liberty* was attacked.

The transfer of the dead and wounded to *America* was completed. Rear Admiral Isaac Campbell Kidd Jr., then assigned in Naples, was appointed by the commander in chief of U.S. Naval Forces Europe to be the investigative officer of the entire incident in accordance with Navy regulations, and he boarded *Liberty* with a small number of officers. On the afternoon of June 9, *Liberty* was under way for Valletta, Malta, with *Davis* and *Massey* continuing their silent and watchful trail astern. The Soviet ships stayed respectfully clear. Valletta was chosen as the port to assess damage and to administer those repairs needed to return to the United States, principally because it was remote and the British still had not left the major shipyard there. On Saturday, June 10, Commander Ralph Hopkins and I officiated at a well-attended memorial service held on the flight deck of *America* for those *Liberty* crewmen who were killed. Also in attendance were forty of their less seriously wounded shipmates.

Over several days C-1A airplanes transferred the wounded to shore airfields where they were later flown to Air Force and Army hospitals in Europe and then returned to the United States. [Subsequently, several authors have written about the attack on *Liberty*. One or two have been highly emotional and have written less than factual books that failed to understand the why or to explore factually the actions of the Sixth Fleet and its aircraft carriers. For those interested in the full story, the most scholarly and factual treatise is the doctoral thesis of Rear Admiral Jay Crystol USNR (ret), currently the chief judge in the federal bankruptcy court in Miami.]

It is amazing how fast a fleet can return to normal operations from a crisis position. We had almost gone to war with the Soviet Union, might have attacked Israel, and had risked war with one or more of the Arab nations. The Mediterranean, locus of conflict for thousands of years, had spawned yet another one. As a result of the Arab-Israeli War, the Suez Canal would remain closed until 1975, and severely bruised international political egos would take a decade to heal to the point of returning to open discourse. On June 19 *America* was headed toward a port visit in Istanbul after thirty days of operations on station south of Crete. Two days earlier I had asked a C-1A pilot to see if Mary was in Athens. I did not know where she might be but surmised that she might have moved closer to the central Mediterranean. In a stroke of pure luck, on the day the pilot flew in, Mary drove up to a hotel in Glyfada, a seacoast suburb of Athens, just as she was paged by the pilot at that hotel. He gave her my message, which simply said, "Go to Istanbul." She, Charles, and Candace—who was following her husband—did just that. They drove through Greece, Yugoslavia, and Turkey and arrived in Istanbul the day that *America* arrived, June 21.

After transiting the Dardanelles and the Sea of Marmara, *America* anchored in the Bosporus at a place that almost seemed to be downtown Istanbul. The Bosporus is the only natural exit or entrance to the Black Sea, and the current from the Black Sea toward the Mediterranean was a respectable and continuous seven knots. I anchored *America* about 100 yards from the European side of the Bosporus and let out 135 fathoms of chain so that the ship rode nicely to the anchor, headed into the current from the Black Sea. *America* returned easily to our port visit routine of a first-evening silent drill and the following social and diplomatic whirl that had accompanied all of our previous port visits.

A number of ship and staff officers and families stayed at the Istanbul Hilton, from which we could see the picturesque minarets of Istanbul's skyline as well as *America* riding well at anchor. With the dress lights, peaked at the top of the mast and strung the 1,000 feet from bow to stern, *America* was the picture of U.S. national power, certainly an impressive peacekeeper.

CHAPTER 6

A New Age, 1976–Present

War in an Unknown Place

Until late April 1982, very few people outside the United Kingdom or Argentina could have answered the question, "Where are the Falkland Islands?", much less the Malvinas, Argentina's name for these obscure outcroppings. The question would have fallen into the same category as contemporary inquiries about Guadalcanal, South Vietnam, or Bosnia-Herzegovina. The answer, depending on who was replying, was down, way down, in the South Atlantic, only a few hundred miles off Argentina, or "down in the bloody coldest, wettest area in the whole bloomin' planet!" Politicians seldom choose paradise in which to fight a war.

The islands—actually two main islands and 200 smaller islands, in an area barely the size of the state of Connecticut—had been home to a hardy band of British citizens for more than a hundred years. But Argentina had contested that claim for nearly as long. Perhaps wanting to focus their countrymen's attention away from the nation's economic woes, government leaders sent a task force including 800 Argentine marines on April 1 to occupy the lightly defended islands. For England, it was not just a matter of invasion, but the fact that British citizens were under an occupying force. Liberation was not only a matter of national pride, but of necessity. By early May, Operation Corporate was underway, and a task force of ships, aircraft, and men was on its long way from England to the forbidding gray ocean of the South Atlantic.

Unique among the aircraft heading that way was the Hawker Siddley Harrier, V/STOL jet fighter. Only one of two such aircraft to attain squadron service—the other being the short-lived Soviet YAK-38, NATO codename "Forger"—the Harrier had been operational for fifteen years. It had not only been flying in squadrons of the Royal Air Force and Royal Navy but had also served with U.S. Marine Corps units since 1971. But for all that time, no Harrier squadron had seen combat.

At one point until the mid-1970s, Britain maintained a small but active traditional aircraft carrier fleet. However, with the stress of economics and the development of the Harrier, it relinquished its remaining carriers for smaller ships that carried combined groups of Harriers and Sea King helicopters.

Two Royal Navy Sea Harrier squadrons participated in the Falklands War, along with one Royal Air Force squadron—Nos. 800 and 801 Naval Air Squadrons, flying Sea Harrier FRS.1s, and the RAF's Nos. 1(F), equipped with Harrier GR.3s. No. 899 and newly formed No. 809 Squadrons contributed pilots, maintainers, and aircraft for the two fleet squadrons. While the GR.3s were dedicated attack versions, the FRS.1 fulfilled a variety of tasks, including fleet defense, ground attack, and air superiority. It was in this latter role that the Sea Harriers—usually called "SHARs," by their crews—would excel, rolling up a 28–0 kill/loss ratio against their Argentine opponents. The 24 SHARs flew from two carriers, HMS *Hermes* and HMS *Invincible*. Another four accompanied the GR.3s aboard the *Atlantic Conveyor*, a container ship in the task force.

Sailing from Portsmouth on April 5, 1982, the task force reached the war zone in time to launch the first attacks on May 1. The Royal Navy Harriers immediately found themselves confronting the Argentine A-4s and Mirage/ Daggers that stalked the Corporate ships, occasionally with some success. ("Dagger" was the Argentine name for the Israeli-built version of the Mirage 5 called the Nesher [Eagle] in Israeli service. Several were sold to Argentina after the Nesher was retired in the late 1970s.) The first scores came on May 1 when a SHAR from 801 Squadron flown by Lieutenant Paul Barton shot down a Mirage III, followed by that Mirage's wingman, destroyed by Lieutenant Steve Thomas. A third kill for the day was registered by RAF exchange pilot Flight Lieutenant Bertie Penfold of No. 800 Squadron over a Dagger.

Lieutenant Commander Nigel David Ward, better known as "Sharkey," was the commanding officer of 801 Squadron aboard *Invincible*. He was the senior SHAR pilot and was an extremely colorful individual and leader. He wrote a paean to his squadron and their mount that sheds a lot of light on how the war was fought from the squadron level. In these excerpts he describes his three kills. The Falklands War is so far the only instance of air-to-air engagements by the Harrier in any service.

A Canadian by birth, Sharkey Ward received his wings in 1969, and flew F-4K Phantoms from the Ark Royal. *By the late 1970s, however, he had transitioned to the Harrier, earning the reputation as the most capable and knowledgeable Sea Harrier pilot. He took command of 801 Squadron in 1981. During the Falklands War, he became the high-scoring pilot with three kills in more than 60 sorties. Promoted to commander and after several late assignments, he retired in 1985 to work in private*

LCdr. Ward at the end of the Falklands deployment. Contrary to U.S. regulations, British aviators are allowed greater latitude regarding facial hair. Note the cloth cover on his helmet visor. (Courtesy of the Royal Navy)

industry. He received the Air Force Cross and Distinguished Service Cross for his wartime service.

From *Sea Harrier Over the Falklands: A Maverick at War*, by Commander Sharkey Ward, RN

The first Harrier loss came near Goose Green on the 27th. Squadron Leader Tubby Iveson had been tasked with providing close support for the Paras. He and his Number Two were to attack Argentine ground forces and Tubby entered into the mission with enthusiasm. The chances of being hit by ground fire were higher on each run, and on the third he was hit and had to leave his aircraft. There was no gain involved, only loss, and now the GR 3 numbers had dwindled to four.

He remained on the ground and on the loose for three days until he was eventually picked up by an Amphibious Force helicopter on the 30th. Although he had not exactly covered himself with glory, he did merit every effort to rescue him. Over-enthusiasm on an operational mission is certainly no crime (although throwing the aircraft away for no return is verging on gross negligence). When Tubby managed to make contact with Harriers and Sea Harriers from the Task Group using his SARBE radio facility, he felt that rescue must be on the way. And so it should have been. But the Flag suspected that his radio calls for help were a 'set-up' by the Argentine forces, that he must have been captured and was being used to drag an unsuspecting rescue helicopter to its destruction.

This conclusion was never accepted by those who had spoken to Tubby from their cockpits in the air. Eventually, it was the Amphibious Force that sent a chopper to retrieve the downed pilot. Without their help he would still probably be yomping around the Falklands, making like a penguin.

Although the 28th saw no losses or claims in the fighter war, I was able to get airborne with the AWI, Morts, for our first CAP mission together. Steve Thomas was still enduring his mandatory rest cure, and it turned out to be fortunate for me that Morts was with me. We each dropped a 1000-pound bomb on Stanley airfield en route to our allotted station. The CAP brought us no trade and so when our fuel state dictated a return to Mother we commenced the climb-out from over the centre of Falkland Sound. As we were going through about 12,000 feet Morts transmitted 'Three bogeys, swept-wing, left 9 o'clock low!'

Morts was to the right and was looking left through his leader. 'Very low, passing underneath us now.'

I inverted my jet and pulled. I levelled the wings in a steep descent and spotted the three bogeys. 'Got them, Morts. Acquiring with Sidewinder.' But for some reason the missiles wouldn't lock.

'Wait, Boss! I think they're GR 3s!'

I had convinced myself that the three were enemy aircraft. But I also knew that Morts, more than anybody, should be able to recognise a GR 3 even from this height and range. I called the control ship, HMS *Minerva*.

'Do you have any friendlies in the area at low level?' If there were any, *Minerva* would know about it.

'Negative. No friendlies in the Sound.'

Just at that moment of distraction, I lost sight of the three swept-wing shapes below. They just disappeared into the multi-coloured background of the water.

'I've lost the fucking things, Morts. Do you hold them?'

'Negative. But I'm sure they were GR 3s.'

I was mad as a hatter, and wasn't thinking straight. I was tired, and 'missing' the enemy jets seemed to drain me of all energy. If I hadn't been so tired I might have considered the line 'better safe than sorry', but I was in no mood for that when I landed on board. The debrief was short and to the point: 'GR 3s, my arse!'

The AWI could see that his Boss was in no state for a constructive discussion and sensibly buttoned his lip. But he was naturally upset at being doubted by the CO, and set to work with the Ops Team to establish exactly what the three jets were. Logically they were not the enemy or they would have attacked something in San Carlos, and they had flown straight past the beach-head. In the early evening he approached me.

'Sir, may I have a word, please?'

I was no longer angry, but was still fretting over the missed 'chance'. 'Yes, Morts, of course. What is it?'

'I've confirmed with *Hermes* that three Harriers did transit Falkland Sound, north to south, at low level at the time of our sighting. They were definitely No 1 Squadron cabs!'

I was shattered. 'Well bugger my old boots! I owe you an apology.' And after a pause, 'I and the Harrier boys owe you our thanks too, Morts. Well done. And please accept my apologies.'

But for Morts's eyesight and aircraft recognition, the Task Group could have been down to one Harrier. No wonder I couldn't get a tone on the Sidewinder. From above, the unannounced Harriers' wings were hiding the hottest part of the jets, the efflux from the nozzles.

'Come on down to the bar. I definitely owe you a beer, and I need a drink too.'

That night I began to realise how tired I was and how tired all those around me were. Wings was almost out on his feet too, but refused to rest. The only man who seemed to be weathering the storm well was JJ. And JJ always slept from midnight to 0600 hours. Only in the event of action stations was he to be disturbed. During the silent hours, his second-in-command, Tony Provest, took the helm. It was good thinking by the Captain, and one reason why he was always able to retain his sense of humour and sense of proportion.

The next day, 29 May, was full of incident, and a costly one for 801 Squadron. The weather over the Task Group was stormy, with 40-knot winds and a heavy sea. Day by day the Flag was moving the Group further east and the transits to CAP stations were becoming longer; approximately 200 miles by this stage.

After the previous day's near-miss and the usual night alert stint, I was no better for wear. I was to fly three CAP missions during the day, and on the first I made a somewhat basic though fortunately uncharacteristic error. Steve Thomas was back on the programme flying as my Number Two, and we launched to bomb Stanley airfield and then take up CAP.

As we ran in towards the airfield target, I carried out the update on my Navhars with a radar fix and prepared to give the drop signal to Steve, who was just behind my wing-line about 20 yards away. It was a beautiful day at altitude; blue sky and hot sun, with the town of Stanley laid out like a model below.

'Now, now, NOW!' I gave the signal and Steve's bomb dropped away towards the target. But there was no characteristic thump from my own aircraft. Suddenly, I realised why. With a great whoosh, my port missile left its rails, searching for a non-existent target and leaving a long trail of white smoke over the airfield and the town. I had failed to switch the armament mode selector from the air-to-air to the air-to-surface mode.

'Oh, fuck, Steve! I've fired a Sidewinder!'

'I can see that, Boss. Glad I'm behind you!' Steve had a very strong, if dry, sense of humour.

We went round again and delivered the second bomb. It seemed to matter little to me that it was on target.

Back on board Wings and the Captain managed to see the funny side of it and I was allowed to get on with the rest of the day's flying without too much ribbing. Between my second and third flight, I was sitting at my desk in my cabin when the phone rang. It was Robin Kent. 'Boss, you'd better get up here. I'm afraid we've just lost another Sea Jet—this time over the side.'

'Is the pilot OK?'

'Affirmative. It's Mike Broadwater and we've got him back on board. He's a bit shaken, that's all.

Up in the crewroom, I listened while Robin explained what had taken place.

'Mike was sitting on deck behind Paul's jet down aft. Both were burning and turning, on the centre-line and ready for launch—so the deck chains had been taken off. It's wet on the deck and you know how slippery it has become recently—like a bloody ice-rink. The ship was in a hard turn to starboard at fairly high speed and so she was heeling to port quite markedly. There was also a 40-knot wind over the deck from starboard, and the deck was heaving up and down badly in the heavy sea. The ship lurched strongly against the sea when without warning the nose of Mike's aircraft swung to the left through 90 degrees and, in spite of the brakes, slid gracefully over the side. He could do absolutely nothing about it and ejected as the nose dropped. He was picked up by the SAR helicopter in less than two minutes!'

Dick Goodenough was on hand to give the final bit of information as to why it had happened. 'The nose-wheel steering is designed so that if too great an outside side-force is pout on it, it will give rather than break. For example, a tractor on the nose-wheel might apply such a force with the nose-wheel steering inadvertently engaged. So it looks as though a combination of the strong wind, the heel of the ship in the turn, and a sudden heave of the deck from the heavy sea took charge of the nose of the aircraft and pointed it at the point of least resistance.'

'Thanks Robin. Thanks Dick. Better get the incident signal drafted while I go and brief Wings. At least they didn't both go over the side.'

It was a pretty unexciting way to lose an aircraft in war, but neither Wings nor the Captain were perturbed about it. We all regretted the loss, but there was nothing more to be said or done about it on board.

I went off to fly my third CAP mission and returned to the ship just in time to attend the evening Command Briefing.

When the briefing was over, we invited JJ to the Wardroom for a quick drink with the boys and he accepted with pleasure. Robin Kent, Ralph Wykes-Snead and others welcomed the Captain and we all relaxed and discussed the events of the day. JJ was obviously enjoying his short break from the bridge and the Ops Room when Dave Braithwaite sauntered over from the bar. At that moment we were chatting about the SHAR going over the side. Suddenly Dave came out with one of the classic lines of the war. Without a by your leave, he looked JJ square in the eye and, in totally inappropriate AWI fashion, bullishly said, 'That'll teach you to treat this ship like a fucking speedboat, eh, Sir?' He was rapidly removed by the Senior Pilot, and when he had gone the Captain had a good laugh.

Steve and I flew the next mission as a pair. There was no trade for us under the now clear blue skies, but we could see that to the south of the Sound HMS *Ardent* had seen more than enough action for the day. She was limping northwards and smoke was definitely coming from more places than her funnel. We were to see more of her on our third and final sortie of the day.

As a squadronmate looks on, a Harrier recovers aboard *Invincible*, May 19, 1982. The Sidewinder on its port rail is missing, indicating the pilot had seen action on this sortie. (Courtesy of the Royal Navy, HMS *Invincible*)

For this final 'hop' we were given the station to the west of San Carlos over the land. We descended from the north-east and set up a low-level race-track patrol in a wide shallow valley. As always, we flew in battle formation—side-by-side and about half a mile apart. When we turned at the end of the race-track pattern, we always turned towards each other in order to ensure that no enemy fighter could approach our partner's 6 o'clock undetected. I had just flown through Steve in the middle of a turn at the southerly end of the race-track when I spotted two triangular shapes approaching down the far side of the valley under the hills from the west. They were moving fast and were definitely Mirages, probably Daggers. I levelled out of the turn and pointed directly at them, increasing power to full throttle as I did so.

'Two Mirages! Head-on to me now, Steve. 1 mile.'

'Passing between them now!' I was lower than the leader and higher than the Number Two as they flashed past each side of my cockpit. They were only about 50 yards apart and at about 100 feet above the deck. As I passed them I pulled hard to the right, slightly nose-high, expecting them still to try to make it through to their target by going left and resuming their track. I craned my neck over my right shoulder but they didn't appear. Instead I could see Steve chasing across the skyline towards the west. My heart suddenly leapt. They are going to stay and fight! Must have turned the other way.

They had turned the other way, but not to fight. They were running for home and hadn't seen Steve at all because their turn placed him squarely in their 6 o'clock. Steve's first missile streaked from under the Sea Harrier's wing. It curved over the tail of the Mirage leaving its characteristic white smoke trail and impacted the spine of the jet behind the cockpit. The pilot must have seen it coming because he had already jettisoned the canopy before the missile arrived; when it did, he ejected. The back half of the delta-winged fighter-bomber disappeared in a great gout of flame before the jet exploded.

I checked Steve's tail was clear but he was far too busy to think of checking my own 6 o'clock. Otherwise he would have seen the third Mirage closing fast on my tail.

Steve was concentrating on tracking the second jet in his sights and he released his second Sidewinder. The missile had a long chase after its target, which was accelerating hard in full burner towards the sanctuary of the west. At missile burn-out the Mirage started to pull up for some clouds. The lethal dot of white continued to track the fighter-bomber and as the jet entered cloud, I clearly saw the missile proximity-fuse under the wing. It was an amazing spectacle.

Adrenalin running high, I glanced round to check the sky about me. Flashing underneath me and just to my right was the beautiful green and brown camouflage of the third Dagger. I broke right and down towards the aircraft's tail, acquired the jet exhaust with the Sidewinder, and released the missile. It reached its target in very quick time and the Dagger disappeared in a ball of flame. Out of the flame ball exploded the broken pieces of the jet, some of which cartwheeled along the ground before coming to rest, no longer recognisable as parts of an aircraft.

Later I was to discover that the third Mirage Dagger had entered the fight from the north and found me in his sights. As he turned towards the west and home he had been firing his guns at me in the turn, but had missed. It was the closest shave that I was to experience.

We were euphorically excited as we found each other visually and joined up as a pair to continue our CAP duties. We had moved a few miles west during the short engagement and now steadied on east for some seconds to regain the correct patrol position. As I was looking towards San Carlos, about

10 miles distant behind the hills, I noticed three seagulls in the sunlight ahead. Were they seagulls?

I called *Brilliant*, 'Do you have any friendlies close to you?'

'Wait!' It was a sharper than usual reply.

A second or two later, *Brilliant* was back on the air. 'Sorry, we've just been strafed by a Mirage. Hit in the Ops Room. Man opposite me is hurt and I think I'm hit in the arm. No, no friendlies close to us.'

Full power again. 'Steve, those aren't seagulls ahead, they're Sky Hawks!' What had looked like white birds were actually attack aircraft that had paused to choose a target. As I spoke the three 'seagulls' stopped orbiting, headed towards the south and descended behind the line of hills. And from my morning flight I knew where they were going.

'They're going for *Ardent!*' I headed flat-out to the south-east, passing over the settlement of Port Howard at over 600 knots and 100 feet.

In quick time I cleared the line of hills to my left and was suddenly over the water of the Sound. Ahead and to the left were the Sky Hawks. To the right was the stricken *Ardent*, billowing smoke like a beacon as she attempted to make her way to San Carlos. I wasn't going to get there in time but I knew that Red Section from *Hermes* should be on CAP on the other side of the water. 'Red Section! Three Sky Hawks, north to south towards *Ardent!* I'm out of range to the west!'

Red Section got the message and appeared as if by magic from above the other bank of the Sound. I saw the smoke of a Sidewinder and the trailing A-4 exploded. The middle aircraft then blew up (a guns kill, so I heard later) and the third jet delivered its bombs into *Ardent* before seeming to clip the mast with its fuselage.

I looked around to see where my Number Two had got to.

'Steve, where are you?' He should have been in battle formation on the beam. No reply. My heart missed several beats. There was only one answer, he must have gone down!

I called *Brilliant*. 'Believe I've lost my Number Two to ground fire. Retracing my track back to the CAP position to make a visual search.' I didn't feel good. My visual search resulted in nothing. But I did hear the tell-tale sound of a pilot's SARBE rescue beacon. Maybe that was Steve? '*Brilliant*, I can't locate my Number Two but have picked up a SARBE signal. Could be him or one of the Mirage pilots. Can you send a helicopter to have a look, please? I'm very short of fuel and must recover to Mother immediately.'

I felt infinitely depressed as I climbed to high level. Losing Steve was a real shock to my system. At 80 miles to run, I called the ship.

'Be advised I am *very* short of fuel. I believe my Number Two has been lost over West Falkland. Commencing cruise descent.'

'Roger, Leader. Copy you are short of fuel. Your Number Two is about to land on. He's been hit but he's OK. Over.'

'Roger, Mother. That is good news. Out.

Invincible could be clearly seen at 60 miles. She was arrowing her way through the water towards me like a speedboat, leaving a great foaming wake. Good for JJ—doesn't want to lose a Sea Jet just for a few pounds of fuel. My spirits had suddenly soared and it felt great to be alive.

I throttled back and didn't need to touch the power again until I was approaching the decel to the hover. On landing with 200 pounds of fuel remaining, I couldn't help thinking what a remarkable little jet the Sea Harrier was. The fuel was right on the button. I had calculated 200 pounds at land-on before leaving San Carlos.

On board, I heard from Steve that he had been hit in the avionics bay by 20-mm machine-gun fire from Port Howard. He had lost his radio, couldn't communicate with me, and thought he might just as well go home. I was too pleased to see him to be angry.

'What was I supposed to think, then?'

'Oh, you were hightailing it after those Sky Hawks, Boss. You can look after yourself and as I didn't have any missiles left I thought the best thing was to get the aircraft back and get it fixed.'

'Steve, that is definitely worth a beer!'

It struck me later that if Red section from *Hermes* had been capping at low level over the sound (where any 801 CAP would have been) instead of at altitude, the Sky Hawks would have had to get through them to get at *Ardent*. The A-4s would not have tangled with the SHARs so *Ardent* would not have been hit again and mortally wounded.

Managing the Store: What It Takes to Run an Aircraft Carrier

As one might imagine, becoming the commanding officer of an aircraft carrier carries tremendous responsibilities that are spread all over the cultural and operational maps. While having to be constantly prepared to sail the ship and its crew into harm's way at a moment's notice, the skipper must also contend with daily routines and problems that are part of life in a small, confined city of five thousand people of widely diverse backgrounds and experience. For most aviators, attaining squadron command of perhaps a dozen expensive aircraft and two hundred people is the pinnacle of their career. Commanding a carrier infinitely expands those responsibilities by quantum leaps.

Newspaper reporter and columnist George C. Wilson got a chance to make a full seven-month deployment with the men of the *John F. Kennedy* (CV-67) in 1983. With the backing of senior Navy officials, including Secretary of the Navy John F. Lehman, Jr., and Chief of Naval Operations Admiral James D. Watkins, Wilson went through survival and physiology training so that he could fly a variety of aircraft in the *Kennedy*'s air wing. For Wilson, it was a dream come true. As a high school senior, he had been accepted for Navy flight training at the end of World War II, but the end of hostilities finished that program. Now, at least in an abbreviated manner, he would finally be part of a Navy carrier's crew as *Kennedy* left Norfolk on September 27, 1983.

The book that came from Wilson's adventure is a fine depiction of life aboard a carrier. At all times, he keeps the human element center stage, describing all phases and levels of the people—officer, enlisted, junior, senior, first-tour, and career sailor. During the cruise, he would see happiness and distress, exhilaration, fear and accomplishment. Men would die, but others would remain to complete their tasks. Something he hadn't expected was an actual combat operation when the *Kennedy* joined the USS *Independence* (CV-62) in a strike against rebel positions in Lebanon in December. *Kennedy* would lose one airplane and its two-man crew on this raid. The pilot of the A-6 would die on the ground of wounds after ejecting, and his bombardier-navigator would be held prisoner for a month before returning.

The excerpt I have chosen offers a close-up view of what a carrier CO thinks and does every day as he deals with people problems aboard his ship, typical of the situation resulting from jamming so many people in such a closely confined space. Although carriers are the largest ships built, the available space for individuals is remarkably tight. Conflicts are bound to develop. It takes a tough, steady hand to resolve issues or redirect a sailor's energies and attitude.

George C. Wilson was beginning Navy flight training when World War II ended. He left the service in 1947 to begin a career of writing about military affairs, which has lasted four decades. He went to Vietnam as a correspondent for the Washington

Post *in 1968 and 1972. He left the* Post *in 1991 to write books. His column, "Talking About Defense," appears in the* National Journal Magazine.

From *SuperCarrier: An Inside Account of Life Aboard the World's Most Powerful Ship, the USS* John F. Kennedy, by George C. Wilson

"This ship, this weapon, this oil tank, this town, this airport, this factory, this everything is unmanageable," I told myself as I marveled at what was going on all around me. The carrier was a million accidents just waiting to happen. I wondered how any one person could run this universe.

I climbed to the bridge many times to talk to Skipper Wheatley in hopes of finding out. He always greeted me warmly and never ducked a question. My first entries into his inner sanctum on the bridge caused some of the officers standing there to stiffen. But Wheatley's warmth eventually thawed everybody. I called him Captain, of course, and he called me George. Even though I wore work khakis, I was never quite in uniform, even on the bridge. I just could not give up wearing Wallabees. The biggest concession I made was to wear black ones.

Wheatley, as skipper of an aircraft carrier, was at once all powerful and powerless. He could fire his officers, send sailors to jail on the ship, work everybody to exhaustion or give slack, or allow a damaged airplane to land on his deck or order it to take its chances and try to reach a land base. He could forbid showers to save water, and issue a thousand other orders to make life better or worse on the ship. In that sense, he was all powerful. But he could not run everything on the ship. He had to trust his subordinates. In that sense, he was powerless. If they did something wrong, even when he was asleep, it would be Wheatley's fault. The captain of the ship was not really all powerful; he was all responsible.

"So how do you manage this unmanageable monster?" I asked Wheatley one day as I stood beside his big upholstered chair on the port side of the bridge. He answered the question as if he were giving one of the management courses he had taken at the Industrial College of the Armed Forces, at George Washington University and at Harvard University on his way up to his present prized command:

"Basically, the thing that has stood me in best stead from all the lessons set forth in management courses I took, was to delegate to subordinates and to trust them—trust them enough to make a mistake. Let them know they can make a mistake, almost to the point of saying, 'If you're not making a mistake, you're not trying.'

"My management challenge becomes one more of motivation and direction than crisis management and involvement in the day-to-day details of what goes on. I don't care how good you are, you can't do it. I've seen some people try that, to run the details themselves, and it's usually counterproductive, regardless of how talented they are."

Through the years, the Navy had developed a management structure for a carrier resembling that of a corporation. Specialized functions are clustered into departments, with a commander or captain in charge of each department. The *Kennedy*'s departments were Administration, Air, Aircraft Intermediate Maintenance, Communications, Dental, Engineering, Medical, Navigation, Operations, Supply, Air Wing, Training, and Safety.

Also, the Navy gave the carrier skipper a vice president to act as his son-of-a-bitch. His deputy, with the title of executive officer, ran around the *Kennedy* yelling at department heads and sailors alike to get the jobs done. The XO was much more visible to the sailors than the remote captain. I asked Wheatley if there was any danger that a super-active executive officer like his own, Captain John Anthony Pieno, would make it seem to the men on board that he, not Wheatly, was running the ship. Wheatley's smile crinkled his face all the way up and under his dark glasses. He took his eyes off the sea dead ahead of the carrier, turned to me and said:

"Very simply, the XO is the man who translates the command goals into reality. I come up with the goals, the ideas. The XO makes them happen. In the final analysis, the man who makes the hard decisions is the captain. You don't get to be a captain by making the easy decisions. I am the man who must punish at mast. The XO may be perceived as being the bad guy because he has got to make things happen. He has to keep the pressure on. The tough decisions have to be made by the man who is responsible."

The skipper of a carrier or any other ship does set the tone, color the personality, and dictate the style and philosophy of his ship. I concluded Wheatley's tone, personality, and style could be summed up in just one word: *correct.* He was proper, but not prim; cordial, but not warm; humorous, but not funny. His philosophy was that of the overachiever—the Middle America conviction that anybody could achieve his goals if he would just set them down and work hard until he reached them. He believed the United States Navy was still one of the places men and women could realize the American Dream of being judged by what they did, not where they came from or what social and political connections they had. He was a zealot when it came to working to get ahead. He told me in one of our chats on the bridge that it infuriated him to see officers or sailors waste their time in the Navy rather than seize its opportunities. I came to learn that this Navy overachiever, this grandson of Tennessee dirt farmers and son of a General Electric plant foreman, this Annapolis graduate—could sound like a missionary, a zealot even, on the subject of making the most of the moment at hand.

Shortly after leaving Norfolk, the sailors who had never sailed on the *Kennedy* before were summoned to the crew's lounge in the stern of the ship for orientation lectures. I decided to attend Wheatley's lecture to the young men to see whether he gave them fire and brimstone or gentle, fatherly advice from "the Old Man." I had been to the lounge many times before to talk to sailors during their off-duty hours. Usually the lounge was heavy with cigarette smoke, especially along the wall where President Kennedy's portrait hung over an artificial fireplace. Card tables were located there. These modern sailors did not play cards much, however. They most often played packaged games of war. This day the lounge was dressed up for the Captain. Its chairs were lined up in neat rows facing the lectern near the door where Wheatley would enter. After I had sat down among the sailors, several eyed me suspiciously. My khakis provoked alerts all down the rows for fear I was a chief petty officer who could give the sailors some grief. Then one of the sailors put his buddies at ease by saying, "He's the guy writing a book about us." The strained expressions turned into smiles and winks. Everyone still likes to see his name in print.

"Attention on deck! Attention on deck!" the chief at the front of the room commanded.

With a scraping of metal chairs, the room full of sailors rose from their seats and stood at the loose attention American sailors have perfected over the years. It is not the rigid attention of the Army airborne or the marines. It is a laid-back attention, as if the sailors were saying to themselves:

"All right, asshole, I'll stand up, but you better have something to say."

I stood up along with the sailors waiting for Wheatley's entrance. I was a khaki blob amid the dark blue dungarees and light blue shirts of the kids who did the heavy lifting on an aircraft carrier. They had already been told that the *Kennedy*, winner of several Navy E's for excellence, was a special ship that would demand their best.

The door snapped open. The young sailors saw their Captain close up for the first time. He was six-foot-three with a thickish, but not fat, build; thinning, reddish-blond hair, and the square chin and clean features of the man in the Arrow shirt ads. He wore a green flight jacket, the equivalent of the varsity sweater for aviators, to remind everybody that he used to fly A-6 bombers before he climbed to the heights of the Navy bureaucracy and won command of an aircraft carrier. He towered over the lectern. He told the sailors to sit down and then caught their attention by almost shouting:

"This is a capital warship, not a love boat."

He went on to explain to the now silent room of sailors:

"We have more destructive force on this ship than the entire Navy carried in World War II.

"We carry two million gallons of jet fuel and two million gallons of ship fuel and bombs, rockets, and bullets. And while we have an automatic sprinkling system in the magazines, we can't lean on them to protect our lives and let somebody smoke a cigarette.

"Every senior petty officer on this ship is a safety officer.

"I want you to stay alert?!"

Changing the tone of his voice from warning to exhortation, Wheatley laid some of that missionary zeal on the confused teenagers sitting before him:

"My personal goal is that every man leaves the Navy with a crow on his shoulder that will represent some skill he can sell on the outside. I don't care whether you stay in the Navy or get out. I have a commitment that you go back as a petty officer and with a skill you can market on the outside. This doesn't happen automatically. Once you've done that, you've got a proven record of achievement. Go crow!

"Another personal reason I want you to advance is that I can't give you a raise. The only way I can give you a raise is to promote you. This is another reason I want you to advance.

"Responsibility." He said the word slowly and paused to let it sink in. "Each of us has responsibility. You have an absolute right to expect me to know my job—to feel confident that I'm not going to do something stupid or dangerous to put your life in jeopardy. What you know or don't know, or do or don't do, affects everybody on this ship. You wouldn't like it if I were high on booze or pills. There is no place on a warship for anybody who diminishes his capability by being on booze or drugs."

As far as getting along on the ship day by day, Wheatley continued in a voice less strident, "a recruit in the Navy only has to do three things. One: be at his place of duty on time; two: do a day's work for a day's pay; three: show mutual respect, not just to the captain, but to those above and below you. And remember that there's no such thing as a stupid question.

"Nobody can guarantee where you're going to be a day from now," Wheatley warned the young sailors as the *Kennedy* steamed south with a schedule that bore no resemblance to where the carrier would actually sail.

"This not knowing causes emotional insecurity. It's one of the personal sacrifices you have to make. But the flexibility is what makes the carrier such a great instrument of national policy.

"Don't ever forget that the only reason this ship is any good is the crew. The difference between a winner and a loser is the people right here in this room."

With that, Wheatley—perhaps to demonstrate that he could relate to the young sailors' personal concerns—asked for questions. He got only a few. Most of the young men were holding back. They saw no reason to risk getting in trouble with The Old Man so early in the cruise by asking stupid questions. The new sailors would be on this boat a long time. Their murmurs and lifted eyebrows to each other suggested that they had concluded the less they saw of this captain over the next seven months, the better off they would be.

Wheatley in his indoctrination address to the new sailors purposely made no attempt to be entertaining or brotherly. We discussed, in one of the many conversations I tape recorded on the bridge, the pros and cons of trying to be popular as a commanding officer. Wheatley had been a bit of a hell-raiser as a young pilot and still loved to party on the beach. But he believed it was a mistake to play to the crowd as a commanding officer, to go out of your way to be popular with your subordinates.

"You find your subordinates have an image of what they want their leader to be," he told me. "And in spite of what people say, they don't want you to be one of the boys. They don't want you to be a clown. They want you to be somebody they can respect—somebody they can look up to because they are in effect saying, 'If I'm successful, I'm going to be him.' So I push real hard both on the officer corps and the enlisted on this concept of mutual respect. I think it's very important."

"But," I pressed, "where is the fun in a job like yours where, as captain, you must distance yourself from the pilots and others with whom you would like to stay social? Where you often must eat alone? Where the demands are endless? Where you are away from your wife and children for months on end when you could be home, where, given your management background, you could be making more money than you are now? Where anything that goes wrong is your fault?"

"It's the unlimited nature of this job," Wheatley replied. "It's got to be the best job in the Navy. There are only thirteen operating carriers in the Navy and I command one of them. That is part of it. Put that together with the great elation one achieves from working with the kind of people I work with, the XO and my department heads. That has a synergistic effect.

"The fact that you say, 'Gee, I'm skipper of a carrier' wears thin after a while. That gets you seven months' deployment away from home. But what is really good is dealing with the individuals you work with, and that includes the sailors. There are some great young people in the Navy."

If that is the up side of commanding an aircraft carrier, of managing the unmanageable, of being captain, mayor, and father confessor for this weapon-town hybrid, what is the downside of trying to be an effective skipper?

Wheatley usually responded to my questions quickly in whole, cogent paragraphs which sounded like written remarks rather than the spontaneous answers that they were. This time he paused and stared out to sea before answering.

"The hardest thing to deal with—and I hope you can find a way to explain it in your book—is the very arduous and deep emotional insecurity that takes hold of so many people on a deployment like this. The never knowing what is going to happen next. Are we going to get home? Are they going to send us through the Suez Canal because something happened in Iran? Is the admiral's staff on board going to be transferred tomorrow? Are we going to have a liberty port? If my wife comes over to see me (when the ship goes into port), am I going to get to see her or are we going to pull out too early for that, like that other ship did? Or like happened to me five years ago? Is there a problem at home, and that's why I'm not getting any mail? Or is the mail just slow?

"People like to be able to count on things. They like to be able to say, 'I'm going to see you in a week,' and count on it. If you could do that and say, 'Hey, this cruise is going to be over on May 1.' If you could believe that in your heart, then you could say it was a cruise."

But in Wheatley's view the *Kennedy* was on a deployment. He corrected anybody who called it a cruise because our duration, liberty ports, and destinations were all uncertain as we steamed toward Rio de Janeiro, despite the seeming specificity of our sailing orders.

I saw Wheatley in many settings as I roamed the ship. At dinner, he could be either the needling extrovert or the warm host. With department head officers, he could be the coach trying to teach. With sailors brought before him for punishment at captain's mast, he could be the ferocious judge or sympathetic captain willing to give the accused a second chance. He was considered a hard ass by the sailors I talked to. They feared going before him, which is the kind of deterrence to stepping out of line which warship commanders like Wheatley try to generate.

One afternoon when I was an observer at captain's mast, a sailor who had been in the Navy for four years without moving up into the petty officer ranks was brought before Wheatley for failing to obey orders of petty officers. He told Wheatley at the outset of this disciplinary proceeding that he was jealous of petty officers who were younger than he. That was why he refused to take orders from them. The sailor was standing at attention on the platform below the ladder leading down from Wheatley's sea cabin off the bridge. Only a narrow lectern separated captain and sailor. The sailor's explanation offended Wheatley. His face reddened as he leaned within inches of the sailor's face and shouted:

"How come you're not a petty officer? You took the test only once in four years. You're not a petty officer because you don't want to be a petty officer. If you wanted to be a petty officer, you could be. Maybe you resent these other men who took the time and effort to study and take the test—who might be junior to you in time but senior to you in rank.

"You ought to be second class by now. Do you think when you get discharged from here and go out you'll be able to pick your own boss? What are you going to do if you've got a younger man, the boss's son, as your boss? Suppose you know more about the job then he does? Are you going to resent him, too?"

"No sir," the sailor answered meekly.

"Then why can't you respect the petty officer who has been appointed over you here? Because you know what's going to happen to you when you're a civilian and you act that way? You're going to get fired! Nobody out there is going to counsel you. You're not going to have a first class petty officer. You're not going to have a chief. You're not going to have a bunch of people who care about you—who are going to sit down and try to improve your performance—to try to keep you out of trouble. You're going to get fired! You're going to be out of a job. It's that simple. Is that what you want?"

"No sir."

"Then why the hell can't you improve your attitude?"

"I'm jealous of the petty officers, sir."

"You don't have to be jealous. You could be a petty officer. You're not a dummy. You're an intelligent, capable man. You could be a barn burner. Why don't you stop quit feeling sorry for yourself and get off your butt and be somebody! You've got the capability. You shouldn't be here at mast. You ought to be one of the leaders. What are you doing—screwing up purposely so you get some attention? I'll guarantee you, you're going to get plenty of attention. You don't have any business being at mast. There's absolutely no reason. You did this deliberately. You know better. There's absolutely no reason for behavior like this out of an individual like you. If you're jealous of people who are senior to you, that's your own fault. You could be senior to them. What are you going to do the first time you have a setback in civilian life—say those bastards are picking on me again and quit?"

"No sir."

"That's what you're saying. The solution starts right here with you. You are a capable, talented man. And you're pissing away your talent, and you ought to be ashamed of yourself for that. You've served honorably for four years. You ought to be a superstar. It makes me sick to see a man like you who has to come to mast. You can't pick your boss, just like I can't pick my boss. You do the best you can. If you suffer a setback, you turn around and keep working. That's the only way you're going to make it in civilian life. You know that. I want you to be a success. I don't want you to be a failure."

Wheatley leaned back from the lectern, brought his head back down close to the stunned sailor and let loose a thunderous summation: "You're not a failure! Quit acting like a failure, God damn it! Act like the man you are! Understand me? Know what I'm saying?"

"Yes sir."

"I find you guilty as charged. I award you a reduction in rate and a $200 fine. On the basis of your superior service, what your superior petty officers say about you, I'm suspending the reduction in rate for six months, which means you don't lose your stripe. Are you going to take that third class exam in March? I don't care if you're getting out of the Navy or not. I want you to take it and pass it."

The chastened sailor mumbled something that sounded like, "Yes sir," saluted and disappeared down the ladder under the landing.

A Helo Crew's War

Helicopters are among the most demanding aircraft to fly. A pair of steady hands, in-depth knowledge of a peculiar set of aerodynamics and of the individual aircraft's systems, as well as a cool head, are prerequisites for a successful career in rotary wing squadrons. The sleeker, more powerful jets may consistently get the glory, but the helicopter lifts its share of the deployment load. Still, most folks would say, besides the occasional rescue missions, which are cause enough for satisfaction, helicopter crews tote mail and passengers, and sometimes chase submarine contacts, so where's the excitement?

Flying from the small flight deck of a surface combatant, i.e., destroyer or cruiser, helicopters certainly require a lot of skill, but, one might ask, where's the mission, especially during a war? There were a few helo crews that saw action during the Gulf War, and a few that even managed to sink a ship or two. British Lynx helicopter crews sank fifteen Iraqi ships during the war. This next excerpt describes the experience and success of one such crew.

Richard Boswell was originally rejected by the RAF and the Navy, but finally was accepted for Navy flight training. He had barely received his wings in September 1990 when he was sent off as part of the buildups in the Gulf following Iraq's invasion of Kuwait in August. Assigned to HMS Manchester, *a Type 42 destroyer, he saw his fair share of action during the six-week war.*

Returning to England after the war, Boswell became an instructor, but was nearly killed in the crash of a light plane he was piloting. The accident ended his Navy career, and he embarked on a lengthy period of rehabilitation. He now flies corporate aircraft and writes as a freelance aviation reporter.

From *Weapons Free: The Story of a Gulf War Helicopter Pilot,* by Richard Boswell

Monday 11 February—NPG

Well, what a day. It started at 0300 with a telephone call saying that an Iraqi fast patrol boat had been found and that we were to come to Alert 15. As our chaff and flare dispenser is not working, Cardiff*'s aircraft was tasked to go north and subsequently hit the target, bringing their total to five.*

We were stood down at 0630 and I immediately went back to bed. I arose again at 1000 and we briefed to go flying at 1100 but we did not actually take off until 1145. It was supposed to be just to clear the deck whilst they received another aircraft but we ended up remaining airborne for an hour whilst we completed a Photex[1] *with* Cardiff. *After a quick fifteen minutes for lunch we were up again, this time with two live missiles, on our way north for a surface search. We landed on the American ship USS* Paul F. Foster, *to refuel before commencing the patrol, and were informed that there was a possible target to the north. After communicating with an American Seahawk, they vectored us in and we fired a missile which was assessed to have hit the target.*

We were then vectored in to another target which we fired at. However, it was assessed as a miss. We have now fired our first shots in anger and undoubtedly killed our first victims of the war. How do I feel?—surprisingly unaffected. During the attack I was scared, afterwards jubilant, now I am back to normal.

We were greeted on our return with half of the ship's company cheering us in the hangar and the Captain with some champagne. It made me appreciate how much the rest of the ship's company are identifying with us and relying upon us to perform for the ship. It was a special

[1] *Photex = a flight to take aerial photographs.*

moment but there is still lots of work to be done so we must not become complacent, but I suppose it still feels good to have pressed home a successful attack.

"A Sea Hawk has been engaged by a small Iraqi fast patrol boat and they want to know if we can assist."

When it came to locating and tracking the enemy, Sea Hawks were very capable helicopters, but bar small calibre machine guns, they were unarmed.

Nick was grinning.

"I've told them that when we land on to refuel, I'll pop down to the control centre to get a tactical update."

I was getting excited. I looked out of the window at the missiles. They looked very menacing.

As soon as we landed on the ship, Nick unstrapped and jumped out. I sat in the cockpit keeping the helicopter running and waving at the guys on the deck like a schoolchild. As they were completing the refuel I saw Nick come running into the hangar. It was the first time I had ever seen him run. As he was strapping back into his seat the American sailor who had been conducting the refuel, came up to my window and waved a jamjar full of fuel at me. I obviously looked confused because he waved it in my face again. Not knowing what was going on, I simply gave him the thumbs up and he wandered off quite content.

"What the fuck was that all about?" I enquired of Nick as he plugged his helmet back into the intercom.

"He's showing you a sample of fuel so you can check that there is no water in it before taking off again."

It was so obvious, I should have known. For the second time in two days I felt very embarrassed.

Nick, however, was very excited.

"The Sea Hawk is tracking the FPB twenty miles to the north. It's already engaged them once so they are keeping well out of its way. I've got a rendezvous position for them. They are going to lead us into the target."

Now I was excited. We might get a chance to fire a missile and there was no fear at this stage. We decided to communicate with the Sea Hawk by the ordinary radio rather than the secure speech. The Iraqi vessel knew that someone was out there as they had already attempted to shoot down the American helicopter. By not using secure speech, at least I would be able to take a much more active part in the proceedings.

Nick called the Sea Hawk, call sign Oceanlord Two-Seven. The reply was instantaneous.

"Roger Three-Six-Zero you're loud and clear. Suggest you RV our position, we'll take you right in."

It was evident that they were old hands at this operation and were completely in control of the situation. They had suggested that we rendezvous in such a way that we would not have to use our radar. It made sense as by now the Iraqis would associate our Sea Spray radar with the Sea Skua missile. To the Iraqis that would mean trouble and we didn't want to show our hand too early. By not transmitting on the radar until the last minute we could, hopefully, lead them into a false sense of security and then attack before they had a chance to jam or decoy our fire control radar.

Nick replied, trying to sound as laid back and as smoothly British as possible. He had never been a *Top Gun* fan.

"Roger Oceanlord Two-Seven. We are presently five miles to the south, with you in two minutes."

I accelerated the helicopter to the maximum permissible speed. I didn't want to give the FPB a chance to get away and so miss another opportunity to launch an attack. I was beginning to feel excited, overlaid with an element of fear. The sight of the Sea Hawk orbiting on the horizon reassured me that things were going to plan. I joined the other helicopter in close formation. One of the crew was leaning out of the back door waving and taking photographs.

"Oceanlord Two-Seven, Navy Three-Six-Zero is now aboard," I transmitted, indicating to the Sea Hawk pilot that we were now in close formation and ready to move off towards the target.

"Roger Three-Six-Zero, contact bears 005 at 15 miles."

The Sea Hawk rolled out and headed to the north. Nick entered the position of the enemy ship in TANS and then we were left with nothing to do so I handed Nick my camera from a pocket in my flight suit.

"Take a quick snap of Oceanlord for me, I'll stick it in my album." It seemed bizarre taking photographs again as we ran in towards the enemy. In close formation I wanted to keep my hands and feet on the controls.

"Contact bears 002 at 12 miles—from Camera Bug, you have weapons free on the target," the Sea Hawk updated the enemy's position. Nick updated it in TANS.

"I'm going to have a quick sweep on radar to confirm," Nick informed me. He set up the radar so that with one sweep he would be able to locate the target. I glanced across at the radar screen as he pressed the transmit button. Exactly where the Sea Hawk had said it would be, a green blob appeared on the illuminated screen. Nick transmitted again.

"Roger Oceanlord Two-Seven we have the contact on radar, many thanks for the service."

"Best of luck Navy Three-Six-Zero." The Sea Hawk broke away and we assumed it was going back to its mother. It felt very lonely again. Nick and I had a quick discussion to decide at what range to fire the missile.

"We will fire closer than usual to maximise the chance of success," Nick commented.

"OK mate, but let's not go inside five miles." We still had no idea of what we were up against except that it was an Iraqi Fast Patrol Boat. Therefore we had no idea as to what type of anti-aircraft guns and missiles she might be armed with. What we did know was that the Iraqis were equipped with SAM 5, a hand-held heat seeking anti-aircraft missile with a range of five miles, and we had to assume that all vessels would be equipped with them. We eventually agreed on a range of six miles.

Nick turned the radar on for another sweep. As he turned it off we both heard it through our headsets. A short bleep, followed by a pause and then another bleep.

"Fire control radar, band three," Nick said almost nonchalantly. My pulse noticeably quickened, I did not need telling. I had heard it as well and had glanced at the screen on our electronic support measure equipment, known as Orange Crop. A series of lights on the third row up indicated that the enemy was looking for us with the radar that controlled the missile system: they must have realised that an attack was imminent. I started a gentle descent. Because radar relies on radio waves, flying lower over the sea minimises your chances of being detected as your radar return becomes confused with the ground 'clutter'. I levelled at fifty feet.

"Ok Dickie, two miles to run." Nick had his radar on all the time now. I was feeling nervous, I could feel my hands becoming very sweaty. I looked down at them: I had a vice-like grip on both the cyclic and collective. I took a deep breath and tried to relax. I remembered my old instructor's words, from when I was learning to hover and tensed up on the controls; "Take a deep breath and try to relax from the arse upwards. If you can unsqueeze your cheeks, you will start to relax". I tried it. It worked, a little.

Bleeeeeeeeeeeeeeeeeeeeeeeeeeeee. The noise was so familiar. I had heard it a hundred times before on the simulator. It sounded like the whine of the television when left on as the transmission ceases. It was the Orange Crop again: we had been locked up. The sound made us both jump. Shit, I must break lock before weapons can be fired at us. I descended even further. I was now flying lower than I had ever dared before. It felt like the wheels should be in the water—this was seriously low. The whining stopped, we had broken lock. Nick offered some encouragement.

"OK mate another mile to run." He didn't comment on the lock up but I could sense the slight apprehension in his voice. We were both scared but controlled. In order to fire the missile we would have to climb again. To attempt to fire at this height would result in the missile simply dropping into the sea.

"Half a mile to run." Nick was now running through the pre-fire checks and activating the missile. I was concentrating very hard on unsqueezing my cheeks and trying to relax.

"Point two of a mile." That was my cue.

"Roger, starting climb," I informed Nick. It all sounded very professional and rehearsed. I was thankful that my body had gone into automatic, overriding the fear and excitement.

I slowed the helicopter down to less than forty knots, I didn't want to continue closing the enemy at almost 150 mph while at this height. As we levelled, I knew that we were sufficiently high to ensure that the missile would have enough room to fall from the weapon station and ignite when fired. The system works thus: the missile fire button is pressed and, within a second, it receives all of the information from the aircraft and internally tests its own systems; it then drops away from the weapon carrier and falls clear, before the burners ignite and it tracks towards the target. This ensures that the helicopter is not engulfed in flames of burning liquid fuel.

I looked across at the radar screen. The little dot indicated that our radar had successfully locked onto the ship. I could hear the bleep, bleep of their radar as they tried to lock us up. It was in rhythm with the heart beat I could feel pounding in my chest. We were now just waiting for the two lights to illuminate on the missile control panel that would indicate the missile was locked onto the target and ready. The two seconds seemed like an eternity. Time was moving at a tenth of its normal speed. Who would win this aerial chess game depended on who would shoot first. We had the advantage as they had yet to lock onto our helicopter again, but the continued regular bleep through the headset indicated that they were frantically trying. Then the 'locked' and 'ready' lights illuminated.

Our missiles had now received all the information from the system computer, knew where to look for the target and wanted to go. Nick pressed the fire button. Nothing happened.

I looked across at Nick as he pressed the fire button again. Still nothing happened. The bleep continued through the headset—they must surely lock up soon. I reached forward to press the fire button on my side. As I moved my arm towards the switch I felt the aircraft twitch laterally as the missile fell from the weapons pylon. Shit. Of course. The missile takes a few seconds to complete an internal self-check before it releases!

Time was moving very slowly indeed. The missile must surely have fallen uselessly into the sea by now. Then it ignited. I had never fired a live Sea Skua before so wasn't ready for the burst of energy, heat and noise as the fuel ignited. It caught me by surprise. I tried to watch but instantly it disappeared into the haze. Almost immediately I turned and descended the helicopter. The helicopter radar must be kept pointing at the target so that the radar waves returning from the contact can guide the missile now thundering towards it. The radar is situated in the front of the helicopter and rotates through ninety degrees either side of the nose, enabling me to turn away from the Iraqi vessel—I did not want to get any closer—while keeping the radar pointing at it. I descended to try and avoid detection by their radar, but we couldn't descend too far and risk losing our own missile.

I glanced in at the missile control panel. The bank of 'time-to-run' lights slowly extinguished as the missile flew towards the ship. Nick counted them out loud.

"Five lights to go . . . four . . . three . . . two . . . one . . . all lights out."

We both looked anxiously in the direction of the ship. With the visibility at less than five kilometres we had never been in visual contact with it, however, exactly when the 'time-to-run' lights in the cockpit indicated impact, an explosion was visible on the horizon, followed shortly afterwards by a thick plume of black smoke.

"Yes!" I punched my clenched fist into the air.

We had achieved our first kill. I felt elated that we had actually achieved the job that we had been dispatched to do. The nerves had gone completely. I relaxed a little and started to shake.

I took a deep breath to calm myself and quickly checked around the cockpit to ensure that all the systems were still working routinely. Good airmanship dictates that this should be done every five minutes or so and I hadn't done it since we left the refuelling ship. While I completed this, Nick reported the engagement back to the controlling authority using secure speech. I was hoping that we would be tasked to close the ship for damage assessment. It's helpful to know how much damage has been done such that you can ascertain whether the target is out of action or not.

Nick finished on the radio and came back on to the intercom.

"There's another contact that they want us to go and attack 25 miles to the north-west. I've been given the position. It's been reported by the SUCAP and believed to be a Boghammer type vessel."

The Boghammer vessels were large military speedboats with four 20 mm anti-aircraft guns fitted to the front. They were a real hazard because of their size; small enough to be very difficult to locate, both with the radar and the naked eye, and yet with enough fire-power to really ruin your day. The standard Sea Skua missile skimmed across the sea at about twelve feet, which meant that the missile would pass right over the top of one without impacting. Fortunately we were carrying a variant of the missile that had been developed especially to counter this type of threat. Its performance parameters had been changed to ensure that it was capable of engaging smaller targets. SUCAP was military jargon for Surface Unit Combat Air Patrol, the American jets assisting in the search for enemy vessels. They were armed with normal free-fall iron bombs. Apparently, hitting a small ship with an unguided bomb proved very difficult and they achieved only very limited success.

This time there was no friendly contact to lead us in. We had no way of knowing how long ago the vessel had been sighted and therefore no indication as to how valid the positional information was. Nick had his radar on pretty much all of the time as we ran in towards the last reported position. He was getting noticeably agitated at his inability to locate any sign of enemy activity amongst the clutter on the radar screen. I kept my eyes firmly glued out of the window, peering into the haze to try and find something visually. We were both concerned that if the intelligence received the previous night was correct, there could well be a number of these vessels around, armed with SAM 5 missiles and with the single objective of shooting down a Lynx helicopter.

Nick spoke again.

"I've think I've got a weak return about seven miles away. It's in the location of the position previously passed."

I glanced across at the radar screen but from where I was sitting I couldn't make out anything.

"OK Nick, I'll hover here and keep pointing towards it." Now it was my turn to sound nonchalant.

Nick was still hunched up over the radar screen, but I could see his lips moving and as he wasn't talking to me, I assumed he was talking on secure speech to Camera Bug. I was right.

"We've got weapons free. Start flying towards it."

Things were a lot calmer and slower with this attack. There was no indication on Orange Crop of any enemy radar activity in the area. We agreed that we would fire at five miles.

Nick counted me down again.

"Seven miles to the target, still a weak return," he informed me. I was flying at fifty feet—no point in flying at five feet when you don't need to.

"Six miles, still a weak return on radar." I glanced at Orange Crop; still no sign of any radar activity from anybody else.

"Five and a half miles." That was my cue. Again, I started to climb and when levelled at optimum height, brought the aircraft into the hover and called 'on condition'. I could tell that Nick was concerned that he might not be able to lock the radar onto the target as it was still such a weak return. In the end he locked on without any problems. With no fire control radar trying to locate us, the 'locked' and 'ready' lights took no time at all to illuminate. The fire button was pressed and this time I was ready for the delay. It seemed very short this time and I was ready for the missile igniting. There was no point in turning away and I held the helicopter in the hover. Once again we watched the 'time-to-run' lights count down. As the last one extinguished I looked into the haze with anticipation: there was no explosion, no plume of black smoke, nothing happened. I felt disappointed. Nick played around with the radar, the contact had been so weak he could not accurately assess whether it had disappeared or not. I looked at the fuel, we were beginning to run low. Nick passed me a heading for the USS *Paul F. Foster* and I turned towards it. After all the excitement of the first attack this had been a complete anticlimax. We discussed the possibility of what might have happened enroute back to the ship.

"It could have been a very small craft and the missile passed straight over their heads," I suggested, "so at least we would have given the rag-heads a shock."

Nick remained less convinced.

"I think we fired at a navigation buoy. There are a lot of them marked on the chart around there."

We were both deflated, but Nick lifted the mood.

"We smoked that first bastard though."

It brought a smile to my face and made me feel good. The American ship began to appear out of the haze.

I landed on the ship and the ground crew commenced the refuel. Noticing the empty weapon pylons they signalled thumbs up inquisitively. We returned the thumbs up and they started jumping and dancing around the deck. They obviously approved of the action. The return flight back to *Manchester* seemed to take forever—in fact it took about forty minutes. The rest of the sortie had taken about three hours but it had felt like ten minutes. It was not until afterwards, when I sat down with Nick, that I had any idea about where we had actually been. It is the Observer's responsibility to look after the navigation, but I used to pride myself in having a good seat-of-the-pants feel as to where we were (extremely difficult when flying over the sea in haze with no landmarks to identify your position). However, on this occasion I really did not have a clue as to where we had been. I had been concentrating so hard on just flying the aircraft safely.

Prowlers in the Gulf

The 1991 Gulf War was a benchmark for naval aviation in several areas, not the least of which was the number of aircraft carriers and related ships involved. Six U.S. carriers and their individual air wings conducted sustained combat operations for six weeks, rivaling the intensity of action at the height of the Vietnam War and the last 18 months of the war in the Pacific. Hundreds of aircraft and a thousand people flew sorties from these six carriers. Several thousand more maintained the aircraft and their carriers to keep up the hectic schedule. It was the first such major exercise of such a large number of carrier resources since 1972.

For most of the flight crews, it was a cold-water initiation into combat. Indeed, some of the very junior aviators had barely finished training at the FRS (fleet replacement squadron) that taught them how to fly the aircraft they would operate in their fleet squadron, i.e., F-14, FA-18, EA-6B.

One of the most important aircraft in the lineup quickly became the Grumman EA-6B Prowler, the electronic jammer that had first seen action in the last year of the Vietnam War. Developed from the A-6 attack bomber, extending the forward fuselage by 40 inches to accommodate two more seats than the bomber's two-man crew, the Prowler quickly established itself as one of the most capable systems against enemy signal and radar. (There was the EA-6A, an A-6 bomber, with its original two crewmen, fitted out with electronic gear, and a huge "football" atop the vertical tail. The Marines had used it in Vietnam to some advantage, especially during the Linebacker campaigns. It had retained the bomber's "Intruder" name, but the EA-6B was such a newer, different platform that it received another name.) To this day, the EA-6B has accumulated hundreds of combat sorties, protecting Navy, Marine, and Air Force, as well as Allied aircraft in countless missions over Bosnia and Iraq. It sounds trite, but many attack crews aver they will not fly unless at least one Prowler is available.

Besides jamming duties, the Prowler also made great use of the AGM-88 HARM, or high-speed, anti-radiation missile, intended to destroy enemy radar sites, especially those connected with surface-to-air missile (SAM) installations. In company with other services, the six Navy and one Marine Corps Prowler squadrons were in constant demand to defend strike groups against Iraqi SAM and flak-site radars. Thus, Prowler crews became some of the most experienced aviators of the American carrier task forces.

Wars generate many memoirs, sometimes written long after the conflicts, but often appearing soon after the fighting has ended and the author has had time to collect his thoughts. The Gulf War saw an especially large number of such first-person accounts written by fairly young participants.

Only 26 when he went to war, Sherman Baldwin was barely out of FRS training when he received orders to join VAQ-136 aboard USS *Midway* (CV-41), one of the Navy's oldest carriers, already on duty in the Persian Gulf as part of Operation Desert Shield. By the end of the war, he had flown 45 missions, quite an initiation to the fleet for a newly-winged jaygee aviator.

His memoir is filled with youthful wonder and occasional annoyance at the system of which he has become a part. The intensity of flying combat, especially at so junior a stage of his career, is especially poignant, as are his desperate attempts to become a good pilot so that the three naval flight officers who make up the rest of his four-man crew will feel confident about flying with him in the dark, moonless nights in the Gulf from one of the Navy's admittedly most unforgiving carriers. Then there is the controlled terror of refueling from an Air Force KC-135, a necessary trial during the Gulf War that every naval aviator had to contend with.

The second excerpt describes a mission against Iraqi flak positions involving a HARM shot.

Sherman Baldwin graduated from Yale University. Following his tour with VAQ-136, he served in several assignments until leaving active duty in 1995. He is currently a venture capitalist focused on information technologies in New York City.

From *Ironclaw: A Navy Carrier Pilot's Gulf War Experience,* by Sherman Baldwin

PART 1

The Iron Maiden

Tonight Cave was in the front seat with me, while Face and Bhagwan were the two ECMOs in the backseat of our Prowler. Face was called *Face* because he liked women and believed they liked him. He had a deep, dark, permanent Mediterranean tan which was fitting for his image. Bhagwan, in contrast, was a short, stocky, and feisty bulldog of a man. His call sign was *Bhagwan* because we thought that if you wrapped a turban around his head, he would look convincingly like a Bedouin with a name like Bhagwan.

Our mission tonight was officially Electronic Surveillance Measures, and we would use the Prowler's sophisticated electronic receivers to detect and locate possible hostile radar emitters in the Gulf of Oman. Face and Bhagwan would search for electronic signals that could be linked to enemy radars that, in turn, would identify an enemy position. The *Midway* was sailing several hundred miles south of the Strait of Hormuz, the narrow body of water that connects the Persian Gulf with the Gulf of Oman, so we did not expect to identify any threats tonight. Actually, the real but unstated mission of the flight was to update the currency of my night-landing qualification, and to see how I would stack up in the constant competition among my fellow pilots that surrounds carrier landings.

It had been fifty-three days since my last night carrier landing in training and I was more than a little nervous. The navy's regulation stated that nugget pilots were allowed a maximum of twenty days between their last night trap in training and their first night trap in their new squadron, if their squadron was at sea. However, the political situation was tense in the Persian Gulf and my squadron needed a new pilot, so the commander of Air Wing Five (known as the CAG, a holdover from when his title was Commander of the Air Group) on the *Midway* had waived this regulation for me. I was glad to be trusted, but that did not make me any less nervous.

Only ten days ago I had completed my shore-based training at Naval Air Station Whidbey Island, Washington, just north of Seattle. Now, on December 10, 1990, I found myself at Saddam Hussein's doorstep four months after the Iraqi Army had invaded Kuwait on August 2. I faced an incredibly steep learning curve. Knowing that the *Midway* would be in the thick of any combat action, I had requested this assignment, but now I was feeling overwhelmed and unsure of myself. At the completion of my two and a half years of flight training I had made a total of twenty daytime carrier landings and only six at night. Not only did I feel inexperienced and awkward in the fast-paced environment of fleet-carrier operations, but there now existed a high probability of combat in the near future. In order to be "combat ready," I knew I needed to increase my pilot proficiency level rapidly, if I hoped to survive. In aviation a pilot's proficiency is perishable over time, meaning that if a pilot does not fly frequently, his skills quickly deteriorate. The great pilots are always a step ahead of every situation in the cockpit and use good judgment to choose among various courses of action. They are proactive, rather than reactive. Being ahead of the aircraft comes from experience and practice. Tonight, being inexperienced and out of practice, I felt slow and reactive, behind the aircraft rather than ahead of it. It was not a good feeling.

"Nav is tight," said Cave as he diligently updated the navigation solution using the Prowler's ground-mapping radar to send position updates to the jet's Inertial Navigation System.

"Roger that," I said. "Hydraulics are good, oil is good, and we're looking at fifteen thousand pounds of gas." We had another forty-five minutes to go until our recovery, when my rusty landing skills would be put to the test. Right now I felt comfortably above my fuel ladder calculations. Projecting forward at the current fuel-flow setting that I had chosen, we would have 11,400 pounds of fuel when the recovery began. That was ample. The Prowler was limited by structural design to a maximum of 8,800 pounds for a carrier landing, so I figured that we might even have to dump some fuel. Fuel is every navy pilot's major worry when the only place to land is on a ship in the middle of the ocean. I began to feel more relaxed. I realized that I was finally starting to think ahead of the aircraft, anticipating the possible sequence of events.

Fortunately, tonight we were within range of a small airfield called Seeb in the United Arab Emirates. The UAE had given the *Midway* permission to use the field for emergencies only. We had been briefed that Seeb was a last-ditch divert airfield because it was a short field without arresting gear to stop our jet, and we were all unfamiliar with it. Since we were roughly one hundred miles from Seeb, it would be a 3.5 bingo. This meant that if we had not landed on the carrier by the time we reached 3,500 pounds of fuel, then we would immediately turn toward Seeb, and commit ourselves to landing with a low fuel caution light at an unfamiliar field. This was a thought that nobody in the crew was excited about. The three ECMOs in my crew all realized that this was my first night trap in the squadron. I'm sure they hoped that I would be able to get aboard without any difficulty, but they also knew that nuggets were unpredictable and often had a rough time with night landings when they first arrived in a fleet squadron. In turn, my crew was ready for anything. During the preflight brief Bhagwan had produced his toothbrush and a clean pair of underwear as testament to the fact that he was prepared in case I was unable to land back aboard the carrier tonight and we were forced to divert to Seeb.

We had flown north for thirty minutes toward the Iranian coast, trying to identify any signals of interest that were being emitted by the Iranian air defense forces. Face and Bhagwan were operating the Prowler's ALQ-99 sur-

Sherman Baldwin during Desert Storm. (Courtesy of the U.S. Navy, Senior Chief Terry A. Cosgrove)

veillance system, which had extremely sensitive receivers able to identify a vast range of electronic signals. The ESM mission was focused primarily on the backseaters. Cave and I were responsible for navigating a specific course that would place us in the optimum position to receive signals intelligence, while Face and Bhagwan worked the system to pick up as many signals as possible. They were probably chatting back and forth about what they were seeing on the system, but I could not hear a word of it because they had the front seat deselected from the intercom so as not to disturb my dialogue with Cave in the front seat regarding the navigation of the mission.

The Prowler's Internal Communications System (ICS) was quite complicated. Usually on missions the backseaters would do most of their talking to each other about the electronic countermeasures (jamming enemy radars) or electronic surveillance (listening to enemy radars) that they were doing. They would normally set up their ICS in the backseat so that they could also hear everything that the frontseaters were saying, but they would need to press a switch to talk to us in the front seat. This created an environment where an insecure pilot might always be wondering what his backseaters were saying about the way he was flying. It did not usually even cross my mind, but I was now the new guy in the squadron, hoping to make a good impression. I could not help wondering what Face and Bhagwan might be saying about me in the backseat. Perhaps they were betting on the odds of my being able to land back onboard the *Midway* tonight. I tried to push such thoughts out of my head and keep my confidence up as we continued to fly through the darkness. "We are showing hardly any activity on the system," said Face.

"Well, keep looking," said Cave. We flew our preplanned route without incident, and about twenty minutes later, I finally heard Cave's voice say in my ear, "We might as well head back to the ship." I nodded in agreement and turned inbound to the ship as Cave began to orchestrate our return. The Tactical Air Navigation (TACAN) system indicated that we were ninety-five miles northeast of the *Midway*. I immediately began to think about the upcoming landing. On our departure from the carrier, we had flown through several layers of clouds at lower altitudes that would be extremely disorienting during the approach to the carrier. Flying in and out of clouds at night was never fun and I was not looking forward to it. After a few minutes of flight toward the carrier, Cave made the first of many standard radio calls. "Strike, Ironclaw 605 is fifty miles to the northeast, state is base plus 8.6." By using the base number from the kneeboard card of the day, which today was four, the ship would know that we had 12,600 pounds of gas.

"Ironclaw 605, Strike, roger. Case III recovery marshal radial is the 090. You're cleared inbound and cleared to switch marshal." I still found the radio dance somewhat confusing and was glad to have Cave, who quickly switched to the marshal frequency where we would be given our holding instructions and other pertinent information about the recovery from the controller aboard the *Midway.* As soon as the frequency was dialed into our radio, we heard some familiar chatter from the other air wing aircraft airborne, preparing for this recovery. There were eight squadrons in the *Midway*'s air wing: three Hornet, two Intruder, one Prowler, one Hawkeye, and one helicopter squadron, for a total of more than sixty aircraft. Because of the small size of the *Midway*'s deck we did not have any F-14 Tomcats or S-3 Vikings in our air wing. We liked to think that our air wing's composition of predominantly Hornets, Intruders, and Prowlers made us the premier attack air wing in the U.S. Navy.

"Marshal, Eagle 510 is checking in, state is 9.0," said the A-6 Intruder's bombardier/navigator (BN), meaning he had nine thousand pounds of gas.

"Eagle 510, marshal, you're cleared to marshal on the 090 radial, angels 13, expect approach time 59, altimeter 30.10." As soon as the BN had read back his aircraft's holding instructions Cave jumped on the frequency. "Marshal, Ironclaw 605 is checking in; state is 12.6."

"Ironclaw 605, marshal, you're cleared to marshal on the 090 radial, angels 14, expect approach time 00, altimeter 30.10."

Cave read back the instructions verbatim as I started a descending turn toward our assigned holding point 29 miles due east of the carrier at 14,000 feet. Before lone, nine aircraft were stacked neatly from 6,000 feet all the way up to 14,000 feet due east of the *Midway.* Each aircraft was separated by 1,000 feet and the holding points were determined by adding the number 15 to the given holding altitude. I was holding at angels 14, or 14,000 feet, so my holding point was 29 miles away from the ship. The lowest aircraft in the stack would fly the first approach starting at time 2152, followed by the aircraft 1,000 feet above him at time 2153, and so on. The aircraft in the marshal stack would continue to fly approaches to the carrier in this way until those of us at the top of the stack had trapped on the carrier's deck.

"Time in fifteen seconds will be 46," said the marshal controller. There was a brief pause and then his voice returned. "Five, four, three, two, one, mark time 46," said the voice of the marshal controller, ensuring that each of the nine aircraft in the marshal holding stack had the correct time so that their approaches would be synchronized. As each aircraft checked in the marshal controller would give their assigned position in the stack.

Each aircraft was expected to commence its approach from the holding point plus or minus five seconds of the given approach time. If you started your approach either earlier or later than five seconds either side of the given time, you were expected to confess over the radio and publicly embarrass yourself. The confession helped the controllers to sequence the jets and ensure that the minimum amount of separation was maintained. The confession also served as a severe form of motivation to the pilots who, to a man, feared nothing more than looking bad in front of their fellow aviators.

Tonight the recovery started out smoothly. The lowest jet in the stack was an F/A-18 Hornet at 6,000 feet. At 2152 I heard, "Dragon 307, commencing, altimeter 30.10." We had eight minutes to go until we would commence our approach to the carrier. The Prowler's holding speed was 250 knots. The technique I used in order to hit the holding point on time was to fly the jet in a six-minute racetrack pattern. At 250 knots and 22 degrees angle of bank, it took the Prowler two minutes to turn 180 degrees. So if I could set myself up heading inbound at the holding point with six minutes to go, then I could fly a two-minute outboard turn, a one-minute outbound leg, a two-minute turn inbound, and then a final one-minute inbound leg, which would place the jet at the holding point exactly on time. The length of the outbound leg could easily be adjusted, depending how much time was remaining. There were now six minutes and thirty-five seconds remaining until the approach time of 2200, and I was on the 090 radial at thirty-two miles. Aside from the timing problem there was also a fuel concern; I still had 11,500 pounds of gas. Flying the approach would require about 800 pounds of gas, so when I commenced my approach in seven minutes, I wanted to have no more than 9,600 pounds in order to land with the maximum allowable limit of 8,800, according to the stress limits of the jet's fuel tanks. I needed to dump gas quickly. "I'm going to dump about two thousand pounds," I announced to my crew, letting them know that I was ahead of the jet. I turned the dumps on as the DME (mileage) indicator in the TACAN read twenty-nine miles, and I commenced my 22-degree angle of bank outbound turn. The clock showed that there were five minutes and forty seconds remaining until my approach time. That translated into a fifty-second outbound leg in order to hit the holding point on time. Cave led me through the challenge-and-reply descent and approach to landing checklists. He turned on the Automatic Carrier Landing System (ACLS) and the Instrument Landing System (ILS), testing each one for proper operation. Both systems seemed to be operating normally, but we would not really know until we were on our final approach.

The word *automatic* in the ACLS system was truly a misnomer. For the Prowler, there was nothing automatic about landing on the *Midway*. On the larger nuclear-powered carriers, some fleet aircraft could be landed, using this system without the pilot touching the controls throughout the approach.

However, the combination of the Prowler's older automatic flight-control system, the *Midway*'s small deck, and its minimal hook to ramp clearance of only ten feet made it a completely manual process for the pilot. Even if it was not truly automatic, the ACLS was still invaluable to all of the *Midway*'s pilots. ACLS was an interactive system between the *Midway* and each aircraft as it flew its approach. The ACLS radar on the carrier could lock on to a jet's radar beacon and then send continuously updated azimuth and glide-slope information to the jet's cockpit. The information was then displayed to the pilot in the form of a vertical and horizontal needle as a background to a small-aircraft symbol. The horizontal needle displayed glide slope and the vertical needle azimuth. The pilot's job was to fly the jet so that the small-aircraft symbol was directly superimposed on the crosshairs by the two needles.

"Checks are complete and I'm securing the dumps. One minute to push and we've got 9,600 pounds of gas. We're in good shape," I said as I smiled under my mask. Everything was going smoothly.

"Delta, delta, all aircraft stand by for new approach times," said the voice of the marshal controller

"Shit," said Cave. The delta call signaled a delay in the recovery and as a result, a delay in our approach time. I wished there were a way to bring back the gas I had just so carefully dumped. The controller now started asking each aircraft its fuel state. "Eagle 510, say your state."

"Each 510 is level angels eleven, state is 7.0," said the Intruder BN ahead of us at 11,000 feet. He had already started his approach to the ship when the delta had been called, so according to procedures, he had leveled off at the next odd altitude after he heard the delta call.

"Ironclaw 605, say your state," said the controller.

"605, state 9.6,." came Cave's terse response. Everyone's thoughts turned to gas. For the moment we were fine, but one could never tell how long the delay might be. "There's no Texaco tonight, but there is an Iron Maiden. Its call sign is Mako 12, at angels twenty-four," said Cave. I cringed. Texaco was the navy term for the carrier-based A-6 tankers which I had learned to tank from in-flight training. Many A-6 pilots on the *Midway* wore Texaco patches on their flight jackets because they provided the air wing with gas. Iron Maiden, on the other hand, was our squadron's nickname for an air force KC-135 tanker. The KC-135 was a converted air force cargo plane that was truly a gas station in the sky. The tanker had earned its nickname because tanking off it was a cruel form of torture that had already broken dozens of our air wing's refueling probes.

"We still have lots of gas to play with," I said cheerfully.

"Yeah, we should be fine," Cave replied. "Don't break out your toothbrush yet, Bhagwan."

It had now been six minutes since the marshal controller had given us the delta call and we were back at our holding point. The weather was disorienting at our holding altitude. There was no discernible horizon, yet I could tell that we were flying in and out of the clouds because of the varying intensity of the reflection of the Prowler's anticollision strobe lights. Delays in a recovery could be caused by a number of different situations. The frustrating aspect of it was that the carrier never seemed to tell you the nature of the delay. *It could be that a few jets had boltered, missing all three of the* Midway's *wires, and the landing pattern around the carrier was now full. Or it could be that a jet has just crashed into the back end of the boat,* I said to myself. I smiled under my oxygen mask, realizing that I had already adopted the naval aviator's habit of referring to the ship as a boat, the stern as the back end, and the bow as the pointy end. It was language used by aviators to annoy the officers in the Surface Warfare community who were cut from the more traditional naval cloth.

The back end of the boat was also called the ramp. When an aircraft crashed into the ramp, it was called a ramp strike. They were rare, but everyone had heard stories of the massive fireballs that would light up the dark night. The ramp at night became the type of a monster that lived in every pilot's nightmares. All navy pilots have had at least one close call with the ramp that they would prefer to forget. The pilots who learned from the encounter forevermore flew on the high side of the glide slope—and those who didn't learn—well, it was just a matter of time. The ramp monster began to creep into my thoughts as we waited to learn our new approach time. My eyes glanced repeatedly at our fuel gauge as it kept getting lower and lower.

Once again the marshal controller asked each aircraft to say his fuel state. And once again we heard, "Eagle 510, state 5.5."

"Ironclaw 605, state 8.0," said Cave. In another ten minutes we would be below our ramp fuel of 7.0, which was the target fuel that Prowlers were expected to land with, according to our air wing's standard operating procedure. As long as we stayed above 4,700 pounds, I would be happy. At 4,700 pounds or less we would be sent to the Iron Maiden, and I really wanted to avoid that at all costs.

After what seemed an eternity, we heard the controller's voice again: "Standby for expected approach times."

"Great," I said. We would be fine if we pushed in the next ten minutes. The time was now 2210.

The controller's voice came over the radio: "Acknowledge your approach time with fuel states. Eagle 510, expect approach time 16."

"510, approach time 16, state is 4.5," repeated the Intruder's BN, who was now getting quite low on gas. It was time to land. Time was becoming critical.

"Ironclaw 605, expect approach time 17."

"Ironclaw 605, approach time 17, state is 7.0" said Cave. "Another six minutes to our approach time. We should call the ball with about 5.8. That's plenty, no problem," said Cave.

"Calling the ball," happened at three quarters of a mile behind the carrier when the pilot transitioned from an instrument approach to a visual approach. It was the most critical part of a night carrier landing. It would take about twenty seconds to fly that final three quarters of a mile. Those twenty seconds were infused with the purest form of survival instinct. Night carrier landings were the practice of overcoming the fear of death that lingered in the back of every pilot's mind.

"The ball" or "meatball" was the nickname given to the navy's Fresnel lens system that offered pilots a visual reference to help them fly a constant glide slope from three quarters of a mile all the way to landing. Five specially cut rectangular lenses of light were stacked vertically in the middle of a horizontal row of green circular lights. The vertical stack of cells projected a yellow "meatball" of light toward an incoming jet. The top four cells were yellow and the bottom cell was red. If a pilot saw the yellow "meatball" higher than the horizontal row of green lights, then his jet was high. If the yellow "meatball" appeared below the row of green lights, then his jet was low. If the "meatball" turned red, then the pilot knew he was dangerously below glide slope and would hit the ramp if he did not make an aggressive power addition. Flying the ball was more of a philosophy or an art than a science. Pilots attached a Zen-like aura to those few who had truly mastered the art.

Tonight I did not feel as if I had the requisite Zen. I was nervous and I prayed that I would not bolter. Bolters were embarrassing, and I was determined not to be embarrassed tonight. However, I realized that boltering was a distinct possibility, and I wanted to have enough gas so that if I did bolter, I could go around again without being forced to go to the Iron Maiden. Bottom line: I needed to land before my jet's fuel gauge indicated 4,700 pounds. As my mind wandered and worried, our expected approach time drew closer.

I needed to focus on the job at hand, which was to hit my holding point on time so that I would get a good start to my approach.

"Two minutes to go," said Cave, as we both closely monitored our progress in the holding pattern. We had ninety degrees of turn left and then we would have a fifty-second inbound leg. The timing problem seemed on track, but I continued to focus on our fuel situation. The gauge indicated about 6,600 pounds, which was about 3,000 pounds less than what I would have liked to have. It would have to be enough. *No bolter tonight*, I told myself.

At time 2216 we heard, "Eagle 510, pushing, altimeter 30.10." The Intruder below us was on his way and we would soon follow in less than a minute. As I rolled out wings level on the inbound course, the timing looked good. The INS indicated a ground speed of 240 knots, and there were four miles to go. I was going to be right on time.

"Ironclaw 605, pushing, altimeter 30.10," said Cave as the second hand passed through the twelve.

"Ironclaw 605, I show you at 29 DME switch button 18," said the marshal controller.

"Switching," said Cave, as I eased back on the throttles to 75 percent RPM, lowered the Prowler's nose ten degrees below the horizon, and extended the speed brakes, which increased the drag on the aircraft and enabled a rapid descent. This maneuver quickly gave the Prowler a 5,000-feet-per-minute rate of descent toward the water below. I felt the rush of speed when I saw with my peripheral vision layers of clouds whipping by the cockpit as we plunged downward. The analog hand of the altimeter unwound quickly as I scanned my instruments to ensure that all systems were normal.

"Approach, Ironclaw 605, checking in at twenty-seven miles."

"Ironclaw 605, continue CV1 approach, cleared to angels 1.2," said the new voice on the radio. The large heavy nose of the Prowler naturally sought the water. The altimeter kept spinning at a quick rate until we passed 5,000 feet and the preset radar altimeter started beeping, warning me to reduce my rate of descent. My thumb pushed the speed-brake switch in, and the Prowler's wingtip speed brakes on both sides closed flush like hands joined at the palms closing together. I then pulled back on the stick in order to reduce my closure with the ocean below.

"Ironclaw 605, platform," said Cave, making the next mandatory radio call as our jet passed through 5,000. Having adjusted my rate of descent, I started to concentrate on the level off.

"One thousand to go," I said over the intercom as I gradually added power and pulled the nose up even farther, easing the descent to 1,200 feet.

At fourteen miles from the *Midway* we were now flying straight and level toward the ramp at 250 knots.

"Ironclaw 605, stay clean through ten, I'll call your dirty up," said the approach controller.

"605," said Cave, acknowledging the call. The normal procedure was to transition to the landing configuration at ten miles, but because the Intruder ahead of us had pushed from a lower altitude than normal, there was a bigger gap between us. The controller wanted me to keep my speed up until eight miles so as to close this gap and expedite the recovery. The transition from 250 knots clean to 130 knots with the landing gear down and the flaps down was a major transition. Eight miles was considered the minimum distance to make a smooth transition.

"Delta, delta," said the voice of the approach controller. "Ironclaw 605, discontinue your approach. Take angels two and continue inbound." Our fuel gauge now indicated 6,000 and it seemed to be decreasing as I glanced at it.

"We have a foul deck for at least ten minutes. Ironclaw 605 estimate state on the ball at time 2230," said the controller.

"Stand by," said Cave.

"Ten minutes at this altitude will be about eight hundred pounds plus the fuel for the approach another eight hundred. We will be at 4.6." I grimaced as I said this figure because it was below 4.7.

"Approach, Ironclaw 605, estimating 4.6 on the ball at time 2230," said Cave matter of factly.

"Ironclaw 605, approach, copy 4.6. Your signal is tank. Mako 12 is overhead at angels 24."

"Damn," I said, as I rammed the throttles to full power to give me the most fuel efficient climb to 24,000 feet. The two Pratt & Whitney P408A engines roared, and the Prowler responded as if it were an angry horse that I had just kicked with the spurs on my boots. We quickly accelerated to .7 indicated mach airspeed, the Prowler's most efficient climb airspeed. *That damned Intruder ahead of us must have broken down in the landing area*, I said to myself. At 16,000 feet the haze layer disappeared below us and visibility drastically improved. The stars were out, and I struggled to pick out the tanker's white light amid the field of constellations. "One thousand feet to go," I said as the altimeter passed through 22,500. "I'll level off at 23,500 until we have the tanker in sight, and then I'll climb to rendezvous."

"Traffic at two o'clock a little high," said Cave. I cranked the jet around to the right to put the nose onto the possible tanker. Off to the right of the nose I saw a white strobe and agreed that it must be the tanker. The trick to any rendezvous is figuring out the aspect and closure rate with the other aircraft. At night, without a fighter's sophisticated air-to-air radar, the Prowler was at a distinct disadvantage. By turning nose to the tanker, I hoped to be able to develop a picture of the relative motion of the tanker to my jet. The white strobe began slowly to rack from right to left across my windscreen. That was good. It was best to join up on the tanker from the inside of its left-hand racetrack pattern. I let the nose of my jet lag behind the white strobe light. This lag increased the rate at which the light passed across my windscreen. I then added power and climbed the final 500 feet to be co-altitude with the Iron Maiden. My airspeed was 350 knots and I expected Mako to be at the standard rendezvous speed of 250.

I was now facing the common tanking dilemma; if the rendezvous was not expeditious, then the length of time would run me short on fuel. Yet, if I was too aggressive in expediting the rendezvous, that too might run us out of fuel because of the high power settings required for maneuvering. I did not want to divert, so I needed to time the rendezvous just right. I now held an aggressive 100 knots of closure on the tanker and knew that I had to be careful. Such a high closure rate at night could easily get out of control. I increased my angle of bank to the left, now putting my jet's nose in front of the tanker. By leading the tanker in this way I was also increasing my closure rate. Afraid of too high a closure rate, I began to ease back on the throttles and decelerate. Now, with a slower airspeed of 300 knots, I felt much more comfortable, yet a left-turn rendezvous was never truly comfortable in the Prowler. The jet's side-by-side seating design, with the pilot on the left, made it incredibly difficult for me to see the tanker on my right side while I was in a left-hand turn. I strained my neck to see the KC-135 over the Prowler's canopy rail on Cave's side of the cockpit.

"Too much closure," said Cave anxiously. The tanker was getting very large very quickly. I pulled the throttles to idle, extended the speed brakes, and lowered the nose to make sure that we did not have a collision. I arrested the closure rate just in time, and, even though it was not a pretty rendezvous, we were now flying off the tanker's left wing.

"Ironclaw 605, port observation, nose cold, switches safe, looking for 5.0," said Cave, who had tanked off the KC-135s many times before and knew what to expect. We ran through the refueling checklist quickly and I selected air-to-air on the Prowler's refueling panel.

"Ironclaw 605, you're cleared in for 5.0," said the tanker pilot.

"Do you want to lower your seat?" asked Cave, knowing that every other pilot in the squadron had learned through experience that it made tanking off the Iron Maiden a lot easier if you lowered your seat.

"No, I'll just leave it like it is," I said with a mild tone of resentment. *Who does he think he is anyway? I'm the pilot, damn it,* I said to myself. I pulled the throttles back and maneuvered the Prowler behind the tanker and took a look at the Iron Maiden's basket for the first time. Our gas gauge now read 4.8, and my flight gloves were soaked with sweat. The leather palms of the Nomex flight gloves felt slippery on the stick's hard black plastic grip. My fingers were clenched around the grip, squeezing it tightly. I needed to relax, but too many things were happening tonight that I had never seen before. I was feeling the stress and knew my crew could tell that I was. If I could not tank successfully, we would have to fly to Seeb. What a nightmare; the embarrassment of not being able to hack it was too great to contemplate. I could imagine Face and Bhagwan taking out their approach plates and control frequencies for the divert field in anticipation of my failure. As a nugget, my every move was watched by everyone in the squadron. If I were to fly in combat, then I would have to be able to tank off these KC-135s at night routinely. I needed to prove myself reliable. I needed to hack it.

The KC-135's refueling basket had a hard steel rim illuminated by small orange lights. The basket was only thirty-six inches in diameter and was connected to a stiff nine-foot reinforced rubber hose by a metal ball joint that swiveled, depending on the position of the basket. The only "night" tanking I had ever done was off an A-6 tanker in training at Naval Air Station Whidbey Island. It had been done one minute after official sunset, and at twenty thousand feet it was still quite light out. At that time my instructor said, "I guarantee you that the first time you tank at night in the fleet it will be pitch black and you will really need the gas." I smiled wryly as I realized how right he had been. From talking to the other pilots in the squadron I knew that there were two obstacles I needed to overcome to tank successfully off the Iron Maiden. The first was the boom operator and the second was bending the hose.

The KC-135, being an air force aircraft, was designed to refuel air force tactical jets and had to be reconfigured to fuel navy jets. In the infinite wisdom of the Department of Defense, the navy and air force accomplish air-to-air refueling with completely contradictory philosophies. In the navy, the receiving aircraft positions itself aft of the tanker's refueling basket and the receiving aircraft's pilot then maneuvers his aircraft's refueling probe into the tanker's basket. In the air force, the receiving aircraft positions itself aft of the tanker and the tanker extends its refueling probe. While the receiving aircraft maintains its position, the tanker's boom operator will fly the tanker's refueling probe into a small basket that is located on the top of the receiving aircraft. On a KC-135 tanker reconfigured for use with navy jets, the tanker's refueling probe is replaced by a basket so that the navy jets can use their refueling probe to "plug" the basket. Problems arise when a navy pilot tries to "plug" the basket and a well-intentioned air force boom operator attempts to guide the basket onto the jet's refueling probe. The result is similar to what happens when one person in a group drops a coin on the floor and two people bend over to pick it up and smack their heads together. Even though both people are trying to accomplish the same thing, they are not coordinated. In the end, one person will bend over and pick up the coin. I hoped that the boom operator would simply let me "pick up the coin" and plug the basket without trying to be too helpful.

The second obstacle was staying in the basket long enough to receive five thousand pounds of gas. The A-6 tanker package had a much longer and more flexible hose that allowed the receiving aircraft to maneuver more freely behind the tanker. The other pilots in the squadron had described how it was necessary to bend the KC-135's hose into an S shape that would allow the pilot to control the swiveling ball joint of the basket. If the hose was not bent properly, then the swiveling ball joint could violently twist around the jet's refueling probe and possibly damage it so that further refueling was impossible. If the hose twisted quickly, then the best course of action was to disengage. However, on disengagement there was always the possibility that the basket's large steel rim could slam down on the nose of the receiving aircraft. With these thoughts swirling in my brain and my palms drenched with sweat, my stomach knotted with fear, I started my approach toward the thirty-six-inch basket.

The Prowler's refueling probe was illuminated by a small red light that was directed upward from the base of the jet's windscreen. My hands felt as if they were shaking, but they were simply making the minute movements necessary to keep the probe moving slowly toward the basket. The basket was not ten feet in front of the probe's tip, and as the distance decreased I watched the basket start to move. Was it my own poor technique or was it the friendly boom operator trying to be helpful? I couldn't tell. I added more power and continued to close on the basket. It began to move to the left and I made the necessary correction. I was almost there when the basket suddenly moved down. The probe tip hit the basket just inside the top of its rim and bent the swivel joint upward. Instead of sliding into the basket, the probe slipped over the top of its rim and the steel basket snapped down and smashed into the nose of the Prowler. "Damn it," I swore over the intercom system. I pulled back on the throttles and winced. I slid back twenty feet behind the Iron Maiden and looked at the nose of my jet.

"No harm done," I said.

"The probe looks all right," said Cave. "Well, let's give it another try." His calm voice belied the fact that our fuel gauge now indicated 4.2. I needed to get in the basket and stay in the basket. After a deep breath and unclenching my fingers several times, I began my second approach. My left hand gently pushed the throttles forward, creating the necessary closure on the basket. Once again the basket stayed steady until the probe was only a few feet away. Now I began to doubt myself. The boom operator was probably doing nothing. Was it just my poor technique that was preventing me from getting into the basket? The frenetic minor corrections I was making to chase the basket were getting smaller and smaller as the probe got closer and closer. At the last instant, the basket started to move down again. I added a handful of power and lowered the nose. The probe slammed home in the center of the basket and the swivel joint whipped the hose around the top of the probe. *Bend the hose, bend the hose*, I told myself. The hose turned and twisted wildly as I tried to stabilize my position and create the bend in the hose that was required. My hands jerked crazily as if I were being electrocuted.

The problem was that I could barely see the basket. The Prowler's canopy has almost as much steel as glass. Since I hadn't lowered my seat, I was forced to lean forward and awkwardly stretch my neck and roll my eyes upward in order to keep sight of the basket and the probe tip. I was leaning so far forward that my chest was right on top of the stick, so it was difficult to make the necessary corrections to keep the probe in the basket. The KC-135 had just reached the end of its holding pattern's straightaway and it started to turn. The angle of bank swung the boom out to the side, and as I tried to make the necessary recorrection my chest got in the way. The oscillations started to become too great. As I leaned back to allow more room for stick movement, I lost sight of the basket behind the canopy's steel frame. I pulled back on the throttles too quickly and could not recorrect in time. The Iron Maiden spit out the probe, and once again I found myself with more sweat and less fuel than I had had a few minutes earlier.

"Let's give it one more try and then we'll divert to Seeb. Our bingo fuel from here is 3.5 and right now we've got 3.8," said Cave.

"Concur. I'm going to lower my seat," I said, recalling Cave's sage advice given fifteen minutes earlier. I could just imagine the twinkle in Cave's eye and the grin on his lips.

The boom swung gently as the lumbering KC-135 continued its turn. I rested and breathed deeply as the tanker turned. It would be much easier to plug and stay plugged once the tanker rolled out of the turn, so I decided to wait. As the tanker rolled out I looked at the gas gage; the needle wavered at 3.7. This was definitely the last chance I would have. The perspective was different now that I had lowered my seat. I felt my neck muscles relax as I was now able to look up comfortably at the basket while I added power and began to close the gap. The new seat position also gave me renewed confidence. The basket started to move again as I closed within ten feet. I pulled some power and stopped the closure. My hands kept moving and I wiggled my toes, which had always helped to relax me I the past. A warm rush of adrenaline pulsed through my veins as I added power. The basket stabilized and the probe struck the center of the basket. The hose quickly bent into an *S* shape, and now that I could see the basket easily, I knew that I could maintain this position.

"We've got good flow," said Cave. My hands kept up the frenetic pace of correction and recorrection. Each power addition would bend the hose more, and each power reduction would bring me closer to losing control of the bend and having the hose wrap wildly around my refueling probe and possibly spit me out again. Finally, after several minutes of torture, the fuel gauge indicated 8.5. "We've got all we need. Great job," said Cave. I knew he was trying to build my confidence because he knew as well as I did that the hard part of the flight was still ahead. I now had to land on the smallest carrier in the fleet.

Part 2

HARM Shoot

"Great, no fighter cover for us," said Gucci, verbalizing what each of us was thinking.

"Ironclaw, Dragon 405, I'm down and returning to Mother. I'll see if they can launch a spare."

"Roger, we're going to press," I said, knowing that all the spares had shut down long ago on the *Midway*'s deck and that even if they could launch one it would be too late to be any good by the time it could catch up to us.

"Good luck, Ironclaw. Dragon 405 is detaching to the right." The Hornet's bright fluorescent strips and red anticollision lights faded quickly in the blackness of the night.

"Let's hope Liberty and AWACS are squared away tonight," I said. The navy's E-2C Hawkeye and the air force's E-3 AWACS were the only aircraft that would be able to provide us with warning of approaching enemy fighters tonight. That was not a fact that filled any of us with great confidence. The guys in the early-warning aircraft quickly reached task saturation on these large, nighttime air strikes, and threat warning quickly dropped a few notches on their priority list. I was resigned to the fact that I would be the pilot of a flying bull's-eye tonight. Our mission called for us to fly a predictable orbit over the oil platforms that might have comm capability with one of several fighter bases in Iraq. The thought did not bring a smile to my face.

We separated from the tanker and flew a 350 heading toward our jamming orbit. "Without an escort I think we should go without our external lights," said Gucci.

"Definitely," I said as my left thumb flicked off the master switch, allowing our jet to blend into the night. Our one advantage was that we knew that at night Iraqi pilots were heavily dependent on ground-controlled radar intercepts and we were also jamming those. Such a radar intercept still demanded the pilot to acquire the target visually in the end game, so whatever we could do to reduce their ability to see us was to our advantage. With our lights out we headed north into enemy territory—alone and afraid.

"Liberty is picture clean," said the anonymous voice of the Hawkeye controller on the strike frequency. This night those words were going to be especially soothing. I couldn't help but feel that I was pilot in command of a flying target just waiting for an Iraqi fighter to thank me for making his first kill of the war so easy.

"Steering is to the outer point of our jamming orbit," said Gucci. "Two minutes to jammers on." Our jamming obscured the strike group from ground-based radars, but it clearly highlighted our general sector in the sky. All it would take is one aggressive fighter flying right down the jamming strobe to ruin our day.

"Guys in back, once you get the jammers set up, I need you to keep up a good lookout. Without an escort we are very vulnerable," I said with a clear tone of concern.

"Understood," said Wolfey, "we'll be looking."

"Jammers on," said Skippy. "We have good power out on two of our pods, but the third is hurting. We're trying to bring it back up."

"OK, let me know when you have it on line," said Gucci. "Tank, we're about one minute out from the first oil platform."

"Roger, I see triple A off the nose." The dark sky was being shredded by the bright lines of 57mm and 76 mm AAA. "Look at those tracers. They're coming up to our altitude," I said.

"Break left, break left," said Gucci. I yanked the stick hard to the left and a stream of tracers lit up the sky where we had been only a second before. Training took over as I pulled hard into the break turn-away from the tracers. At that moment I saw a bright flash in my peripheral vision and felt the jet shudder around me. My immediate thought was, *Oh my God, we've been hit.* I reversed my break turn and pulled even harder. I pulled as hard as I could and saw another bright, blinding explosion aft near the tail section, and as I increased the pull I felt even more vibrations from the aircraft. My heart racing, I pushed the throttles as hard as I could and pulled on the stick even harder.

"Altitude," piped up Gucci.

"Got it," I said as I reduced the pull, leveled the wings, and tried to start climbing out of the hole I had fallen into. My maneuvering had been reckless and costly. We had lost five thousand feet in the various defensive maneuvers I had made and now we were flying in the very heart of the AAA fire that we had been trying to avoid. Climbing uphill the Prowler had about as much maneuverability as a Mack truck crossing the continental divide. My left arm was locked and the throttles were pressed to the stops as I tried to squeeze every ounce of thrust out of the Pratt & Whitney engines underneath me. We managed to avoid the AAA somehow, and when I finally got up to twenty thousand feet, I realized that I was behind on the timeline for the HARM launch for the mission. Maintaining max thrust we accelerated toward our launch point.

"Liberty is picture clean," came the voice of our eyes-in-the-sky broadcasting over strike frequency. We were glad to hear that there seemed to be no Iraqi fighter activity. My mind quickly shifted back from the tactical picture to flying my jet.

"The jet seems to be handling OK, but I thought we had been hit," I said.

"Me, too," said Gucci. "I released two chaff-and-flare programs to help us out," he said over the intercom as he continued to work on the navigation solution.

"You released two programs? Oh my God!" I said.

"What's wrong?" asked Gucci.

"What I thought was flak exploding near us was just you releasing chaff-and-flare programs without me knowing it." My inexperience and Gucci's nerves led us to the embarrassed realization that in the darkness and confusion of the night we had been trying to avoid our own chaff and flares. What had felt like explosions was airframe buffet attributable to the strength of my own pull. I had yanked the Prowler into heavy buffet dangerously close to stall speed.

"Hell, that's embarrassing," I said, realizing the mistake was now behind us. "But now we have to launch this missile on time."

"Steering is to the launch point and if we maintain this speed we will be five seconds late to our point," said Gucci.

My arm remained locked at the elbow as I tried to squeeze out all the thrust from our engines. It felt like the Prowler still had some acceleration left and I wanted to launch this HARM on time. My eyes darted back and forth from the steering needle, to the time to go on the computer, to the clock. All the parameters were closing in and coming to a climax.

"Missile on Station Two is designated. You have your needles," said Gucci. They were the same needles I referenced for night landings, but now they were linked to the HARM control panel. I pulled the jet around to the right and then back to the left once the needles had centered up. As the horizontal needle began to come down I prepared to pull the trigger. I judged the rate of movement of the horizontal bar and squeezed the trigger just prior to its meeting the horizon. I closed my eyes to avoid being blinded by the brilliance of the HARM's rocket motor in the darkness. As soon as the missile left the rail I keyed the mike and once again broadcast the warning to my fellow strikers: "Ironclaw magnum."

"Right on time," said Gucci. "Let's turn outbound." I pulled the Prowler around to a southeasterly heading and we continued our jamming as we headed outbound from the target.

"Liberty's picture clean," came the reassuring voice from the E-2C Hawkeye. A few minutes passed in silence, which was then broken by good news.

"Hammer flight feet wet," pronounced the Hornet strike lead.

"Thunder flight feet wet," echoed the Intruder lead. It appeared as though the Iraqis had missed everyone again. I couldn't believe it, but I relaxed and climbed to the RTF altitude and slowed to the RTF airspeed for the boring yet satisfying flight back to the ship.

We had recovered from our mistake and had salvaged the mission. After performing a death spiral into the heart of the oil platform's AAA envelope we had been able to speed up enough to hit our launch point on time. We were less optimistic that we had hit the target, however, because the signal had disappeared before the missile had impacted. That fact meant that it was more than likely that the HARM missile we had fired, no longer having a radar signal to home in on, had missed its intended target. This time the Iraqi radar operators had survived, but our mission was still successful because we had forced them to turn off their radar and without it, they were blind, with no way to shoot down our striking aircraft.

Bailout in the Balkans

Long a hotbed of internal rivalries, the Balkans have been the background for events leading up to several conflicts, not the least of which was World War I. A young assassin's bullet in Sarajevo ignited a new level of warfare that spread across the globe for four bloody years. Even during the next world war 20 years later, the only thing that brought the feuding factions in the region together was their common hatred of German occupation. Communism brought a measure of stability to the Balkans under somewhat benevolent dictator Josip Broz, also known as Marshal Tito. However, after his death in 1980 and the demise of the Soviet Union in 1991, the area slid back into the myriad civil conflicts that finally boiled over in open war that engulfed not only the land but the people.

European leaders tried to contain the war for four years. The U.S. eventually brokered a cease-fire with the 1995 Dayton Peace Accords, which ended the fighting in Bosnia and Croatia for a while. But by 1997, rising tumult, fed by the festering hatred the communities had for each other, again erupted into all-out war. Eventually, and somewhat unexpectedly, NATO, which had slumbered for so long after World War II, awoke to a major war in its own backyard. The enemy was not the predicted Soviet Union and its minion states of the Warsaw Pact, but a relatively small area split asunder by people who simply abhorred each other and would go to nearly any lengths to kill their neighbors.

At first, in the mid-1990s, Britain and the United States, occasionally supplemented by France and Italy, had tried to protect the general population as well as the crews of the airlift mounted to bring in supplies and peacekeeping personnel. The Adriatic Sea has not seen many carrier operations, even during World War II. The presence of American, British, and French ships launching aircraft on periodic reconnaissance and strike missions gave this prolonged conflict a unique aspect.

In April 1994, Serbian forces assaulted the Bosnian Muslim town of Gorazde in southeastern Bosnia-Herzegovina, south of Sarajevo. The engagement turned into a major battle, with Allied forces quickly entering the fray. American F-16s and FA-18s struck the Serbs in what has become noted as the first attacks mounted by NATO in its 40-year existence. The fighting was not one-sided. A French reconnaissance Étendard IVP took a hit from an enemy missile, and a Royal Navy Sea Harrier FRS.1 was shot down on April 16. The British pilot, Lieutenant Nick Richardson of 801 Naval Air Squadron, had launched from the carrier HMS *Ark Royal* on a reconnaissance mission to find Serb targets.

Ejecting from his stricken Harrier, Lieutenant Richardson soon found himself in considerable danger as he began trying to evade capture. His book is unusual for several reasons, including having been written by a Harrier pilot engaged in one of the 20th century's last wars. While there have been many books written by British naval aviators from World War II, little has come from subsequent generations; thus, this book is all the more welcome, particularly because it has a happy ending.

Lieutenant Richardson describes the launch from *Ark Royal* and the shootdown.

After his experience, Lieutenant Richardson served a two-year exchange tour with the RAF, being promoted to lieutenant commander. In 1997, he returned to the Fleet Air Arm and served as an instructor with 899 Naval Air Squadron, the Sea Harrier OCU (operational conversion unit), the British equivalent of the American FRS. Retiring in January 2000, he remains in the Royal Naval Reserve as a Harrier aviator. He currently lives in Saudi Arabia where he is an instructor with the Royal Saudi Arabian Air Force.

From *No Escape Zone,* by Nick Richardson

The yellowcoat's hand changed from a balled fist into the signal that gave me my instruction to roll: fingers and thumb of the right hand splayed wide, the gesture held there in freeze-frame for a good long moment.

I felt *Ark Royal* heave as another big wave started to ride the length of her. The swell merged uncomfortably with the steel claw that was already raking its way through my guts. No matter how many times I launched off the deck of a ship, each time seemed like the first. Controlled bloody mayhem, rendered more interesting by a two-minute infusion of high-octane adrenalin.

The Sea Harrier's Pegasus engine was already belting out tons of thrust, but still I toggled the throttle for more. The aircraft bucked against the brakes and held there, its entire rear half hanging over the fan tail, with nothing but the height equivalent of a six-storey flock of flats between me and the surface of the Adriatic. A little extra power never hurts at this stage if you want to keep from deep-sixing over the side.

Out of the corner of my eye I caught a glimpse of angry grey eddies and white foam whipped up by *Ark Royal*'s powerful screws, then the rear of the ship started top haul back up again. Never mind the yellowcoat. This was the real signal to move.

I released the brake and eased the Sea Harrier forward.

The yellowcoat gave me a thumbs-up and pointed to the flight deck officer, the FDO, distinguishable by his white vest; my next point of contact.

I manoeuvred the aircraft down the tramlines, reached the 500ft marker and stopped. My wingman was visible in my mirror a respectable distance behind me.

Five hundred feet between you and the ramp exit feels like nothing when you're sitting in a fully loaded aircraft—a thin skin of aluminium wrapped around several tons of fuel, a quaking power-plant and several mission quid's worth of electronics. On this occasion, because of the high threat level, I had the additional weight of a 1,000lb bomb to contend with. Why we bothered, I couldn't really fathom. No one had dropped a bomb on the Serbs yet. And if the UN carried on the way it was going—moving the goalposts every time Karadzic and his cronies pretended to step into line with UN resolutions—no one ever would.

The FDO was standing 30 feet to my right. He was holding his red flag in the air and the green one down low, his eyes fixed intently on the lights by the bridge that would tell him when he was clear to launch. Seconds from the signal, there was still one last big check I had to do. If the Pegasus was going to fail on me, I needed to know now.

I rotated the engine nozzles all the way down and spun up the power. The Pegasus's fan is as wide as a big family saloon car and the noise that it throws out is deafening. Even with his ear-defenders on, the FDO winched visibly.

He was so close I could see the tautness of his expression and rivulets of spray on his face. The poor sod had been out here long enough for a thin crust of salt to form like fine powder on his cheeks and in little drifts in the corners of his mouth. Even though I was only seconds away from being launched across the tops of the waves, I knew where I'd rather be.

The rule book said you needed to be registering 100 per cent or more as you thundered down the deck. After one second of roll, though, I was committed to launching whether I got full power or not. A carrier deck is a place of simple truths.

The light on my instrument panel told me I was riding a good engine. I brought the throttle back to idle, rotated the nozzles aft, and then, almost immediately, shoved the power back up to 55 per cent. Ahead of me, I could see the bow starting to fall and the first licks of ocean swelling beyond the grey metal of the ship. Come on, I found myself willing the FDO as I risked a last glance at my flickering engine instruments. Let's light the bloody candle.

I got my wish.

I had a momentary impression of the lights changing below the bridge, then the FDO brought up his green flag. I slammed the throttle forward. There was a seemingly interminable pause as the Pegasus fought to reach full power. The sound of its screaming machinery filled the cockpit. I stood on the brakes and felt the vibration transmit through my feet into my body. The aircraft wanted to go, but not yet, I told it, not yet.

I shoved the throttle to the stops and the power hit 100 per cent. The aircraft started to skid across the deck. I released the brakes and felt a giant boot in the back as 21,000 lb of thrust shot out of the nozzles and propelled the Sea Harrier forward.

The FDO and the bridge disappeared in a sickening streak of colour on the periphery of my vision.

A quick glance at the rpms, a minute adjustment on the rudder bars to keep the aircraft straight and I was heading for *Ark Royal*'s ski-jump at the speed of heat. For a brief moment it filled my vision: a grey mountain, almost indistinguishable from the sea beyond. Then I shot over the ramp, gasping as the aircraft, no longer supported by the deck, lurched towards the waves.

Before it lost all its ballistic energy, I rotated the nozzles 35 degrees and felt the cushioning downward thrust of the Pegasus as I clawed for airspeed.

Only when I heard the ker-klunk of the undercarriage as it folded into the belly of the plane did I relax my grip on the stick and tilt the Sea Harrier towards the Dalmatian coastline.

Twenty minutes later, a crackle in my headset signalled I had a message inbound. The voice of the AWACS controller steadied on the ether.

'Vixen Two-Three, this is Magic from Chariot. Proceed to Italy and contact Fortune Zero-Five on TAD Three.'

I acknowledged before I had absorbed all the information. My sixth sense must have already got the gist of it, though, because the hairs on the back of my neck were standing up. Something big was going down and we'd just

Nick Richardson in full flying kit in front of a Sea Harrier. (Courtesy of Nick Richardson)

flown into the thick of it. The exercise had switched to something infinitely more deadly.

I checked my map. 'Italy'—the codename for the besieged town of Gorazde—was 20 miles to the south-east of my present position, 10,000ft over the battered ruins of Sarajevo.

I banked the Sea Harrier and reached for the bundle of OS maps tucked under my left leg.

Keeping one eye on my instruments, I picked out the main map from the fan of charts, shoved those I didn't need back under my thigh, then did a swift bit of one-handed origami to ensure that the folds of the main map took me away from the Sarajevo area and out over Gorazde.

My last move in this cockpit version of Twister was to open up my notebook—an aircrew companion known as the 'green brain'—at the page containing the authentication codes the forward air controller, the FAC, had to provide to demonstrate that he was genuine.

I looked up and saw the weather closing in. The cloud cover over Sarajevo had been intermittent, but the base of the dotted cumulus stacks had steadied at around 12,000ft. As I rocketed towards Gorazde, I found myself reluctantly forcing the nose of the aircraft down towards the 10,000ft mark. To see the target, I needed to stay below the angry wisps of grey steam vapour that seemed to be bunching over the target area.

I cursed under my breath. I was heading for SAM City and there was not a damned thing I could do about it.

Before I had taken off, I had refamiliarised myself with the locations of Serbian SA-2 and SA-6 missile batteries. These were ancient but formidable Soviet-supplied surface-to-air missile systems—SAMs—with operating altitudes of 90,000ft and 50,000ft respectively. Because our intelligence was up to speed on the positions of these weapons it was easy enough to steer clear of them, but what the intel guys couldn't plot, because there were just too many of them, were the Serbs' man-portable air defence systems, or MANPADS, shoulder-launched missiles. On top of that, there was the Triple-A, anti-aircraft artillery. That was bloody everywhere, too.

Normally, neither the MANPADS nor the Triple-A bothered us that much, because we spent most of our time above their effective height range of 10,000ft. But as I watched my altimeter dip below the magic safe base height mark, I kissed all that goodbye. I'd just crossed into a very dangerous patch of sky.

The green brain contained the codes that hooked me into the new frequency. Trying to read this shit as I'm thudding through the choppy air spinning up from the mountains below is like trying to read a telephone directory while driving round a pot-holed version of the M25 at 120mph. At last the numbers swam into view. I reached up and twisted five dials low down on my left-hand side.

'Vixen Two-Three. TAD Three. Go.' I checked with my wingman that he was tuned in as we readied ourselves at the top of the switchback ride.

I'm a family man with three kids. I pictured them back home, savouring the sunshine of a warm, mid-April afternoon, or dodging the showers on a shopping run in the local town. This isn't really happening, the voice in my head attempted. It's just another alert. It'll all be over by the time you get there, mate. You wait. You'll see.

Sixty seconds to target.

'Vixen Two-three,' I checked with the wingman.

'Two-Four,' his reply crackled back.

As a pair, a fighting unit, we're locked and loaded. Now to make contact with the FAC.

'Fortune Zero-Five, this is Vixen Two-Three.'

There was a brief pause, then a voice burst in my ears: 'Vixen Two-Three, this is Fortune Zero-Five. You're loud and clear.'

At the first attempt, I failed to get the necessary authentication off the guy. As I was wondering what the problem was, I heard what at first I took to be an irregular jamming signal, like a series of thumps. Then it started to dawn on me what was happening.

'We haven't got time for this shite, mate!' the voice on the ground yelled in between more artillery bursts. 'We're getting bloody shelled here!'

Through the head-up display, the HUD, I cold now see a pall of smoke between the ground and the cloud base ahead of me.

'Authenticate, X-ray Yankee,' I insisted.

Three or four seconds ground by. The airspeed indicator was clipping 440 knots. The tension was killing me.

Suddenly, there was a crackle in my headset. 'Bravo. It's Bravo.'

I heard several more bursts of shellfire. Then the FAC said: 'We know there's a tank or two above the ridgeline to the north of Gorazde. That's what you're gonna take out, mate. All right?'

I grabbed the map and scoured the topography. A moment later I found it, a sharp divide between two alpine faces running north-south about 10 miles north of the besieged Muslim enclave. I was now down to only 8,000ft, right in the heart of the MANPADS and Triple-A envelope.

I shot over Gorazde, banking the Sea Harrier to the left. Columns of smoke were rising from the houses below, but in the still air it could have been wood smoke. It was not what I imagined at all. From my vantage-point, there was little discernible damage.

The ridgeline suddenly veered towards me. For a brief couple of seconds, it undulated and coiled below the aircraft. And then it was gone. I was out over the mountains again and pulling into a 'dumb-bell' manoeuvre that would bring me back again, this time from another direction.

As I hauled back on the stick, feeling the gs wrenching at my oxygen mask and sucking my guts into my boots, my brain tried to review every nook and cranny of the densely wooded topography I'd just seen.

'Vixen Two-Four, spot anything?' I asked.

He was positioned 500 yards behind me and a little to the right, in strike formation. I was about 500 feet below the base of the clouds. He was a little higher. As wingman, his job was to watch out for me. Over this place, that meant keeping his eyes peeled for SAMs.

'Negative.'

I heard the disappointment in his voice. The FAC must have caught it, too. 'We're pretty bloody sure there are two tanks down there, mate,' he yelled, hope in his voice. 'Do you see 'em?'

The ridgeline loomed large in the HUD again. I banked the aircraft and peered hard past the canopy glare. My eyes watered with the effort.

'Come on!' the FAC yelled, incredulous when I told him I'd found nothing, 'you must have seen 'em.'

I elected to give it one last try. This time I pressed even lower, scanning the terrain feverishly as I tore towards the ridgeline. For a moment I could see nothing but trees. Then, quite unexpectedly, a plume of smoke broke through the branches.

Instinctively, I made a minute course correction towards it and shot overhead. As I did so, I caught the unmistakable outline of an olive-green main battle tank. And then I spotted another. The second vehicle was on the move, sending a stream of thick, clogging exhaust into the air.

'Tally!' I yelled. 'Two T-55s.'

'Tally!' the wingman responded. He'd seen them, too.

'That's your target,' the FAC announced drily.

As we started pulling round, I asked to be 'cleared live' by Vicenza. This was the authorisation I had to have to drop the bomb. In the meantime I armed the fuse. I was in the middle of this instinctive routine when I heard Vixen Two-Four shout a warning.

'Flares!'

As I hit the countermeasures button to release the flares, I threw my head around, wondering if I'd catch a glimpse of the sliver-thin frame of the projectile as it slammed into the aircraft. Instead, I saw a trail of smoke rising vertically from the ground and disappearing into the cloudbase to my right. My relief was tempered by the realisation that the missile had passed between our two aircraft before I'd had a chance to react.

The voice of the FAC was back in my headset. 'Come on, man. We're getting shelled to shit down here. Do something, for Christ's sake!'

I switched the HUD to ground-attack mode, lined up on the ridgeline and rolled the Sea Harrier into a dive. The tanks slid neatly into the middle of the sight. I pressed the accept button, waiting for the radar to range the distance between me and the tanks, but nothing happened. The diamond symbol that had flashed up with unfailing regularity every time I'd done it on exercise failed on the one occasion I really needed it to materialise.

I hauled back on the stick and pulled into a 5g, 30-degree climb. I went into another dive, but the same thing happened again. Then Vixen Two-Four had a go. The same thing happened to him.

I looked at my gauges. My fuel state was pretty iffy, but I was determined to have one last go. I approached from the north and went into the dive. Once again, the tanks were lined up perfectly in my sights. And once again I failed to get a radar lock.

I was pulling out of the dive, 30 degrees nose-up and feeling sick about the whole thing, when there was a massive bang beneath the aircraft and a violent jolt upwards. For a moment, the force that propelled me against my straps threatened to rip my shoulders off. Then, as a black curtain snapped around my head, there was a fearsome ripping noise, as if my brain and body had just parted company. I found myself falling into a cold place with a beguiling absence of sound. I was still falling when a flash of searingly bright light cut through the darkness, accompanied by an angry, grating noise that seemed to emanate from somewhere behind my eyes.

And then, suddenly, I was staring at a wall of flashing cockpit lights with the master warning horn going off in my ears. I'd been out for less than a second, but it felt as if I'd been to the end of the universe and back.

The Sea Harrier was upside down and hurtling towards the ground.

Recce Over Kosovo

France had one small carrier immediately before World War II, the *Bearn*, converted from a battleship under construction during World War I. The ship saw no action against the invading German army. There were ambitious plans for other conversions as well as new ships, but these were quickly put aside in the face of the German Blitzkrieg of September 1939. After the war, with the loan of several small former British and American carriers, a few of which saw action in Southeast Asia in the late 1940s and early 1950s, France embarked on its own carrier program, building two small ships, roughly comparable to the American *Essex* class.

Built in the 1950s, the *Clemenceau* and *Foch* served long and well, although again, they saw little combat action for most of their careers. Like many such weapons, however, their presence was often enough to convey French intentions to various areas of the world. They occasionally participated in American operations in the eastern Mediterranean in the 1980s, sending reconnaissance missions over rebel positions in Lebanon. During the 1991 Gulf War, the *Clemenceau* played a support role, transporting much-needed aircraft to shore bases.

As the situation in the former Yugoslavia deteriorated in the early 1990s, NATO's concern generated a growing presence of ground, sea, and air forces in the Balkans, staging from several airfields in Italy and neighboring countries, or from several aircraft carriers from Britain and the U.S., and finally from France.

Short campaigns against Serbian forces yielded little. It wasn't until March 1999, with the beginning of Operation Allied Force, that a three-month-long sustained series of attacks against the Serbs, centered around the small southeastern province of Kosovo, bordering Albania and Macedonia, showed any effect. Most NATO countries sent contingents of their air forces to supplement the main forces of American and British aircraft. Aerial action included several kills by USAF fighters as well as one by a Royal Netherlands Air Force F-16, the first kill by a Dutch pilot since World War II.

Clemenceau had been retired in 1997. Scheduled for retirement in 2000, but held in service because of problems with the nuclear-powered carrier *Charles de Gaulle*, the *Foch* was on station in the Adriatic, sending strikes and reconnaissance flights over the embattled province. One of its embarked squadrons, or "flottilles," was 16F with Étendard IVPM photo-reconnaissance jets, reminiscent of the U.S. Navy's RF-8 Crusader, and nearly as old, having first flown in 1960.

From January 1955, 16F became the Aeronavale's sole dedicated reconnaissance squadron, again similar to the U.S. Navy's two light-photo squadrons, VFP-62 and VFP-63. The squadron had transitioned to the Étendard IVP as far back as 1969 and had participated in every carrier deployment since then. In anticipation of the oncoming Rafale shipboard fighter and the *de Gaulle*, the remaining operator of the F-8E(FN) Crusader, 12F, retired its veteran "Crouzes" and decommissioned in December 1999, following the conclusion of Allied Force. Recce squadron 16F retired its Étendard IVPMs and decommissioned in July 2000. *Foch* decommissioned the following November, and began a new career as the Brazilian carrier *Sao Paolo*. Her departure leaves the new nuclear carrier *Charles de Gaulle* as the only French aircraft carrier.

French participation in operations in Bosnia and Kosovo included an active role by Aeronavale carrier resources, including 16F's Étendard IVPMs, modernized IVPs. Enseigne de Vaisseau 1st Classe (Lieutenant, j.g.) Emmanuel Delin accumulated 136 missions over Bosnia and Kosovo. In the following excerpt, he describes a reconnaissance mission over Kosovo during Allied Force.

Author and photographer Alexandre Paringaux has assembled some of the most striking photographs of modern carrier action I have seen. These pictures are well supported by a text equally full of action and the experience of carrier aviators. It would be great to see an English edition. I happened to find this wonderful book through retired Captain Robert Feuilloy, himself a former Étendard pilot and instructor, as well as an exchange pilot with VA-46, flying A-7Bs aboard USS *John F. Kennedy* 1975–1977. A special thanks to Stephane Nicolaou, who translated EV1 Delin's story. I especially recommend this chapter to former U.S. Navy and Marine Corps RF-8 Crusader drivers.

From *Le Porte-Avions Foch,* by Alexander Paringaux

Translated by Stephane Nicolaou

Right behind the catapult, I take advantage of the last moments of quiet to recall the last few hours. The ritual never changes: the release, late in the evening, of the targets tasked by NATO; the long, fastidious preparation; the briefing with our SEAD [suppression of enemy air defenses, usually U.S. EA-6B Prowlers (Ed.)] crews based in Italy, who will protect us during our ingress tomorrow. Then, a few hours of sleep before settling the last-minute details on the carrier's position early in the morning, and the weather report.

At last, "Bravo" time is coming. The "chien jaune" ["yellow dog," meaning the yellow-shirted flight-deck director] shows me the way to the catapult through a bunch of men wearing a wide variety of colored jerseys. They move around my airplane like a well-rehearsed ballet. The green flag is held high: full power, last check of the weight board, a salute, and go!

Two seconds of acceleration with a glance at the airspeed indicator as I leave the deck at more than 150 knots. That's good. The plane is airborne. My wingman is ready to launch, too. I give my bird back its life. I clean up, then turn toward the rendezvous point with our SEM [Super Étendard IVM, serving as a tanker], which will nurse us before going feet dry over the Alba-

Two IVPMs of 16F during operations in the Adriatic. (Courtesy of Alexandre Paringaux)

nian coast. It's right there—circuit, contact, no problem to tank. As my wingman refuels, I start talking to the AWACS, which will keep a sharp lookout for us during the mission. It gives me the positions of all active SAM sites, and I quickly locate them on my charts.

We leave our tanker to cross Albania, then Macedonia. Armament, EW, camera, everything is ready. Now, my Étendard is ready. The Kosovo Mountains appear. We descend to our working altitude, the lowest point of the mission, and also the most dangerous. That's why we have bodyguards, SEAD escorts. Today, we have one U.S. Navy EA-6B and two USAF F-16CJs on our wing. Radio checks confirm their presence, but we won't see them because they fly at a higher altitude. I know they will watch out for us, making their flight coincide with ours, following precisely the information I sent them last night.

I keep my speed at about 550 knots. It's part of our life insurance. Then, I give my "go" signal at the push point. Our run starts now. So begins this special feeling, mixing attention, fear and care, plus a desire to make this mission succeed.

Now it goes! First targets—some hangars at Pristina airfield that were bombed last night, and a few miles further, an artillery position, still active. I read a chart, then switch to another. I prepare my run-in, scrutinize the ground to be ready if it turns bad. A glance at my wingman. He's staying at an ideal position, a little higher and behind me. His main role is to warn me about any SAM threat, especially the IR-guided missiles, as no signal will ring in my cockpit if those are launched.

The landscape rushes by. The airfield is near. The three SA-6 sites active in the area are reason for caution, but right now, everything seems quiet. Only some smoke here and there testifies to recent activity of NATO planes or Serbian slaughter in Kosovar villages. I select the cameras, roll the airplane, aim and shoot. Lights are flashing, needles are turning, it's in the box! Now. We're approaching the gun positions. I roll on the other side, changing the cameras. Over, then we are flying to the next targets. They are going by one after the other, from bridges to factories, radar positions to EW installations. Yet, we are still in Serbia.

The second part of the mission is an immediate post-strike over an armament stock attacked by F-15s, F-16s, and Tornados. Near the target, I'm surprised not to see heavy smoke, although I'm right on time. Suddenly, a series of explosions makes my doubts vanish. The wind is not in the right direction. I have to change my penetration axis to avoid the dense curtain of smoke that could degrade my photos.

EV1 Delin in the cockpit of his Étendard IVPM. (Courtesy of Alexandre Paringaux)

We are leaving the building in flames and the deep part of Serbia behind to fly over a final target when a high-pitched sounds rings in my helmet. The low-frequency flash confirms that radar has locked us up at three o'clock. I call for an escape maneuver and launch IR decoy flares. The deceiver is now emitting signals. A white plume crosses over us and plunges toward the threat. That's our SEAD planes that have launched a HARM to destroy the radar station. The code word to leave the area comes over the radio. Full speed puts us out of SAM range, going to a higher altitude in the direction of Macedonia, sending reports to the AWACS. We've taken pictures of all the targets except the last.

Flying back, we leave Serbia and Kosovar territories, crossing back through Macedonia and Albania down to the Adriatic Sea to find the carrier. There it is. Ready for the racetrack, I concentrate for the last time. I recover with an OK 3-wire.

The weapon-system crew is already running to bring the film to the photo lab. In less than one hour, they will digitally transmit via satellite, and the pictures will be available at NATO headquarters.

I am climbing down from my Étendard IVP, which, despite its age, has contributed its small part to the success of the operation.

Women in the Cockpits

The Navy, indeed all U.S. military services, experienced a tremendous sociological change following the 1991 Tailhook scandal. Much has been written and discussed in every type of forum about this far-reaching event. But while Tailhook '91 generated a major escalation in how and where women served in the military, the gender gap had been slowly closing since the 1970s. And truth be told, women had been in aviation since the days of ballooning.

The decades between the world wars, a time that saw goals achieved then surpassed over and over, also saw the rise of a cadre of female aviators such as Amelia Earhart, Jackie Cochran, Anne Lindberg, and Ruth Nichols who challenged their male counterparts for the headlines. Their wings took them everywhere on the planet, everywhere, that is, except the ready room of a military squadron. During World War II, the only example of designated female aviators and flight crews flying combat was in the Soviet Union. In the hard-pressed Soviet air force, women flew every type of frontline aircraft—bombers, fighters, transports—often seeing intense, sustained combat . . . and dying just as the men in the units. One squadron even boasted an 11-kill ace, Lilya Litvak, eventually killed in action in 1943.

After the war, although the Russians retained some of their women aviators, no other country permitted the crossing of such a barrier. Women's liberation, the demonstrations of the 1960s, and a new awareness by women of their place outside the kitchen, though strong, sometimes violent, could not penetrate the wall of men-only cockpits. However, things began changing by the 1970s.

When Lieutenant, j.g. Barbara Ann Allen received her Navy Wings of Gold on February 22, 1974, the wall was breached, and what at first was a trickle became a steady stream, not a flood, but a constant flow of young women intent on flying. The firsts kept coming:

- First woman to qualify aboard a carrier—Lieutenant Donna L. Spruill, June 20, 1979
- First female flight surgeons—Lieutenant Jane O. McWilliams and Lieutenant Victoria M. Voge, designated on December 20, 1973
- First woman to carrier qualify in an F-14—Lieutenant Kara Hultgreen, July 31, 1994
- Women allowed to fly aircraft in combat—July 31, 1991: The U.S. Senate votes to overturn a 43-year-old law prohibiting women from flying in combat

Sadly, only three months after her milestone qualifying flight, Lieutenant Hultgreen died trying to eject from her F-14 while attempting to land aboard USS *Abraham Lincoln* (CVN-72) on October 25, 1994. She had not been the first female naval aviator to die during a flight, but her death and the mishap that led to it raised a storm of questions about women flying high-performance aircraft and the training they went through to do it.

Loree Draude Hirschman was an S-3 pilot in the same air wing as Kara Hultgreen. She had qualified aboard *Lincoln* a few weeks before and was ashore when Lieutenant Hultgreen crashed. She writes about her flying in this excerpt, and when she heard about the F-14 mishap.

Receiving her wings in 1992, Lieutenant Commander Hirschman made two deployments with VS-29, flying the S-3B Viking. During her squadron tour, she became the first woman training-qualified landing signal officer in the Navy. She left the Navy in 1999 to pursue a master's degree. She works as a management consultant in the San Francisco Bay area.

Her book about her experiences as a naval aviator is one of several that have appeared in recent years. A few of these have been somewhat heavy-handed, even vitriolic, railing at the system and the intensity of the problems the authors encountered during the squadron tours. Loree Draude Hirschman's treatment is more even-handed and mature, and is a good account of one young woman's experience as a military aviator.

From *She's Just Another Navy Pilot: An Aviator's Sea Journal,* by Loree Draude Hirschman and Dave Hirschman

25 October 1994 started as the kind of day that made me feel lucky to be a navy pilot.

I climbed into the cockpit of an S-3 Viking at the naval air station at North Island in San Diego, started the plane's two jet engines, and prepared to fly the airplane and three other crew members to the USS *Abraham Lincoln.* The nuclear-powered aircraft carrier that soon would become our home was waiting about fifty miles offshore. This was a warmup for the ship's air wing, a chance for aviators to practice landing and taking off from the moving deck of the aircraft carrier in the open sea. In six months our squadron was scheduled to fly aboard the *Lincoln* and steam across the Pacific Ocean, through the South China Sea and Indian Ocean and into the Persian Gulf. There we would enforce the "no-fly zone" over Iraq and monitor the volatile military situation in the Middle East.

Our cruise would make history, too. The *Lincoln* was about to become the first West Coast aircraft carrier to go to sea with female crew members. I was among seventeen female aviators, and the only woman piloting S-3 Vikings—twin-engine jets that hunt submarines, drop bombs, and refuel other planes in flight.

The morning was cool and clear with steady trade winds and streaks of high stratus clouds that looked like huge white brush strokes on a deep blue canvas. I took off from the west-facing runway at North Island, turned south over San Diego harbor to avoid overflying the residential areas of Point

Loma, then joined up with another S-3 headed toward the *Lincoln.* We climbed together to fifteen thousand feet and circled in loose formation about thirty miles south of the aircraft carrier.

Given its role as the first Pacific Fleet aircraft carrier to carry female pilots, people throughout the navy jokingly referred to the *Lincoln* as the *Babe-raham Lincoln.* The genesis of the nickname was the movie *Wayne's World* ("She's such a babe that if she was president, she'd be Babe-raham Lincoln"). I got a chuckle out of the moniker and even used it in letters to my parents.

The air was smooth at our altitude, and that made it easy to keep abreast of the other S-3 as we made lazy circles in the sky. Our Vikings, pejoratively known as "Hoovers" for the vacuumlike sucking sound their twin turbo-fan engines make, are hardly the glamour planes of the fleet. Compared with the supersonic single-seat F/A-18 Hornets I had flown in a previous assignment, or the sleek afterburning F-14 Tomcats made famous by the movie *Top Gun,* Vikings are slow and ponderous. But in the air, with the landing gear tucked neatly into the wheel wells and the flaps retracted, even plodding Vikings look graceful.

Through my plane's large canopy I had an excellent view of the lead aircraft. Like most carrier-based jets, it was painted a dull gray. The drab color did a good job of obscuring the many fuel leaks that always seemed to streak along the undersides of the wings and line the fuselages of our aging planes. The navy had been flying Vikings since the mid-1970s, and some of the planes at North Island had been in service since I was in elementary school.

I kept my head turned to the left and studied the lead S-3. Like all the planes in our squadron, VS-29, it had the silhouette of an ancient Viking ship painted in black on the tail. Most of the tall-masted ship with the billowing sail covered the vertical stabilizer, and the rear of the boat stretched onto the rudder. But to me the emblem resembled a cartoon rubber ducky, and the side-to-side movements of the plane's rudder sometimes made it look as though the little duck was waddling through the sky. Even though I was becoming more tense as we approached the ship, the comical rubber ducky made me smile.

I had finished my final carrier landing qualifications in an S-3 a month before, and this was going to be my first time aboard the *Lincoln* as a full-fledged member of the Pacific Fleet. After four college years in the Reserve Officer Training Corps, three years of navy flight training, and two years in a land-based composite squadron, I was finally about to start doing the job my country had prepared me to do: I was going to sea with a fleet squadron.

We were part of a five-thousand-member team aboard a technologically advanced aircraft carrier. And that ship was at the heart of a floating arsenal, one of the most powerful naval battle groups ever assembled. Our ship alone had more firepower than the entire U.S. Navy had delivered in all of World War II.

As an added bonus, I was going to be part of navy history: the first group of female aviators on the West Coast to deploy at sea as fully qualified combat pilots. Women in the military—and especially in military aviation—had overcome tremendous obstacles to get to this point. My bookshelves at home were stocked with accounts of their heroism. Many military women had given up hope that this day would ever come. But thanks to their sacrifices and dedication during more than fifty years of military flying, nineteen of us were about to fulfill their dream. We were going to be allowed to go as far as our ability and imagination would take us. The artificial gender barrier was gone, and we would be allowed to succeed or fail on our own merits.

As usual, when our S-3s arrived at the holding pattern about twenty miles

Lt. Loree Hirschman stands with her husband, Lt. Harry Hirschman, before their respective aircraft. Harry had flown to North Island from his home field at LeMoore in his FA-18C to visit Loree, who was instructing at VS-41. (Courtesy of Kathleen Spane)

from the ship, air-traffic controllers instructed us to stay at a high altitude. Vikings are relatively fuel-efficient, and we could wait in a holding pattern while the gas-guzzling F-14 Tomcat fighters, which had to get on the deck quickly, were guided in. Next came the F/A-18 Hornet fighter/bombers, then the two-seat A-6 Intruder bombers and four-seat EA-6B radar-jamming planes.

Finally it was our turn to begin our approach, and my adrenaline surged as the *Lincoln*'s air-traffic controllers told us to drop down to a lower altitude. Even though I had already performed dozens of carrier landings, each one was an exciting challenge. The goal was to fly a perfect approach at exactly the right airspeed, then hit a spot on the deck while holding the plane at precisely the right attitude. There are hundreds of variables, so no two landings are exactly alike. Each approach and landing is graded, and the results are posted in every squadron's "ready room" for all to see.

Carrier landings differentiate navy and marine aviators from their counterparts in the other services, and we measure ourselves by how well we perform this critical task. No matter how good we are at bombing, dogfighting, hunting submarines, or refueling other aircraft, all of that is meaningless unless we can be counted on to bring our airplanes back to the ship at any time of day and in any kind of weather.

During my previous trips to "the boat," there had been several delays that kept planes in the holding pattern longer than expected. The deck crews were new to their jobs, and so were many of the pilots. It wasn't uncommon for operations to lag a few minutes behind schedule. But on this day, as we kept circling in seemingly endless lefthand patterns, the situation was becoming absurd. Our planes had plenty of fuel, so we were capable of staying aloft for hours; time wasn't a problem. But why had the ship's air-traffic controllers brought us down to a lower altitude if the deck wasn't clear to come aboard? It didn't make any sense. We orbited so long that my neck was getting sore from constantly looking left toward the lead airplane.

Then, at last, we were told to approach the ship for landing. There was a thin layer of clouds over the ship, and I followed the lead airplane through the wispy white cloud. We came out of the clouds about two thousand feet above the blue ocean surface.

My mouth went dry, and my heart was beating so hard that I could hear it thumping through the earphones inside my helmet. I tried to calm myself down by concentrating o the procedures as I entered the pattern for my first fleet carrier landing, or "trap." My limbs felt as though they had electricity running through them, and my stomach tightened. It was like a bad case of stage fright. This was going to be my first time aboard the *Lincoln*, and I wanted to make a good impression.

I brought my S-3 into tight formation with the lead aircraft, and together we streaked toward the ship at about 400 miles an hour, just eight hundred feet above the ocean. Whitecaps from the wind-blown ocean passed by us in a blur, and the choppy air near the water's surface made it feel as though we were racing cars down a rutted country road. As soon as the *Lincoln* disappeared beneath us, the lead plane banked hard left and began slowing down in preparation for landing.

I continued straight ahead about one mile, then swung my aircraft onto its side in a tight left turn, yanked the two throttle levers back to idle, and extended the speed brakes. The g-forces of the hard turn pressed us all down into our ejection seats at three times our normal weight, and I glanced at the airspeed indicator as it unwound. At 186 knots, my left hand found the round landing gear lever on the instrument panel and pulled it sharply downward. A series of mechanical groans and clanks followed as the hydraulic system pushed the wheels into the slipstream and locked them into place.

My goal was to touch down exactly forty-five seconds after the plane in front of me. That's the minimum amount of time the deck crew needed to move one plane out of the landing area and prepare for the next arrival. By evenly spacing our arrivals at the shortest possible intervals, aviators can reduce the amount of time the ship is required to steam into the wind—a predictable situation that makes the ship and its crew especially vulnerable to enemy attack.

The *Lincoln* was about a mile off my left wing and the lead plane was making its final approach to the steel deck when I grabbed the square flap handle with my left hand and pulled it all the way back to the landing position. My plane was traveling at 156 knots when I tugged on the lever, and it decelerated as though I had tossed out an anchor. In four seconds the flaps reached their full travel of 65 degrees.

I pushed the noise of the airplane down to maintain airspeed, and I tried to anticipate the altitudes at certain key points in the arrival pattern. As soon as we were abeam the ship's arresting wires, I banked the plane 25 degrees to the left and began a descending turn. I focused intently on the instrument panel. With 90 degrees of turn to go, my S-3 was 450 feet above the water. As I rolled into the "groove" on final approach, the ship's frothy white wake was 325 feet below. So far so good.

The naval flight officer, or "NFO," in the right seat made the "ball call" as the "meatball," an automated lighting device that showed the plane's position relative to the deck, came clearly into view.

"Seven-oh-five, Viking, ball," he said in a calm, clear voice. "Four point-oh," he added, letting air-traffic controllers know that we had a little under two tons of fuel remaining.

Everything looked right. The lights on the meatball were centered, just as they should have been, and I was properly aligned with the white painted center stripe on the black deck. My eyes quickly shifted between the gauges on the instrument panel and the rapidly changing view outside. Airspeed was 100 knots, fine. The rate of descent was roughly 600 feet a minute, good. I glanced at the numbers and, by force of habit, reviewed my procedures out loud: "Meatball, lineup, angle of attack."

Under my breath, I admonished myself not to screw up.

From this point on there was no reason to look at the instrument panel. The final portion of any carrier landing is purely visual, and I kept my eyes on the panoramic picture in front of me. The ball started to sink, indicating that I was getting too low, so I added a touch of power. But that changed the aircraft's pitch, so I pushed the nose down to compensate. Then the airspeed increased along with the rate of descent. I was flying like a rookie reacting to the ball instead of using the airplane to actively place it where I wanted it to be.

Suddenly we slammed against the steel deck with a loud screech, and the forged metal hook dangling beneath our plane's tail caught the third of four arresting cables. Maybe it was beginner's luck, but I had caught the target wire. We came to an abrupt stop with the nose wheel only a few yards from the edge of the deck. I could feel the rolling motion of waves crashing against the ship's hull directly beneath us.

There was no time to savor the successful trap, however. We knew that the next plane was rolling onto the groove just forty-five seconds behind us, and we had to raise the tailhook and clear the landing area quickly. The flight officer in the right seat pulled the lever that unlocked and folded the wings, and in my side mirror I could see them folding into a giant X above us. I added power and steered through a narrow gap between several parked airplanes as we rolled to catapult number one at the bow of the ship. Everything was hurried, but nothing could be overlooked. We unfolded the wings, then extended the launch bar on the front landing gear after I taxied the airplane to the catapult.

As soon as I got the hand signal from the launch director, I brought both engines up to full power. The jet engines whined, and the airplane rocked and shook against its restraints, then squatted down as the catapult went into tension. A pair of deck hands scurried in front of the plane, then ducked down to check the nose wheel and make sure the aircraft was properly attached to the catapult. Moments later, they scrambled away to safety.

I clicked the launch bar switch to "retract" so that it would raise up automatically and not interfere with the landing gear at the end of the catapult stroke. Next I moved the joystick all the way left and right, forward and aft, then around in a circle to make sure that the movement was free and unobstructed. I pushed the rudder pedals firmly back and forth with my feet. Everything felt normal. Then I scanned the gauges a final time to make sure that the jet engines were up to speed and none of the warning or caution lights was on. A few seconds later, the checklist was complete.

I saluted the yellow-shirted catapult officer and watched him slowly turn his head to look up and down the four-hundred-foot length of the catapult, making sure that nothing was in our way. Then I took a deep breath and prepared for the sudden acceleration of a catapult shot. As soon as the catapult officer was satisfied that all was clear, he bent down and gently touched the deck with is left hand, his long-sleeved jersey rippling in the stiff wind. At that moment, the "shooter" behind him hit the button that sent us on our way like a rock from a slingshot. In three breathtaking seconds, we lunged forward from zero to 130 miles an hour. The bow of the ship passed underneath us in a blur.

As our Viking began climbing into the salty air, the furious noises and strange sensations of the catapult faded away. We were flying again. The smooth, comforting sounds of the two jet engines and the wind passing over the wings filled the cockpit. The plane was under my control, and the horizon before us was wide and expansive.

I couldn't wait to do it again.

I made four traps and four "cat shots," as the catapult launches are known, before being told to fly home to nearby North Island. The short trip back to the coast gave me a chance to calm down, and landing on the eight-thousand-foot runway that runs parallel to the beach on Coronado Island seemed like child's play compared with setting the plane down on the ship. By the time our S-3 rolled to a stop in front of the hangar at our base and we shut the engines down, I was relaxed and happy with my day's work. My passes hadn't been perfect, but each of them had brought us aboard the ship safely. The day's events had increased my confidence that I could perform the demanding tasks required of navy pilots, and that with experience I would become a valued team member.

I bounded upstairs to our squadron ready room with the rest of the crew. The four of us were still talking and laughing when someone asked if we had heard about an F-14 crash. Crash rumors circulate through navy squadrons with amazing speed, and they are seldom unfounded.

A chill ran through me. There were three female officers in VF-213, the F-14 squadron assigned to the *Lincoln.* Two—Lt. Kara Hultgreen and Lt. Carey Lohrenz—were pilots, and a third, Lt. Christina Taylor, was a radar intercept officer, or "RIO." I had worked with Christina in a previous assignment at another squadron, and I prayed that she wasn't involved. But no one had much solid information. All anyone knew for sure was that an F-14 Tomcat had gone down a few minutes before we had arrived at the ship. That explained our long delay in the arrival pattern before landing.

The next morning the *San Diego Union-Tribune* printed a little blurb about the accident on the front page. The article said that an F-14 Tomcat had crashed while attempting to land on the *Lincoln.* The RIO in the back seat of the two-person fighter had been plucked from the 63-degree water and treated for mild hypothermia. But the pilot was gone—lost at sea. The paper said the pilot's name was being withheld pending notification of "his" next of kin. I was slightly relieved by the pronoun, because the only people I knew in that squadron were women; if it had been one of them, I assumed, the paper would have written "her" next of kin. I went to the VS-29 ready room and prepared to fly back out to the *Lincoln* for several more days at sea.

INDEX

THE EDITOR

Peter B. Mersky graduated from the Rhode Island School of Design in 1967, and was commissioned an ensign in the U.S. Naval Reserve in 1968. He served in a variety of assignments on active duty in the Navy and Naval Reserve, retiring as a commander in 1992. He has written a dozen books and more than a hundred magazine articles on Navy and Marine Corps aviation, as well as a ground-breaking account on Israeli fighter aces. Since 1982, he has reviewed more than four hundred books in his regular column for *Naval Aviation News*. He retired from the Naval Safety Center in 2000 after serving for more than sixteen years as the assistant editor, then editor of *Approach*, the Navy and Marine Corps aviation safety magazine.